Photosynthesis Bibliography

volume 2 1971

References no. 9088 - 12069 / AAC - ZUR

Editors Z. Šesták & J. Čatský

Springer-Science+Business Media, B.V. 1975

ISBN 978-90-6193-040-2   ISBN 978-94-017-2636-8 (eBook)
DOI 10.1007/978-94-017-2636-8

PREFACE

The bibliography includes papers in all fields of photosynthesis research - from studies of model biochemical and biophysical systems of the photosynthesis mechanism to primary production studied so-called growth analysis. In addition to papers devoted entirely to photosynthesis, papers on other topics are included if they contain data on photosynthetic activity, photorespiration, chloroplast structure, chlorophyll and carotenoid synthesis and destruction, etc., or if they contain valuable methodological information (measurement of selected environmental factors, leaf area, etc.). In many branches it has been very difficult to define the limits of interest for photosynthesis researchers. This problem has arisen e.g. in topics dealing with the transport of gases, where - in addition to the papers on CO_2 transfer - some papers on water vapour transfer are included, these being of general application. On the other hand, many papers dealing with the anatomy and physiology of stomata have been omitted, if the aspect of carbon dioxide or water vapour exchange has not been discussed.

To maximize the value of the bibliography the references are arranged alphabetically by author's names, and each volume is provided with three indexes. The authors' index to this volume contains all names of authors, co-authors and editors.

The subject index covers only primary items chosen according to their interest for photosynthesis researchers. In this volume its preparation was based on the paper titles, key words and abstracts.

In the plant index, only important crop plants and selected plant types and groups are indexed.

Cumulative indexes accompany every fifth volume, i.e. Vols. 1, 6, 11, etc.

We have tried to cover fully the relevant papers which have appeared in the most important scientific periodicals and books. Articles published in local journals, mimeographed booklets, abstracts of theses and of symposia contributions, etc., were chosen mostly from reprints and lists of publications received direct from the authors.

Since some 3000 relevant papers are currently published every year and included in this bibliography, and since the majority of citations have been checked with the originals, collecting and preparing for publication such a large amount of material would habe been impossible without the collaboration of the authors of the relevant publications. The courtesy of those authors who have already supplied us with reprints and lists of their publications is highly appreciated.

We acknowledge with thanks the cooperation of our colleagues from the Department of Physiology of Photosynthesis and Water Relations of the Institute of Experimental Botany of the Czechoslovak Academy of Sciences in Prague, especially Mrs. DRAHOMÍRA TĚŽKÁ who helped in preparing the card material, retyping the manuscript and preparing the author index, Dr. INGRID TICHÁ who helped with completing and checking the references and reading the proofs, and Mr. PETR ZÁZVORKA who supplied us with rare periodicals. In addition, the librarian of the Institute, Mrs. ALENA ŠTĚTINOVÁ, and Ing. ZDENA MITSCHKOVÁ helped us with checking the references.

Dr. Z. ŠESTÁK and Dr. J. ČATSKÝ
Institute of Experimental Botany
Czechoslovak Academy of Sciences
Flemingovo nám. 2

160 00 PRAHA 6

Czechoslovakia

INSTRUCTIONS FOR USE

All references are arranged alphabetically according to the authors' names and the year of publication. They are numbered and these numbers are used in the indexes. In case of a book title, the number is preceded by B.

The references contain the original unshortened title of the paper (book). English, French, and German titles are cited in the original language. Titles in other languages are supplemented with a translation in English (sometimes using the title of the respective English abstract or a shortened title with omitted deadweight words). Titles of Japanese, Chinese etc. papers are given in English translation only. The journals' names are abbreviated mainly according to the "Style Manual for Biological Journals" (Second Edition, Amer. Institute of Biological Sciences, Washington, D.C. 1964), e.g.:

Abhandlungen	chimicus	Husbandry	Publication
Abstract	chinese	imperial	Publishers
Abteilung	Chromatography	Industry	quantitative
Academy	Commission	inorganic	Quarterly
Acta	Communication	Institute	Radiation
Africa	comparative	international	Radiobiology
agricultural	Comptes rendus	Investigation	RasteniǐT
Agriculture	Conference	italian	Report
Agronomy	Congress	Jahrbuch	Research
Akademie	Contribution	japanese	Review
Algology	Cytochemistry	Japan	royal
allgemeine	Cytology	Journal	russian
american	czechoslovak	Klasse	russkiǐT
America	Dendrology	Laboratory	Science
analytical	Department	Landwirtschaft	Section
Anatomy	Deutschland	Limnology	Series
angewandte	Disease	Magazin	Society
Annals	Dissertation	marine	sovetskiǐT
annual	Doklady	Mathematics	soviet
anorganisch (-nic)	Ecology	Microbiology	special
applied	Education	miscellaneous	SSSR
Arbeit	Embryology	molecular	Station
Archiv	Encyclopedia	Monograph	Supplement
Atmosphere	Engineer	moskovskiǐT	Survey
atomic	Enzymology	Mycology	Symposium
Australia	european	national	technical
Beiheft	experimental	natural	Technology
Belgique	Experiment	Naturforschung	Tijdschrift
Bericht	Faculty	neerlandicus	Transaction
biochemical	Federation	Netherlands	Travail (-aux)
Biochemistry	Fizika	New Zealand	tropical
biokhimicheskiǐT	Fiziologiya	nuclear	ukrainian
Biokhimiya	Forestry	Oceanography	UK
biological	Forschung	Optics	US, USA
Biology	Foundation	organic	USSR
biophysical	France	original	University
Biophysics	Gazette	Pathology	végetale
Bodenkunde	general	Pflanzen-	Virology
bolgarskiǐT	genetical	Philosophy	Virusforschung
botanical	Genetics	physical	Volume
Botany	Gesellschaft	Physics	Weekblad
british	Giornale	physiological	Wetenschappen
Bulletin	helveticus	Physiology	Wissenschaft
Canada	Histochemistry	Phytopathology	Zeitschrift
central	Histology	Plant (-arum)	Zeitung
chemical	Horticulture	polish	Zentralblatt
Chemistry	hungaricus	Proceedings	Zhurnal

The numbers at the end of each reference of a journal article denote: volume (issue): first page - last page, year of publication. The number of issue is given only in journals where each issue is paginated separately.

Book titles are cited according to the title page, not to the book jacket or cover (if the names of the editors are not given on the title page, they are not cited in the reference). The publishing house, place and year of publication are included.

Brackets at the end of the reference give bibliographic details and explanations to the contents, not given in the original. The following abbreviations are used most often:

ab	abstract	Latv.	Latvian
Belorus.	Belorussian	Lithu.	Lithuanian
Bulg.	Bulgarian	Norweg.	Norwegian
Car	carotenoids	PC	paper chromatography
CC	column chromatography	PhAR	photosynthetically active radiation
Chin.	Chinese		
Chl	chlorophyll	Pol.	Polish
Croat.	Croatian	Ps	photosynthesis
E	English	R	Russian
F	French	Roum.	Roumanian
G	German	Span.	Spanish
GC	gas chromatography	Swed.	Swedish
Hung.	Hungarian	TLC	thin-layer chromatography
IRGA	infra-red gas analyser	Tr	transpiration
Ital.	Italian	Ukr.	Ukrainian
Jap.	Japanese	Uz.	Uzbeg

The transliteration of Cyrillic characters is in accordance with the BSI-ASA/SC-Z39 draft table, i.e.:

Translit.	Cyrill.	Translit.	Cyrill.
a	а	r	р
b	б	s	с
ch	ч	sh	ш
d	д	shch	щ
e	е	t	т
f	ф	ts	ц
g	г	u	у
i	и	v	в
ī	й	y	ы
k	к	ya	я
kh	х	yu	ю
l	л	z	з
m	м	zh	ж
n	н	"	ъ
o	о	'	ь
p	п		

Several exceptions apply for Ukrainian and Belorussian:

	Translit.		Cyrill.
Ukrainian:	y	=	и
	i	=	i
	ī	=	ī
Belorussian:	ŭ	=	y

Authors' names are presented in spelling used in the original paper. If this spelling does not correspond to the original spelling used by the author (e.g. Russian papers of English authors), one spelling is referred to the other in the Authors' Index.

Printers' errors in the original papers are marked by underlining the respective words (letters).

Errata to Volume 1. Part 1.

Ref. no.	For	Read
282	rastenīĭ plants [In R.]	rastenīĭ sorgo sorghum plants [In R, ab: Tadzh.]
2611	[In R.]	[In R, ab: Tadzh.]
2614	63-67	63-66
2956	[In R.]	[In R, ab: Tadzh.]
3477	246	245
3777, 4042, 7853	[In R.]	delete title in Russian [In Georg., ab: R.]
3864	[In R.]	[In R, ab: Tadzh.]
4847	Altaya Altai [In R.]	Alaya Alai [In R, ab: Tadzh.]

Part 2.

5774	PosvyaschennoĬ	PosvyashchennoĬ
5775	[In R.]	[In R, ab: Tadzh.]
6548	sulphanate	sulphonate
7242	3:	1967 (3):
7853	TAKAISHVILI	TAKAÏSHVILI
8196	GIRAULT, J.M.	GIRAULT, G.

p. 538 - 9th line from the top
NIKOLAYEVA, A.A.   NIKOLAYEVA, M.K.
p. 544 - 30th line from the bottom
RODZHERS   ROGERS
p. 544 - 24th line from the bottom
ROGERS   RODZHERS
p. 583 - 28th line from the top
4741   4721

9088 - AACH, H.G., FRANCK, U.F., BAUER, R.: Vergleich der Chlorophyllfluoreszenz-Induktion von Photosynthesemutanten. - Naturwissenschaften *58*: 525, 1971.

9089 - AASE, J.K.: Growth, water use, and energy balance comparisons between iso-genic lines of barley. - Agron. J. *63*: 425-428, 1971. [Chl.]

9090 - ABDUL-BAKI, A.A.: Biochemical differences between embryonic axes from green and sun-bleached lima bean seeds: Synthesis of carbohydrates, proteins, and lipids. - J. amer. Soc. hort. Sci. *96*: 266-270, 1971. [Chl.]

9091 - ABDUL-BAKI, A.A.: Changes in chlorophylls and carotene contents of green and bleached lima bean seeds during development and maturation. - J. amer. Soc. hort. Sci. *96*: 576-580, 1971.

9092 - ABDULLAEV, Kh. A., USMANOV, P.D., TAGEEVA, S.V., NASYROV, Yu.S.: Struktura i funktsiya khloroplastov pigmentnykh mutantov *Pisum sativum* i *Arabidopsis thaliana* (L) HEYNH. [Structure and function of chloroplasts of pigment mu-tants of *Pisum sativum* and *Arabidopsis thaliana* (L) HEYNH.] - In: NASYROV, Yu.S. (ed.): Geneticheskie Aspekty Fotosinteza. Pp. 77-106. Donish, Dushanbe 1971. [In R.]

9093 - ABDURAKHMANOVA, Z.N., BELAN, N.F., NASYROV, Yu.S.: Fotosinteticheskaya assimilyatsiya $C^{14}O_2$ i metabolizm ugleroda u vytsvetayushchego mutanta *Ara-bidopsis thaliana* (L). [Photosynthetic assimilation of $^{14}CO_2$ and carbon metabolism in a bleaching mutant of *Arabidopsis thaliana* (L.)] - In: NASYROV, Yu.S. (ed.): Geneticheskie Aspekty Fotosinteza. Pp. 143-158. Donish, Dushan-be 1971. [In R.]

9094 - ABDURAKHMANOVA, Z.N., KHODZHAEVA, R.M., BELAN, N.F., NASYROV, Yu.S.: Sootnos-henie puteT assimilyatsii ugleroda v svyazi s aktivnost'yu fotosintetiches-kogo apparata. [Interrelation of carbon fixation pathways in connection with the activity of the photosynthetic apparatus.] -In: ZALENSKIĬ, O.V. (ed.): Fotosintez i Ispol'zovanie SolnechnoT Energii. Pp. 156-160. Nauka, Leningrad 1971. [In R, ab:E.]

9095 - ABOLINA, G.I., RIKHSIBAEV, N.: Vliyanie razlichnykh doz i srokov primeneniya NRV i MU na urozhaT i kachestvo kartofelya. [Effect of different doses and rates of use of petroleum growth substance and trace-nutrient fertilizer on the yield and quality of potatoes.] - In: NRV v Sel'skom KhozyaTstve. Pp. 77-83. ELM, Baku 1971. [Ps, Chl: in R.]

9096 - ABOLINA, G.I., USMANBEKOV, N.: Vliyanie razlichnykh doz mineral'nykh udobre-niT, NRV, MU na razvitie kornevoT sistemy, urozhaTnost'i kachestvo kartofe-lya v usloviyakh serozemov Uzbekistana. [Effect of different doses of min-eral fertilizers, petroleum growth substance, and trace-nutrient fertilizer on the development of the root system, the yield, and quality of potatoes in Uzbeg SSR sierozems.] - In: NRV v Sel'skom KhozyaTstve. Pp. 84-87. ELM, Baku 1971. [Chl; in R.]

*9097 - ABRAMYAN, A.G.: Zavisimost´ obnovleniya khlorofillo-lipoproteidnogo komplek-sa ot funktsional´nogo sostoyaniya korne-listovoT svyazi u rasteniT. [Depen-dence of restoration of chlorophyll-lipoprotein complex on the functional state of the root-leaf relation in plants.] - In: Ontogenez Vysshikh Raste-niT. Pp. 274-279. Izd. Akad. Nauk Arm. SSR, Erevan 1970. [In R.]

9098 - ABRAROV, A.A., KARIEV, A.U.: Stimulyatsiya biosinteza khlorofilla i belka pod vliyaniem khlorkholinkhlorida. [Stimulation of biosyntheses of chloro-phyll and protein affected by CCC.] - Sel´.-khoz. Biol. *6*: 358-361, 1971. [In R, ab: E.]

9099 - ACOCK, B., THORNLEY, J.H.M., WARREN WILSON, J.: Photosynthesis and energy conversion. - In: WAREING, P.F., COOPER, J.P. (ed.): Potential Crop Produc-tion. A Case Study. Pp. 43-75. Heinemann Educational Books, London 1971.

9100 - ADACHI, M., HAMADA, K.: [Joint action of photosynthesis inhibiting herbicides.] - Zasso Kenkyu *1971* (12): 59-64, 1971. [In Jap., ab: E.]

9101 - ADAMS, M.S.: Effects of drying at three temperatures on carbon dioxide ex-change of *Cladonia rangiferina* (L.) WIGG. - Photosynthetica *5*: 124-127, 1971.

9102 - ADAMS, M.S.: Temperature response of carbon dioxide exchange of *Cladonia ran-*

giferina from the Wisconsin Pine Barrens, and comparison with an alpine population. - Amer. Midland Natur. *86*: 224-227, 1971.

9103 - ADAMS, M.S., LOUCKS, O.L.: Summer air temperatures as a factor affecting net photosynthesis and distribution of eastern hemlock (*Tsuga canadensis* L. (CARRIERE)) in south-western Wisconsin. - Amer. Midland Natur. *85*: 1-10, 1971.

9104 - ADEDIPE, N.O., HUNT, L.A., FLETCHER, R.A.: Effects of benzyladenine on photosynthesis, growth and senescence of the bean plant. - Physiol. Plant. *25*: 151-153, 1971.

9105 - AEROV, I.L.: Spektroskopicheskie osobennosti list'ev rasteniǐ v sine-fioletovoǐ oblasti spektra. [Spectroscopic features of plant leaves in the blue-violet spectral region.] - Zh. prikl. Spektroskop. *15*: 272-277, 1971. [In R.]

9106 - AFANAS'EVA, T.A.: Vnekornevoe vnesenie NRV v usloviyakh Zapolyar'ya. [Extra-root application of petroleum growth substance beyond the polar circle.] - In: NRV v Sel´-skom Khozyaǐstve. Pp. 109-118. ELM, Baku 1971. [Ps; in R.]

9107 - AGATA, W., KUBOTA, F., KAMATA, E.: [Dry matter production of forage plants. 1. Comparison of the amount of dry matter calculated from CO_2 balance with one observed in forage plant populations.] - J. jap. Soc. Grassland Sci. *17*: 223-228, 1971. [In Jap., ab: E.]

9108 - AGATA, W., KUBOTA, F., KAMATA, E.: [Dry matter production of forage plants. 3. Influences of light extinction coefficient (k) on dry matter production and optimum frequency of cutting in forage plant populations.] - J. jap. Grassland Sci. *17*: 235-242, 1971. [In Jap., ab: E.]

9109 - AGEEVA, O.G., LYUTOVA, M. I.: Vliyanie teplovogo zakalivaniya gorokha *Pisum sativum* L. na fotosintez i fotokhimicheskie reaktsii. [Effect of heat hardening of pea on photosynthesis and photochemical reactions.] - Bot. Zh. *56*: 1365-1373, 1971. [In R.]

*9110 - AKHRAMOVICH, N.I.: Fraktsionirovanie pigmentnogo fonda yachmenya s pomoshch'yu dezoksikholata. [Fractionation of barley pigments using desoxycholate.] - In: Tezisy IV Nauchnoǐ Konferentsii Molodykh Uchenykh po Sovremennym Problemam Biologii. Pp. 58-60. Minsk 1970. [In R.]

9111 - AKOYUNOGLOU, G., MICHALOPOULOS, G.: The relation between the phytylation and the 682 → 672 nm shift *in vivo* of chlorophyll *a*. - Physiol. Plant *25*: 324-329, 1971.

9112 - AKSENOVA, E.I.: O sravnitel'nom aspekte ispol´zovaniya khlorofil´nogo metoda. [Comparative aspect of using the chlorophyll method.] - Gidrobiol. Zh. *7* (4): 105-107, 1971. [In R.]

9113 - AKULOVA, E.A., MUKHIN, E.N.: O veshchestvakh zelenykh list´ev - vozmozhnykh regulyatorakh svetovykh reaktsiǐ fotosinteza. [Green-leaf substances as possible regulators of photosynthetic light reactions.] - Dokl. Akad. Nauk SSSR *198*: 956-958, 1971. [In R.]

9114 - AKULOVICH, N.K., RASKIN, V.I.: Formirovanie protokhlorofill-golokhroma v etiolirovannykh list'yakh i ego fotoprevrashchenie v khlorofill-golokhrom. [Formation of protochlorophyll-holochrome in etiolated leaves and its phototransformation in chlorophyll-holochrome.] - In: Problemy Biosinteza Khlorofillov. Pp. 5-52. Nauka i Tekhnika, Minsk 1971. [In R.]

9115 - AKULOVICH, N.K., RASKIN, V.I., ORLOVSKAYA, K.I., GODNEV, T.N.: Rol' korotkovolnovoǐ formy protokhlorofillida v protsesse khlorofilloobrazovaniya v etiolirovannȳkh list'yakh lipy. [Role of a short-wave form of protochlorophyllide in chlorophyll formation in etiolated linden leaves.] - Dokl. Akad. Nauk Belorus. SSR *15*: 1038-1040, 1971. [In R.]

9116 - AL-ANI, T.A., BIERHUIZEN, J.F.: Stomatal resistance, transpiration, and relative water content as influenced by soil moisture stress. - Acta bot. neerl. *20*: 318-326, 1971.

9117 - ALBERDA, T.: Potential production of grassland. - In: WAREING, P.F., COOPER, J.P. (ed.): Potential Crop Production. A Case Study. Pp. 159-171. Heinemann Educational Books, London 1971.

9118 - **ALDERFER, R.G., GATES, D.M:** Energy exchange in plant canopies. - Ecology *52*: 855-861, 1971.

9119 - **ALEKPEROVA, M.S.:** DeĬstvie NRV v gerbitsidnykh dozakh na rost, razvitie i fiziologo-biokhimicheskie protsessy khlopchatnika. [Action of petroleum growth substance in herbicide doses on the growth, development, and physiological and biochemical processes of cotton.] - In: NRV v Sel'skom KhozyaĬstve. Pp. 339-342. ELM, Baku 1971. [Chl; in R.]

9120 - **ALIEV, E.A., PIONTKOVSKIĬ, V.I., KADYSH, A.G.:** Vliyanie uglekisloĬ podkormki rastenii v gidroponnykh teplitsakh na urozhaĬ ogurtsov. [Effect of CO_2 supply to plants in hydroponic greenhouses on cucumber yield.] - Fiziol. Biokhim. kul't. Rast. *3*: 430-433, 1971. [Growth analysis; in R.]

9121 - **ALIEV, S.A., RZAEV, N.M.:** Vliyanie udobreniĬ na fotosintez, radiatsionnyĬ rezhim i transpiratsiyu khlopchatnika. [Effect of fertilizers on photosynthesis, radiation regime, and transpiration of cotton.] - Agrokhimiya *1971* (9): 88-92, 1971. [In R.]

9122 - **ALIEVA, S.A., TAGEEVA, S.V., TAIRBEKOV, M.G., KASATKINA, V.S., VAGABOVA, M.E.:** Strukturnoe i funktsional'noe sostoyanie khloroplastov v zavisimosti ot vodnogo rezhima rasteniĬ. [Structural and functional state of chloroplasts in relation to plant water relations.] - Fiziol. Rast. *18*: 494-500, 1971. [Ps, Chl; in R, ab: E.]

9123 - **ALIEVA, S.A., TAIRBEKOV, M.G., KASATKINA, V.S., TAGEEVA, S.V.:** Svyaz' fotofosforilirovaniya s ul'trastrukturnoĬ organizatsieĬ i mekhano-khimicheskimi svoĬstvami khloroplastov. [Relation between photophosphorylation and the ultrastructural organization and mechano-chemical properties of chloroplasts.] - Dokl. Akad. Nauk SSSR *197*: 1189-1192, 1971. [In R.]

9124 - **ALLEN, C.F., GOOD, P.:** Acyl lipids in photosynthetic systems. - In: COLOWICK, S.P., KAPLAN, N.O. (ed.): Methods in Enzymology. Vol. 23. Pp. 523-547. Academic Press, New York - London 1971.

9125 - **ALLEN, H.L.:** Primary productivity, chemo-organotrophy, and nutritional interactions of epiphytic algae and bacteria on macrophytes in the littoral of a lake. - Ecol. Monogr. *41*: 97-127, 1971. [Chl, Car.]

9126 - **ALLEN, L.H. Jr.:** Variations in carbon dioxide concentration over an agricultural field. - Agr. Meteorol. *8*: 5-24, 1971.

9127 - **ALLEN, L.H. Jr., JENSEN, S.E., LEMON, E.R.:** Plant response to carbon dioxide enrichment under field conditions: A simulation. - Science *173*: 256-258, 1971.

9128 - **ALLEN, M.B.:** High-latitude phytoplankton. - Annu. Rev. Ecol. Systematics *2*: 261-276, 1971. [Also production.]

9129 - **ALLISON, J.C.S.:** Analysis of growth and yield of inbred and crossbred maize. - Ann. appl. Biol. *68*: 81-92, 1971.

9130 - **ALSCHER, R.G., HARDY, S.I., CASTELFRANCO, P.A.:** Gas exchange and possible hormonal effects on greening. - Plant Physiol. *47* (Suppl.): 44, 1971.

*9131 - **AL'SHEVSKIĬ, N.G.:** DeĬstvie bornykh i mednykh mikroudobreniĬ na urozhaĬ i biokhimicheskiĬ sostav sakharnoĬ svekly. [Effect of boron and copper microfertilizers on the yield and biochemical composition of sugar beet.] - In: Mikroelementy v Sel'skom KhozyaĬstve i Meditsine. Vol. 3. Pp. 90-96. Naukova Dumka, Kiev 1967. [Chl, Car; in R.]

9132 - **AMBLER, R.P.:** The amino acid sequence of cytochrome *c*-551.5 (cytochrome c_7) from the green photosynthetic bacterium *Chloropseudomonas ethylica*. - FEBS Lett. *18*: 351-353, 1971.

9133 - **AMESZ, J., KRAAN, G.P.B.:** Fluoreszenz und Photochemie in der Photosynthese. - Umschau *71*: 715, 1971.

9134 - **AMESZ, J., VISSER, J.W.M.:** Light-induced shifts in pigment absorption in green, red and blue-green algae. - Biochim. biophys. Acta *234*: 62-69, 1971.

9135 - **AMESZ, J., VISSER, J.W.M., DIRKS, M.P., van den ENGH, G.J.:** Kinetics of plastoquinone and other intermediates of the photosynthetic chain. - In: BRODA, E., LOCKER, A., SPRINGER-LEDERER, H. (ed.): Proceedings of the First European Biophysics Congress. Vol. 4. Pp. 73-77. Verlag Wiener Med. Akad., Vienna 1971.

9136 - AMIR, S., REINHOLD, L.: Interaction between K-deficiency and light in ^{14}C-sucrose translocation in bean plants. - Physiol. Plant. *24*: 226-231, 1971.

*9137 - AMIRDZHANOV, A.G.: Pokazateli struktury vinogradnogo kusta v svyazi s ego produktivnost'yu. [Characteristics of the structure of grape bushes in relation to their productivity.] - Sel'skokhoz. Biol. *2*: 365-371, 1967. [Growth analysis; in R, ab: E.]

9138 - AMIRDZHANOV, A.G.: Opyt polucheniya urozhaya vinograda po zadannoĭ programme. [Grape yield programming.] - Sel'-skokhoz. Biol. *6*: 688-697, 1971. [Growth analysis; in R, ab: E.]

9139 - ANALYTIS, S., KRANZ, J., STUMPF, A.: Eine Methode zur Berechnung der Blattfläche. - Angew. Bot. *45*: 111-114, 1971.

9140 - ANDERSEN, K.S.: Ribosome synthesis in greening primary leaves of bean seedlings (*Phaseolus vulgaris*). - Biochem. Physiol. Pflanzen *162*: 245-264, 1971.

9141 - ANDERSEN, K.S., SMILLIE, R.M., BISHOP, D.G.: Photosynthetic electron transfer in plants containing agranal chloroplasts. - Proc. aust. biochem. Soc. *4*: 23, 1971.

9142 - ANDERSEN, W.R., CRIDDLE, R.S.: Control of ribulose diphosphate carboxylase activity in green plants. - Amer. J.Bot *58*: 477, 1971.

*9143 - ANDERSON, D.E., TOLBERT, N.E.: Phosphoglycolate phosphatase. - In: COLOWICK, S.P., KAPLAN, N.O. (ed.): Methods in Enzymology. Vol. 9. Pp. 646-650. Academic Press, New York - London 1966.

*9144 - ANDERSON, G.C., ZEUTSCHEL, R.P.: Release of dissolved organic matter by marine phytoplankton in coastal and offshore areas of the Northeast Pacific Ocean. - Limnol. Oceanogr. *15*: 402-407, 1970. [Chl, liguid scintillation counting method.]

9145 - ANDERSON, J.M., BOARDMAN, N.K., SPENCER, D.: Phosphorylation by intact bundle sheath chloroplasts from maize. - Biochim. biophys. Acta *245*: 253-258, 1971.

9146 - ANDERSON, J.M., WOO, K.C., BOARDMAN, N.K.: Photochemical properties of mesophyll and bundle sheath chloroplasts from C$_4$ plants. - In: HATCH, M.D., OSMOND, C.B., SLATYER, R.O. (ed.): Photosynthesis and Photorespiration. Pp. 353-360. Wiley-Interscience, New York - London - Sydney - Toronto 1971.

9147 - ANDERSON, J.M., WOO, K.C., BOARDMAN, N.K.: Photochemical systems in mesophyll and bundle sheath chloroplasts of C$_4$ plants. - Biochim. biophys. Acta *245*: 398-408, 1971.

9148 - ANDERSON, L.E.: Chloroplast and cytoplasmic enzymes. II. Pea leaf triose phosphate isomerases. - Biochim. biophys. Acta *235*: 237-244, 1971.

9149 - ANDERSON, L.E.: Chloroplast and cytoplasmic enzymes. III. Pea leaf ribose 5-phosphate isomerases. - Biochim. biophys. Acta *235*: 245-249, 1971.

9150 - ANDERSON, M.C.: Light measurement in photosynthesis research. - In: HATCH, M.D., OSMOND, C.B., SLATYER, R.O. (ed.): Photosynthesis and Photorespiration. Pp. 551-553. Wiley-Interscience, New York - London - Sydney - Toronto 1971.

9151 - ANDERSON, M.C.: Radiation and crop structure. - In: ŠESTÁK, Z., ČATSKÝ, J., JARVIS, P.G. (ed.): Plant Photosynthetic Production: Manual of Methods. Pp. 412-466. Dr. W. Junk N.V. Publ., The Hague 1971.

9152 - ANDERSSON, F.: Methods and preliminary results of estimation of biomass and primary production in a south Swedish mixed deciduous woodland. - In: DUVIGNEAUD, P. (ed.): Productivity of Forest Ecosystems. Pp. 281-288. Unesco, Paris 1971.

9153 - ANDREEVA, R.A., KOMLEVA, V.P.: Vliyanie obrabotki kornevoĭ sistemy tomatov geteroauksinom na sintez khlorofilla v list'yakh i zhelezoporfirinovykh soedineniĭ v kornyakh. [Effect of heteroauxin treatment of tomato root system on the synthesis of chlorophyll in leaves and heme porphyrins in roots.] - Fiziol. Rast. *18*: 209-211, 1971. [In R.]

*B9154 - ANDREEVA, T.F.: Fotosintez i Azotnyĭ Obmen List'ev. [Photosynthesis and Nitrogen Metabolism of Leaves.] - Nauka, Moskva 1969. [In R.]

9155 - ANDREEVA, T.F., AVDEEVA, T.A., VLASOVA, M.P., NGUEN-TCHYU-TCHYOK, NICHIPO-
       ROVICH, A.A.: Vliyanie azotnoge pitaniya rasteniĭ na strukturu i funktsiyu
       fotosinteticheskogo apparata. [Effect of nitrogen nutrition of plants on
       the structure and function of the photosynthetic apparatus.] - Fiziol. Rast.
       18: 701-707, 1971. [In R, ab: E.]

9156 - ANDREWS, T.J., HATCH, M.D.: Activity and properties of ribulosediphosphate
       carboxylase from plants with the C_4-dicarboxylic acid pathway of photosyn-
       thesis. - Phytochemistry 10: 9-15, 1971.

9157 - ANDREWS, T.J., JOHNSON, H.S., SLACK, C.R., HATCH, M.D.: Malic enzyme and
       aminotransferases in relation to 3-phosphoglycerate formation in plants with
       the C_4-dicarboxylic acid pathway of photosynthesis. - Phytochemistry 10:
       2005-2013, 1971.

9158 - ANDREWS, T.J., LORIMER, G.H., TOLBERT, N.E.: Incorporation of molecular oxy-
       gen into glycine and serine during photorespiration in spinach leaves. -
       Biochemistry 10: 4777-4782, 1971.

9159 - ANDRIANOV, V.K., BULYCHEV, A.A., KURELLA, G.A., LITVIN, F.F.: O svyazi foto-
       indutsirovannykh izmeneniĭ potentsiala pokoya s nalichiem khloroplastov v
       kletkakh Nitella. [Connection of the photoinduced changes in resting poten-
       tial with the presence of chloroplasts in Nitella cells.] - Biofizika 15:
       190-191, 1970. [In R, ab: E.]

9160 - ANIKUSHKIN, N.F.: O razvitii khloroplastov semyadoleĭ eli Picea excelsa LINK
       na svetu i v temnote. [Development of cotyledon chloroplasts in Picia excel-
       sa LINK in the light and dark.] - Bot. Zh. 56: 1687-1689, 1971. [In R.]

9161 - ANTOSZEWSKI, R., MIKA, A.: Translocation of some assimilates from the sink
       to the donor in apple tree. - Biol. Plant. 13: 43-49, 1971.

9162 - ANTSUPOVA, L.V.: Sezonnyĭ ritm pigmentnogo sostava planktona severo-zapadnoĭ
       chasti Chernogo morya. [Seasonal rhythm of the pigment composition of plank-
       ton in the northwestern part of the Black Sea.] - Biol. Morya 22: 115-129,
       1971. [In R.]

9163 - D'AOUST, A., BATE, G.C., CANVIN, D.T.: Effect of O_2 concentration and ref-
       erence cell CO_2 content on the calibration of infrared CO_2 gas analyzers.
       - Can. J.Вот. 49: 317-319, 1971.

1964 - AQUINO, O., MAKKINK, G.F.: Influence of photoperiod and water supply on pro-
       duction of seed and dry matter in three varieties of soybean, Glycine max
       (L.) MERR. - Neth. J. agr. Sci. 19: 168-175, 1971. [Ps.]

9165 - ARGYROUDI-AKOYUNOGLOU, J.H., AKOYUNOGLOU, G.: Analysis of chlorophyll and
       other porphyrin "hydrolysates" in the amino acid analyser. - Photosynthetica
       5: 153-159, 1971.

9166 - ARGYROUDI-AKOYUNOGLOU, J.H., FELEKI, Z., AKOYUNOGLOU, G.: Formation of two
       chlorophyll-protein complexes during greening of etiolated bean leaves. -
       Biochem. biophys. Res. Commun. 45: 606-614, 1971.

*9167 - ARKHIPOV, V.N.: Issledovanie fotoprovodimosti plenok khloroplastov. [Photo-
        conductivity of chloroplast layers.] - Nauch. Dokl. vyssh. Shkoly, biol.
        Nauki 13 (3): 135, 1970. [In R.]

*9168 - ARKHIPOV, V.N.: Vliyanie nekotorykh faktorov na elektricheskie svoĭstva ple-
        nok khloroplastov. [Effect of some factors on the electrical properties of
        chloroplast membrane.] - Vestn. mosk. Univ., Ser. 6 - Biol. Pochvoved. 25
        (5): 93-95, 1970. [In R.]

9169 - ARMSTRONG, J.J., SURZYCKI, S.J., MOLL, B., LEVINE, R.P.: Genetic transcrip-
       tion and translation specifying chloroplast components in Chlamydomonas
       reinhardi. - Biochemistry 10: 692-701, 1971. [Chl.]

*9170 - ARNAUTOVA, A.I.: Ob izvlekaemosti khlorofilla a malopolyarnym rastvoritelem
        iz raznovozrastnoĭ khvoi Picea pungens v razlichnye periody goda. [Extract-
        ability of chlorophyll a by a solvent of low polarity from Picea pungens nee-
        dles during the year.] - In: Tezisy IV Nauchnoĭ Konferentsii Molodykh Uchen-
        ykh po Sovremennym Problemam Biologii. Pp. 2-3. Minsk 1970. [In R.]

*9171 - ARNAUTOVA, A.I., MEL'NIKOVA, L.M., KHODASEVICH, E.V., GODNEV, T.N.: K metodike vydeleniya i radiokhimicheskoǐ ochistki pigmentov khvoǐnykh. [Isolation and radiochemical purification of conifer pigments.] - In: Fiziologo-biokhimicheskie Issledovaniya Rasteniǐ. Pp. 204-209. Nauka i Tekhnika, Minsk 1970. [In R.]

9172 - ARNDT, U.: Konzentrationsänderungen bei Blattfarbstoffen unter dem Einfluss von Luftverunreinigungen. Ein Diskussionsbeitrag zur Pigmentanalyse. - Environ. Pollut. 2: 37-48, 1971. [Chl, Car.]

9173 - ARNOLD, W., AZZI, J.: The mechanism of delayed light production by photosynthetic organisms and a new effect of electric fields on chloroplasts. - Photochem. Photobiol. 14: 233-240, 1971.

9174 - ARNOLD, W.A., AZZI, J.: Electric field and chloroplast membranes. - In: MANSON, L.A. (ed.): Biomembranes. Vol. 2. Pp. 189-191. Plenum Press, New York - London 1971.

9175 - ARNON, D.I.: The light reactions of photosynthesis. - Proc. nat. Acad. Sci. U.S.A. 68: 2883-2892, 1971.

9176 - ARNON, D.I., KNAFF, D.B., McSWAIN, B.D., CHAIN, R.K., TSUJIMOTO, H.Y.: Three light reactions and the two photosystems of plant photosynthesis. - Photochem. Photobiol. 14: 397-425, 1971.

9177 - ARNON, D.I., KNAFF, D.B., McSWAIN, B.D., TSUJIMOTO, H.Y., CHAIN, R.K., MALKIN, R., BEARDEN, A.J.: Photosynthetic electron transport and phosphorylation induced by three light reactions in chloroplasts. - In: QUAGLIARIELLO, E., PAPA, S., ROSSI, C.S. (ed.): Energy Transduction in Respiration and Photosynthesis. Pp. 237-268. Adriatica Editrice, Bari 1971.

9178 - ARNTZEN, C.J., DILLEY, R.A., NEUMANN, J.: Localization of phosphorylation and proton transport activities in chloroplasts. - Plant Physiol. 47 (Suppl.): 9, 1971.

9179 - ARNTZEN, C.J., DILLEY, R.A., NEUMANN, J.: Localization of photophosphorylation and proton transport activities in various regions of the chloroplast lamellae. - Biochim. biophys. Acta 245: 409-424, 1971.

9180 - ARNTZEN, C.J., NEUMANN, J., DILLEY, R.A.: Inhibition of electron-transport in chloroplasts by a quinone analogue: evidence for two sites of $DPIPH_2$ oxidation. - J. Bioenerg. 2: 73-83, 1971.

9181 - ARONOFF, S., ELLSWORTH, R.K., WICKLIFF, J.: A metabolic pathway for chlorophyll a in *Chlorella*. - Bot. Rev. 37: 263-293, 1971.

9182 - ARONOFF, S., HOULSON, P.R., ELLSWORTH, R.K.; Investigations on the biogenesis of chlorophyll a. V - Ordering of submutants of ultraviolet chlorophyll mutants of *Chlorella*. - Photosynthetica 5: 166-169, 1971.

9183 - ARONSON, R.B., van SLYKE, D.D.: Manometric determination of CO_2 combined with scintillation counting of C-14. - Anal. Biochem. 41: 173-188, 1971.

9184 - ARTAMKINA, I.Yu., KUTYURIN, V.M.: Vliyanie vody na okislitel'no-vosstanovitel'nye svoǐstva khlorofilla i ego analogov. [Effect of water on redox properties of chlorophyll and its analogues.] - Dokl. Akad. Nauk SSSR 196: 980-983, 1971. [In R.]

*B9185 - ARTSIKHOVSKAYA, N.V.: Fotosintez. Ukazatel' Otechestvennoǐ i Inostrannoǐ Literatury. Tom II. 1958-1962. Chast' 1. [International Bibliography of Photosynthesis. Vol. II. 1958-1962. Part 1.] - Izd. mosk. Univ., Moskva 1968.

*B9186 - ARTSIKHOVSKAYA, N.V.: Fotosintez. Ukazatel' Otechestvennoǐ i Inostrannoǐ Literatury. Tom II. 1958-1962. Chast' 2. [International Bibliography of Photosynthesis. Vol II. 1958-1962. Part 2.] - Izd. mosk. Univ., Moskva 1968.

*B9187 - ARTSIKHOVSKAYA, N.V.: Predmetnyǐ Ukazatel' k Bibliografii "Fotosintez". (K tomam I i II). [Subject Index of the Bibliography "Photosynthesis". (Volumes I and II).] - Izd. mosk. Univ., Moskva 1970.

*9188 - ARUGA, Y.: [Technical problems for measuring primary production in the sea and inland waters and the data reported from various areas.] - Bull. Plankton Soc. Jap. 15: 19-22, 1968. [In Jap., ab: E.]

*9189 - ARUGA, Y., ICHIMURA, S.: Characteristics of photosynthesis of phytoplankton and primary production in the Kuroshio. - Bull. Misaki mar. biol. Inst. Kyoto Univ. *1968* (12): 3-20, 1968.

9190 - ASADA, K., TAKAHASHI, M.: Effect of potassium chloride on photosystem II of spinach chloroplasts. - In: HATCH, M.D., OSMOND, C.B., SLATYER, R.O. (ed.): Photosynthesis and Photorespiration. Pp. 387-393. Wiley-Interscience, New York - London - Sydney - Toronto 1971.

9191 - ASADA, K., TAKAHASHI, M.: Purification and properties of cytochrome *c* and two peroxidases from spinach leaves. - Plant Cell Physiol. *12*: 361-375, 1971.

9192 - ASADA, K., TAKAHASHI, M.: Conservation of electron transport and energy transfer reactions of spinach chloroplasts in glycerol. - Plant Cell Physiol. *12*: 709-715, 1971.

*9193 - ASANA, R.D., CHATTOPADHYAY, N.C.: Relation of some growth attributes of tall and dwarf wheat (*Triticum aestivum* L.) varieties to their productivity. - Indian J. agr. Sci. *40*: 309-317, 1970. [Growth analysis.]

9194 - ASEEVA, I.V., PINEVICH, T.G., DOBROVOL'SKAYA, T.G.: Pigmenty kokkovykh form mikroorganizmov iz litofil'nykh lishanTnikov. [Pigments of coccous forms of microorganisms from lithophilic lichens.} - Nauch. Dokl. vyssh. Shkoly, biol. Nauki *14* (5): 91-95, 1971. [Car; in R.]

9195 - ASHOUR, N.I.: Comparative studies on the effect of low supply of nitrogen or phosphorus on photosynthetic pigments and nitrogen contents in sunflower plants. - Advanc. Front. Plant Sci. *28*: 285-289, 1971.

9196 - ASHOUR, N.I., THALOOTH, A.T.: Effect of putrescine on growth and photosynthetic pigments of broad bean plants grown under chloride salinization conditions. - Biochem. Physiol. Pflanzen *162*: 203-208, 1971.

9197 - ASLAM, M.: Nitrate assimilation in barley and corn: The interaction of respiration and photosynthesis in the induction of nitrate reductase activity. - Diss. Abstr. Int. B *32*: 659-B - 660-B, 1971.

9198 - ASROROV, K.A.: Koeffitsient ispol´zovaniya sveta posevom khlopchatnika v zavisimosti ot ob"ema plodorodnogo sloya pochvy. [Coefficient of radiant energy utilization by cotton crop in relation to the volume of fertile soil layer.] - In: ZALENSKIĬ, O.V. (ed.): Fotosintez i Ispol'zovanie SolnechnoĬ Energii. Pp. 39-44. Nauka, Leningrad 1971. [In R, ab: E.]

9199 - ATANASIU, L.: Photosynthesis and respiration in some lichens in relation to winter low temperatures. - Rev. roum. Biol. - Sér. Bot. *16*: 105-110, 1971.

9200 - ATANASIU, L.: Transformări de energie în fotosinteză. [Energy changes in photosynthesis.] - Natura (Bucureşti) *23* (6): 3-10, 1971. [In Roum.]

9201 - ATAULLAEV, N.A., KHODZHAEV, Sh., ABIDOV, A., RAKHIMOVA, R.: Vliyanie NRV na izmenenie fiziologo-biokhimicheskikh protsessov v rasteniyakh v usloviyakh Uzbekistana. [Effect of petroleum growth substance on a change in physiological and biochemical processes in plants in Uzbekistan.] - In: NRV v Sel'-kom KhozyaĬstve. Pp. 363-380. ELM, Baku 1971. [Chl; in R.]

9202 - ATKINS, C.A., CANVIN, D.T.: Analysis of ^{14}C-labeled acidic photosynthetic products by ion-exchange chromatography. - Photosynthetica *5*: 341-351, 1971.

9203 - ATKINS, C.A., CANVIN, D.T.: Photosynthesis and CO_2 evolution by leaf discs: gas exchange, extraction, and ion-exchange fractionation of ^{14}C-labeled photosynthetic products. - Can. J. Bot. *49*: 1225-1234, 1971.

9204 - ATKINS, C.A., CANVIN, D.T., FOCK, H.: Intermediary metabolism of photosynthesis in relation to carbon dioxide evolution in sunflower. - In: HATCH, M.D., OSMOND, C.B., SLATYER, R.O. (ed.): Photosynthesis and Photorespiration. Pp. 497-505. Wiley-Interscience, New York - London - Sydney - Toronto 1971.

9205 - ATKINS, C.A., GRAHAM, D.: Light-induced pH changes by cells of *Chlamydomonas reinhardii*: Dependence on CO_2 uptake. - Biochim. biophys. Acta *226*: 481-485, 1971.

9206 - ATKINS, C.A., PATTERSON, B.D., GRAHAM, D.: Isozymes of carbonic anhydrase in plants. - Proc. aust. biochem. Soc. *4*: 10, 1971.

9207 - ATTILA, A.: Dünnschichtchromatographische Untersuchungen über die Farbverän-

derungen an grünen Erbsen während der Konservierung. - Ind. Obst-Gemüsever-
wertung *56*: 335-337, 1971. [Chl.]

9208 - AVADHANI, P.N., OSMOND, C.B., TAN, K.K.: Crassulacean acid metabolism and
the C_4 pathway of photosynthesis in succulent plants. - In: HATCH, M.D.,
OSMOND, C.B., SLATYER, R.O. (ed.): Photosynthesis and Photorespiration. Pp.
288-293. Wiley-Interscience, New York - London - Sydney - Toronto 1971.

9209 - AVRON, M.: Biochemistry of photophosphorylation. - In: GIBBS, M. (ed.):
Structure and Function of Chloroplasts. Pp. 149-167. Springer-Verlag, Berlin -
Heidelberg - New York 1971.

9210 - AVRON, M., SHNEYOUR, A.: On the site of action of plastocyanin in isolated
chloroplasts. - Biochim. biophys. Acta *226*: 498-500, 1971.

*9211 - AXELSSON, B., GÄRDEFORS, D., LOHM, U., PERSSON, T., TENOW, O., WALLIN, L.:
Components of variance and the cost of a sampling programme concerning biomass of hazel *Corylus avellana* L. available to leaf-eating insects. - Oikos
21: 203-207, 1970.

*9212 - AZMI, A.R., OMROD, D.P.: Carbon dioxide compensation values in some members
of the genus *Oryza*. - Pak. J. Bot. *2*: 21-24, 1970.

9213 - AZMI, A.R., ORMROD, D.P.: Rates of net carbon dioxide assimilation in rice
as influenced by growth temperature and photoperiod. - Agron. J. *63*: 543-
546, 1971.

9214 - BABADZHANOVA, M.A., DOMAN, N.G., CHEKINA, N.G.: Vydelenie i ochistka preparatov karboksilazy ribulezodifosfata iz list'ev gorokha i fasoli v atmos-
fere inertnykh gazov. [Isolation and purification of ribulose diphosphate
carboxylase preparations from bean and pea leaves in the atmosphere of
inert gas.] - In: ZALENSKII, O.V. (ed.): Fotosintez i Ispol'zovanie Solnechnoi Energii. Pp. 222-226. Nauka, Leningrad 1971. [In R, ab: E.]

*9215 - BABADZHANOVA, M.A., GORENKOVA, L.G., KHAITOVA, L.T.: Fiksatsiya $C^{14}O_2$ fer-
mentnymi preparatami, vydelennymi iz list'ev razlichnykh mutantov gorokha i
rezushki Talya (*Arabidopsis thaliana*). [$^{14}CO_2$ fixation by enzymatic preparations isolated from leaves of various mutants of pea and *Arabidopsis thaliana*.] - In: Tezisy Dokladov Vtorogo Vsesoyuznogo Biokhimicheskogo S"ezda,
Sektsiya "Problemy Fotosinteza". P. 49. FAN, Tashkent 1969. [In R.]

9216 - BABADZHANOVA, M.A., KHAITOVA, L.T.: Aktivnost' fermentov karboksilirovaniya
i potentsial'naya intensivnost' fotosinteza u auksotrofnogo mutanta *Arabidopsis thaliana* (L) HEYNH. [Activity of carboxylating enzymes and potential
photosynthetic rate in auxotrophic mutant *Arabidopsis thaliana* (L.) HEYNH.]
- In: NASYROV, Yu.S. (ed.): Geneticheskie Aspekty Fotosinteza. Pp. 133-142.
Donish, Dushanbe 1971. [In R, ab: E.]

9217 - BABADZHANOVA, M.A., KHAITOVA, L.T., GORENKOVA, L.G.: Potentsial'naya inten-
sivnost´ fotosinteza i aktivnost' fermentov karboksilirovaniya u iskhodnykh
i mutantnykh form rasteniT. [Potential photosynthetic rate and activity of
carboxylating enzymes in normal and mutant forms of plants.] - Dokl. Akad.
Nauk Tadzh. SSR *14*: (4): 74-77, 1971. [In R, ab: Tadzh.]

9218 - BABADZHANOVA, M.A., KHAITOVA, L.T., KAS'YANENKO, A.G.: Detstvie leTtsina na
potentsial'nuyu intensivnost' fotosinteza i aktivnost' fermentov, otvetst-
vennykh za fiksatsiyu CO_2 u iskhodnoT i mutantnoT form *Arabidopsis thaliana*
(L.) HEYNH. [Effect of leucine on potential photosynthetic rate and activity
of enzymes of CO_2 fixation in normal and mutant forms of *Arabidopsis thaliana*
(L.) HEYNH.] - Dokl. Akad. Nauk Tadzh. SSR *14* (5): 50-52, 1971. [In R, ab:
Tadzh.]

9219 - BABADZHANOVA, M.A., KHAITOVA, L.T., NASYROV, Yu.S.: Vliyanie kinetina na
fiksatsiyu $C^{14}O_2$ beskletochnymi preparatami *Arabidopsis thaliana* (L.)
HEYNH. [Effect of kinetin on $^{14}CO_2$ fixation by cell-free preparations of
Arabidopsis thaliana (L.) HEYNH.] - Dokl. Akad. Nauk Tadzh. SSR *14* (8): 62-
64, 1971. [In R, ab: Tadzh.]

9220 - BABAEVA, T.N.: Zavisimost' intensivnosti fotosinteza mutantov *Arabidopsis
thaliana* ot kontsentratsii uglekisloty. [Dependence of photosynthetic rate
of *Arabidopsis thaliana* mutants on CO_2 concentration.] - In: ZALENSKII, O.V.

(ed.): Fotosintez i Ispol´zovanie SolnechnoĬ Energii. Pp. 254-257. Nauka, Leningrad 1971. [In R, ab: E.]

9221 - BABAEVA, T.N., NIKITINA, A.N.: VlIyanie kinetina na fotosinteticheskiĬ apparat mutantov gorokha. [Effect of kinetin on the photosynthetic apparatus of pea mutants.] - In: NASYROV, Yu.S. (ed.): Geneticheskie Aspekty Fotosinteza. Pp. 195-206. Donish, Dushanbe 1971. [In R, ab: E.]

9222 - BABUK, V.I.: Vliyanie udobreniĬ na ploshchad' listovoĬ poverkhnosti, intensivnost' nakopleniya sukhogo veshchestva v list'yakh i soderzhanie uglevodov v otvodkakh i kornyakh matochnykh rasteniĬ podvoya yabloni EM IV. [Effect of fertilizers on leaf area, rate of dry matter accumulation in leaves and carbohydrate content in root scions and roots of the mother plants of apple rootstock EM IV.] - Tr. kishinev. sel'skokhoz. Inst. *69* (Plodovodstvo): 40-47, 1970. [In R.]

*9223 - BABUSHKIN, L.N.: PolevoĬ pribor dlya kolichestvennykh sravnitel´nykh izmereniĬ intensivnosti fotosinteza. [Field apparatus for quantitative comparative determinations of photosynthetic rate.] - In: Tezisy Dokladov Vsesoyuznogo Soveshchaniya po Unifikatsii Metodov i Priborov dlya Massovykh IzmereniĬ Intensivnosti Fotosinteza. Pp. 3-5. Nauch. issled. Inst. Rastenievod. N.I. Vavilova, Leningrad - Pushkin 1970. [In R.]

9224 - BACCARINI-MELANDRI, A., MELANDRI, B.A.: Partial resolution of the photophosphorylating system of *Rhodopseudomonas capsulata*. - In: COLOWICK, S.P., KAPLAN, N.O. (ed.): Methods in Enzymology. Vol. 23. Pp. 556-561. Academic Press, New York - London 1971.

9225 - BACHOFEN, R.: Die Wirkung von Makrotetralidantibiotika auf photosynthetische Reaktionen in Spinatchloroplasten. - Experientia *27*: 770-772, 1971.

9226 - BACK, A., RICHMOND, A.E.: Interrelations between gibberellic acid, cytokinins and abscisic acid in retarding leaf senescence. - Physiol. Plant. *24*: 76-79, 1971. [Chl.]

9227 - BADDELEY, M.S.: 1. Biochemical aspects of senescence. - In: PREECE, T.F., DICKINSON, C.H. (ed.): Ecology of Leaf Surface Micro-organisms. Pp. 415-429. Academic Press, London - New York 1971. [Chl, chloroplast.]

9228 - BAGAUTDINOVA, R.I.: Zavisimost' mezhdu intensivnost'yu fotosinteza, raspredeleniem assimilyatov i produktivnost'yu u sortov soi i kartofelya. [Photosynthetic rate, photosynthate distribution and productivity in soybean and potato cultivars.] - In: ZALENSKIĬ, O.V. (ed.): Fotosintez i Ispol'zovanie SolnechnoĬ Energii. Pp. 116-122. Nauka, Leningrad 1971. [In R, ab: E.]

9229 - BAGAUTDINOVA, R.I., BORZENKOVA, R.A.: Napravlennost' pervichnogo sinteza uglevodov u kartofelya. [Direction of primary carbohydrate synthesis in potato.] - Nauch. Dokl. vyssh. Shkoly, biol. Nauki *14* (2): 64-69, 1971. [In R.]

9230 - BAGGE, P., LEHMUSLUOTO, P.O.: Phytoplankton primary production in some Finnish coastal areas in relation to pollution. - Merentutkimuslait. Julk./ Havsforskningsinst. Skr. *235*: 3-18, 1971.

9231 - BAHL, J.: Sur la préparation de suspensions de chloroplastes isolés de *Spirodela polyrrhiza* (L) SCHLEIDEN. - Compt. rend. Acad. Sci. Paris, Sér. D *272*: 2185-2187, 1971. [Chl.]

9232 - BAHL, J., LECHEVALIER, D., MONÉGER, R.: Sur les lipides chloroplastiques des frondes de la *Spirodela polyrrhiza* (L) SCHLEIDEN cultivées sur milieu minéral et sur milieu saccharosé. - Compt. rend. Acad. Sci. Paris, Sér. D *272*: 2320-2323, 1971. [Chl.]

9233 - BAILEY, J.L., DOWNTON, W.J.S., MÄSIAR, E.: The proteins of photosystems I and II in mesophyll and bundle sheath chloroplasts of *Sorghum bicolor*. - In: HATCH, M.D., OSMOND, C.B., SLATYER, R.O. (ed.): Photosynthesis and Photorespiration. Pp. 382-386. Wiley-Interscience, New York - London - Sydney - Toronto 1971.

*9234 - BAKER, A.L., TOLBERT, N.E.: Glycolate oxidase (ferredoxin-containing form). - In: COLOWICK, S.P., KAPLAN, N.O. (ed.): Methods in Enzymology. Vol. 9. Pp. 338-342. Academic Press, New York - London 1966.

*9235 - BAKER, A.L., TOLBERT, N.E.: Purification and some properties of an alternate form of glycolate oxidase. - Biochim. biophys. Acta 131: 179-187, 1967.

*9236 - BALCEREK, W.: Metoda do szybkiego i seryjnego oznaczania karotenów w zielonkach i suszach roślin pastewnych. [Method for rapid and serial determination of carotene in fresh and dried feed crops.] - Zesz. nauk. Wyzsz. Szk. roln. Szczecin 30: 3-23, 1969. [In Pol., ab: E.]

9237 - BALDRY, C.W., BUCKE, C., COOMBS, J.: Progressive release of carboxylating enzymes during mechanical grinding of sugar cane leaves. - Planta 97: 310-319, 1971.

9238 - BALDY, C.M.: Analyse de la photosynthèse du maïs dans les condition naturelles par une méthode gravimétrique. - Oecol. Plant. 6: 101-113, 1971.

9239 - BALDY, C.M., LE BUHAN, J.-P.: Répartition de la photosynthèse nette dans les feuilles de tabac. - Photosynthetica 5: 421-423, 1971.

9240 - BALLARD, R.: Interrelationships between site factors and productivity of radiata pine at Riverhead Forest, New Zealand. - Plant Soil 35: 371-380, 1971.

9241 - BALTSCHEFFSKY, H., BALTSCHEFFSKY, M., THORE, A.: Energy conversion reactions in bacterial photosynthesis. - In: SANADI, D.R. (ed.): Current Topics in Bioenergetics. Vol. 4. Pp. 273-325. Academic Press, New York - London 1971.

9242 - BALTSCHEFFSKY, M.: Carotenoids as endogenous indicators of the energized state in chromatophores. - In: QUAGLIARIELLO, E., PAPA, S., ROSSI, C.S. (ed.): Energy Transduction in Respiration and Photosynthesis. Pp. 639-648. Adratica Editrice, Bari 1971.

*9243 - BANKIN, M.P.: Opredlenie sutochnoĭ produktivnosti kartofelya kombinirovannym metodom. [Determination of diurnal productivity in potato by a combined method.] - In: Tezisy Dokladov Vsesoyuznogo Soveshchaniya po Unifikatsii Metodov i Priborov dlya Massovykh Izmereniĭ Intensivnosti Fotosinteza. Pp. 6-10. Nauch. Issled. Inst. Rastenievod. N.I. Vavilova, Leningrad - Pushkin 1970. [In R.]

9244 - BANKIN, M.P.: Nepreryvnoe opredelenie gazoobmena kartofel'nogo rasteniya v estestvennykh usloviyakh. [Continuous determination of gas exchange of the potato plant in natural conditions.] - Vestn. Leningrad. Univ., Ser. biol. 26 (1): 154-155, 1971. [In R, ab: E.]

9245 - BAR-AKIVA, A.: Functional aspects of mineral nutrients in use for the evaluation of plant nutrient requirement. - In: SAMISH, R.M. (ed.): Recent Advances in Plant Nutrition. Vol. 1. Pp. 115-142. Gordon and Breach Sci. Publ., New York - London - Paris 1971 [Chl.]

*9246 - BAR-AKIVA, A., LAVON, R.: Carbonic anhydrase activity as an indicator of zinc deficiency in citrus leaves. - J. hort. Sci. 44: 359 - 362, 1969.

*9247 - BARALDI, D.: Effect of gamma irradiation on carotenoid synthesis and colour development in tomatoes. - Agrochimica, Atti Simp. int. 8: 556 - 562, 1971.

9248 - BARALDI, D.: Separation of pigments extracted from tomato by thin layer chromatography. - Agrochimica 15: 371 - 376, 1971.

*9249 - BARASHKOV, B.I., BORISOV, A.Yu., GODIK, V.I., SAMUILOV, V.D., CHIBISOV, A. K.: Uchastie tripletnykh sostoyaniĭ v bakterial'nom fotosinteze. [Participation of triplet states in bacterial photosynthesis.] - Nach. Dokl. vyssh. Shkoly, biol. Nauki 13 (3): 135-136, 1970. [In R.]

9250 - BARBER, J., VARLEY, W.J.: Millisecond delayed light as an indicator of electrical gradients across chloroplast thylakoids. - Nature - new Biol. 234: 188 - 189, 1971.

*9251 - BARBOÏ, N.I., DILUNG, I.I.: Issledovanie fotokhimicheskogo okisleniya khlorofilla i ryada ego proizvodnykh elektrometricheskimi metodami. [Electrometric study of photochemical oxidation of chlorophyll and of some of its derivatives.] - Biofizika 14: 980-985, 1969. [In R, ab: E.]

*9252 - BARBOÏ, N.I., DILUNG, I.I.: Issledovanie prirody fotogal'vanicheskogo effekta, voznikayushchego pri fotoreaktsii khlorofilla s nitrosoedineniyami. [The nature of photogalvanic effect initiated in the photoreaction of chlo-

rophyll with nitro compounds]   - Biofizika *15*: 608-611, 1970. [In R, ab:
E.]

9253 - BARDEN, J.A.: Factors affecting the determination of net photosynthesis of
apple leaves. - HortScience *6*: 448-451, 1971.

9254 - BARR, R., CRANE, F.L.: Quinones in algae and higher plants. - In: COLOWICK,
S.P., KAPLAN, N.O. (ed.): Methods in Enzymology. Vol. 23. Pp. 372-408. Academic Press, New York - London 1971.

*9255 - BARR, R., HALL, J.D., CRANE, F.L., AL-ABBAS, H.: Lipophilic quinones in mi-
neral-deficient maize leaves. - Proc. Indiana Acad. Sci. *80*: 130-139, 1970.
[Chl.]

9256 - de BARREIRO, O.C.: Sulphydryl groups and the *in vitro* enzymatic synthesis
of 5-amino-laevulinic acid and porphobilinogen in *Rhodopseudomonas spheroides*. - FEBS Lett. *17*: 257-260, 1971.

9257 - BARRETT, J., JEFFREY, S.W.: A note on the occurrence of chlorophyllase in
marine algae. - J. exp. mar. Biol. Ecol. *7*: 255-262, 1971.

*9258 - BARSKIĬ, E.A., NAZARENKO, A.V., SAMUILOV, V.D.: Zavisimost'fiksatsii C^{14}-
bikarbonata ekstraktami *Rhodospirillum rubrum* ot istochnikov energii. [De-
pendence of fixation of ^{14}C-bicarbonate by *Rhodospirillum rubrum* extracts
on energy sources.] - Nauch. Dokl. vyssh. Shkoly, biol. Nauki *13* (3): 141,
1970. [In R.]

9259 - BARSKY, E.L., BORISOV, A.Yu.: Determination of the quantum yields of the
primary photosynthesis events and the photosynthetic unit types in purple
bacteria. - J. Bioenerg. *2*: 275-281, 1971.

*9260 - BARTA, A.L.: Characterization of photosynthesis in hardening winter wheat.
- Diss. Abstr. Int. B *30*: 4876-B, 1970.

9261 - BARTELS, F.: Strukturelle Veränderungen der Chloroplasten-Thylakoide in
Palisadenzellen von *Peperomia metallica* im Licht-Dunkelwechsel.. - Protoplasma *72*: 27-41, 1971.

9262 - BARTSCH, R.G.: Cytochromes: bacterial. - In: COLOWICK, S.P., KAPLAN, N.O.
(ed.): Methods in Enzymology. Vol. 23. Pp. 344-363. Academic Press, New
York - London 1971.

9263 - BARTSCH, R.G.: High potential iron proteins: bacterial. - In: COLOWICK,
S.P., KAPLAN, N.O. (ed.): Methods in Enzymology. Vol. 23. Pp. 644-649.
Academic Press, New York - London 1971.

9264 - BARTSCH, R.G., KAKUNO, T., HORIO, T., KAMEN, M.D.: Preparation and properties of *Rhodospirillum rubrum* cytochromes c_2, cc', and $b_{557.5}$, and
flavin mononucleotide protein. - J. biol. Chem. *246*: 4489 - 4496, 1971.

*9265 - BARUFFINI, A., BORGNA, P., PAGANI, G.: Acidi succinanilici e derivati.
[Succinanilic acids and derivatives.] - Farmaco, Ed. Sci. *22*: 781-795,
1967. [Ps; in Ital, ab: E.]

9266 - BASSHAM, J.A.: Photosynthetic carbon metabolism. - Proc. nat. Acad. Sci.
U.S.A. *68*: 2877-2882, 1971.

9267 - BASSHAM, J.A.: Regulation of metabolite transport through the outer membrane of photosynthesizing isolated spinach chloroplasts. - In: QUAGLIA-
RIELLO, E., PAPA, S., ROSSI, C.S. (ed.): Energy Transduction in Respiration and Photosynthesis. Pp. 869-882. Adriatica Editrice, Bari 1971.

9268 - BASSHAM, J.A.: The control of photosynthetic carbon metabolism. - Science
172: 526-534, 1971.

9269 - BASZYŃSKI, T.: Distribution of plastid quinones and pigments along the
leaf in different stages of plastid development of *Zea mays* L. seedlings.
- Ann. Univ. Mariae Curie-Skłodowska, Sect. C *26*: 187-198, 1971.

9270 - BASZYNSKI, I., BRAND, J., KROGMANN, D.W., CRANE, F.L.: Plastocyanin participation in chloroplast Photosystem I. - Biochim. biophys. Acta *234*: 537-
540, 1971.

9271 - BASZYŃSKI, T., DUDZIAK, B., ARNOLD, D.: Light-stimulated synthesis of plastid benzoquinones and pigments in cells of *Euglena gracilis* in absence of

photosynthesis. - Ann. Univ. Mariae Curie-Skłodowska, Sect. C *26*: 199-206, 1971.

9272 - BATALOV, R.B., KVITKO, K.V.: Induktsirovanie letal'nykh mutatsiĭ i ihk ge-netisheskiĭ analiz u *Arabidopsis thaliana*. [Induction of lethal mutations of *Arabidopsis thaliana* and their genetic analysis]. - In: ZALENSKIĬ, O.V. (ed.): Fotosintez i Ispol'zovanie Solnechnoĭ Energii. Pp. 257-260. Nauka, Leningrad 1971. [In R, ab: E.]

9273 - BATE, G.C., CANVIN, D.T.: A gas-exchange system for measuring the produc-tivity of plant populations in controlled environments. - Can. J. Bot. *49*: 601-608, 1971.

9274 - BATE, G.C., CANVIN, D.T.: The effect of some environmental factors on the growth of young aspen trees (*Populus tremuloides*) in controlled environments. - Can. J. Bot. *49*: 1443-1453, 1971. [Ps.]

9275 - BAUDOUIN, M.F., SCOPPA, P.: Fluorometric determination of chlorophyll *a* in the presence of pheopigments: effect of the half-value width of the excita-tion beam. - Mar. Biol. *10*: 66-69, 1971.

9276 - BAUER, A., SCHAEFER, C., ERISMANN, K.H.: Zur Frage der Stickstoffinkorpora-tion in Aminosäuren und Proteine bei *Lemna minor* unter Photosynthesebedin-gungen. - Verh. Schweiz. naturforsch. Ges. *151*: 107-109, 1971.

9277 - BAUERNFEIND, J.C., BRUBACHER, G.B., KLÄUI, H.M., MARUSICH, W.L.: Use of carotenoids. - In: ISLER, O. (ed.): Carotenoids. Pp. 743-770. Birkhäuser Verlag, Basel - Stuttgart 1971.

9278 - BAUMAN, A.J., BOETTGER, H.G., KELLY, A.M., CAMERON, R.E., YOKOYAMA, H.: Isolation and characterization of keto-carotenoids from the neutral extract of algal mat communities of a desert soil. - Europe. J. Biochem. *22*: 287-293, 1971.

*9279 - BAVRINA, T.V.: Izmenenie v soderzhanii plastidnykh pigmentov v svyazi s fotoperiodizmom i ontogenezom rasteniĭ. [Changes in content of plastid pigments in relation to photoperiod and plant ontogenesis.] - In: Onto-genez Vysshikh Rasteniĭ. Pp. 141-147. Izdat. Akad. Nauk Arm. SSR, Erevan 1970. [In R.]

*9280 - BAVRINA, T.V.: Vliyanie dliny dnya na fotokhimicheskuyu aktivnost' khloro-plastov list'ev rasteniĭ razlichnykh fotoperiodicheskikh grupp. [Effect of daylength on the photochemical activity of chloroplasts from plants of dif-ferent photoperiodical groups.] - Dokl. Akad. Nauk SSSR *192*: 451-454, 1970. [In R.]

9281 - BAZZAZ, F.A., BLISS, L.C.: Net primary production of herbs in a central Illinois deciduous forest. - Bull. Torrey bot. Club *98*: 90-94, 1971.

9282 - BAZZAZ, F.A., DUSEK, D.: Photosynthesis, respiration, transpiration, and Δ_9THC content of tropical and temperate populations of *Cannabis sativa*. - Amer. J. Bot. *58*: 462, 1971.

*9283 - BAZZAZ, F.A., PAOLILLO, D.J. Jr., JAGELS, R.H.: Photosynthesis and respi-ration of forest and alpine populations of *Polytrichum juniperinum*. - Bryologist *73*: 579-585, 1970.

9284 - BAZZAZ, M., GOVINDJEE: Studies on the photocemical activities and spectral characteristics of bundle sheath and mesophyll chloroplasts in *Zea mays*. - Plant Physiol. *47* (Suppl.): 32, 1971.

9285 - BEALE, S.I.: Studies on the biosynthesis and metabolism of δ-aminolevulinic acid in *Chlorella*. - Plant Physiol. *48*: 316-319, 1971. [Chl.]

9286 - BEALE, S.I., APPLEMAN, D.: Chlorophyll synthesis in *Chlorella*: Regulation by degree of light limitation of growth. - Plant Physiol. *47*: 230-235, 1971.

9287 - BEARD, B.H.: Chlorophyll mutation from recurrent X-irradiation of flax seed. - Crop Sci. *11*: 317-319, 1971.

9288 - BECK, E., STRANSKY, H., FÜRBRINGER, M.: Synthesis of hamamelose-diphospha-te by isolated spinach chloroplasts. - FEBS Lett. *13*: 229-234, 1971.

9289 - BEEVERS, H.: Comparative biochemistry of microbodies (glyoxysomes, peroxisomes). - In: HATCH, M.D., OSMOND, C.B., SLATYER, R.O. (ed.): Photosynthesis and Photorespiration. Pp. 483-493. Wiley-Interscience, New York - London - Sydney - Toronto 1971.

9290 - BEEVERS, H.: Photorespiration: assessment. - In: HATCH, M.D., OSMOND, C.B., SLATYER, R.O. (ed.): Photosynthesis and Photorespiration. Pp. 541-543. Wiley-Interscience, New York - London - Sydney - Toronto 1971.

9291 - BEHNKE, H.-D.: Zum Feinbau der Siebröhren-Plastiden von *Aristolochia* und *Asarum* (*Aristolochiaceae*). - Planta *97*: 62-69, 1971.

9292 - BEHNKE, H.-D.: Über den Feinbau verdickter (nacré) Wände und der Plastiden in den Siebröhren von *Annona* und *Myristica*. - Protoplasma *72*: 69-78, 1971.

9293 - BEKINA, R.M., KRASNOVSKIĬ, A.A.: Khranenie izolirovannykh fotosinteziruyushchikh struktur bez izmeneniya ikh osnovnykh fotokhimicheskikh funktsiĭ. [Storage of isolated photosynthetically active structures without changes of their basic photochemical functions.] - In: ZALENSKIĬ, O.V. (ed.): Fotosintez i ispol'zovanie SolnechnoĬ Energii. Pp. 145-149. Nauka, Leningrad 1971. [In R, ab: E.]

*9294 - BEKOV, S.M.: Effektivnost' primeneniya gerbitsidov na oroshaemykh vinogradnikakh Checheno-Ingushetii. [Effectivity of herbicide use on the irrigated vine-yards of Checheno-Ingushetium.] - In: Primenenie Pestitsidov v Sel'skom KhozyaĬstve. Pp. 65-70. Min. sel'. Khoz. SSSR; mosk. sel´.-khoz. Akad. Timiryazeva, Moskva 1970. [Ps; in R.]

9295 - BELL, J.F.: Sampling with unequal probabilities to estimate cubic-foot volume growth for immature Douglas-fir on permanent sample plots. - Diss. Abstr. int. B *31*: 7018-B-7019-B, 1971.

9296 - BELL, L.N.: Primenenie fotokalorimetrii dlya issledovaniya energetiki fotosinteza. [Use of photocalorimetry for studying energetics of photosynthesis.] - In: Biofizicheskie Metody v Fiziologii RasteniĬ. Pp. 106-128. Nauka, Moskva 1971. [In R.]

9297 - BELL, L.N., SHUVALOVA, N.P.: Blue radiant energy storage uncoupled from oxygen evolution. - Photosynthetica *5*: 113-123, 1971.

9298 - BELLIN, J.S., GERGEL, A.: Photosensitized degradation of *C*-phycocyanin. - Photochem. Photobiol. *13*: 399-409, 1971.

*9299 - BELOUS, A.I., KARTASHEV, L.N.: Kompleksnoe ispol'zovanie stimulyatorov i ingibitorov rosta pri vnekornevoĬ podkormke sakharnoĬ svekly. [Complex use of growth stimulators and inhibitors during the spray dressing of sugar beets.] - Tr. khar'kov. sel'skokhoz. Inst. *90* (Issledovaniya po Fiziologii i Biokhimii RasteniĬ) :16-20, 1970. [Ps, Chl; in R.]

9300 - BELOZEROVA, L.S., LUKASHOVA, G.N.: DeĬstvie ortofenantrolina na fotosintez i prevrashchenie organicheskikh kislot v list'yakh *Bryophyllum daigremontianum* na svetu. [Effect of *o*-phenanthroline on photosynthesis and organic acid metabolism in leaves of *Bryophyllum daigremontianum* exposed to light.] - Vestn. leningrad. Univ., Ser. biol. *21*: 108-112, 1971. [In R, ab: E.]

9301 - BELYAKOV, V.A., EVSTIGNEEV, V.B.: Khemilyuminestsentsiya pri reaktsiyakh vosstanovleniya fotookislennogo khlorofilla. [Chemiluminescence in the reduction reactions of photooxidized ichlorophyll.] - Biofizika *16*: 544-546, 1971. [In R, ab: E.]

9302 - BEN AZIZ, A., GROSSMAN, S., ASCARELLI, I., BUDOWSKI, P.: Carotene-bleaching activities of lipoxygenase and heme proteins as studied by a direct spectrophotometric method. - Phytochemistry *10*: 1445-1452, 1971.

9303 - BEN-AZIZ, A., GROSSMAN, S., BUDOWSKI, P., ASCARELLI, I.: Enzymic oxidation of carotene and linoleate by alfalfa: Properties of active fractions. - Phytochemistry *10*: 1823-1829, 1971.

9304 - BENDALL, D.S., DAVENPORT, H.E., HILL, R.: Cytochrome components in chloroplasts of the higher plants. - In: COLOWICK, S.P., KAPLAN, N.O. (ed.): Methods in Enzymology. Vol. 23. Pp. 327-344. Academic Press, New York - London 1971.

9305 - BENDALL, D.S., SOFROVÁ, D.: Reactions at 77 ^{0}K in Photosystem 2 of green plants. - Biochem. biophys. Acta *234*: 371-380, 1971.

9306   BENDER, M.M.: Variations in the $^{13}C/^{12}C$ ratios of plants in relation to the pathway of photosynthetic carbon dioxide fixation. - Phytochemistry *10*: 1239-1244, 1971.

9307 - BENEDICT, C.R., KETRING, D.L.: The genetic control of chloroplast development. - Plant Physiol. *47* (Suppl.): 44, 1971.

9308 - BEN-HAYYIM, G., AVRON, M.: Light distribution and electron donation in the Z scheme. - Photochem. Photobiol. *14*: 389-396, 1971.

9309 - BENNETT, A., BOGORAD, L.: Properties of subunits and aggregates of blue-green algal biliproteins. - Biochemistry *10*: 3625-3634, 1971.

9310 - BENNETT, A., BOGORAD, L.: Studies on the structure of *c*-phycoerythrin. - Plant Physiol. *47* (Suppl.): 46, 1971.

9311 - BENNETT, J., SCOTT, K.J.: Quantitative staining of fraction I protein in polyacrylamide gels using Coomassie brilliant blue. - Anal. Biochem. *43*: 173-182, 1971.

9312 - BENNOUN, P.: Étude de la luminescence des chloroplastes d´Épinards en présence d´hydroxylamine et de 3-(3,4-dichlorophényl)-1,1-diméthylurée (DCMU). - Compt. rend. Acad. Sci. Paris, Sér. D *273*: 2654-2657, 1971.

9313 - BENNUN, A.: Hypothesis for coupling energy transduction with ATP synthesis or ATP hydrolysis. - Nature - new Biol. *233*: 5-8, 1971.

9314 - BENNUN, A.: Properties of chloroplast's coupling factor-1 and a hypothesis for a mechanism of energy transduction. - In: BRODA, E., LOCKER, A., SPRINGER-LEDERER, H. (ed.): Proceedings of the First European Biophysics Congress. Vol. 4. Pp. 85-91. Verlag Wiener med. Akad., Wien 1971.

9315 - BENSON, A.A.: Lipids of chloroplasts. - In: GIBBS, M. (ed.): Structure and Function of Chloroplasts. Pp. 129-148. Springer-Verlag, Berlin - Heidelberg - New York 1971.

9316 - BENSON, A.A., GEE, R.W., JI, T.-H., BOWES, G.W.: Lipid-protein interactions in chloroplast lamellar membrane as bases for reconstitution and biosynthesis. - In: BOARDMAN, N.K., LINNANE, A.W., SMILLIE, R.M. (ed.): Autonomy and Biogenesis of Mitochondria and Chloroplasts. Pp. 18-26. North-Holland Publ. Comp., Amsterdam - London 1971.

9317 - BENZING, D.H., RENFROW, A.: Significance of the patterns of CO_2 exchange to the ecology and phylogeny of the *Tillandsioideae (Bromeliaceae)*. - Bull. Torrey bot. Club *98*: 322-327, 1971. [Ps.]

9318 - BENZING, D.H., RENFROW, A.: The significance of photosynthetic efficiency to habitat preference and phylogeny among tillandsioid bromeliads. - Bot. Gaz. *132*: 19-30, 1971.

9319 - BERDAHL, J.D.: The effect of leaf area on grain yield, photosynthesis, and light penetration in barley. - Diss. Abstr. int. B *31*: 5757-B-5758-B, 1971.

9320 - BERKALOFF, C., DEROCHE, M.-E.: Pertes chlorophylliennes subies, par les chloroplastes frais ou lyophilisés de feuille de Blé au cours des diverses étapes préparatoires à l'observation en microscopie électronique. - Compt. rend. Acad. Sci. Paris, Sér. D *273*: 1367-1370, 1971.

*9321 - BERLAND, B.R., BONIN, D.J., DAUMAS, R.A., LABORDE, P.L., MAESTRINI, S.Y.: Variations du comportement physiologique de l'algue *Monallantus salina* (Xanthophycée) en culture. - Mar. Biol. *7*: 82-92, 1970. [Ps, Chl.]

*9322 - BERMANN, B., GINZO, H.D., SORIANO, A.: Eco-fisiología del maíz I. Relaciones entre la economía del agua y el crecimiento, en plantas de maíz con riego y sin riego. [Eco-physiology of maize I. Relation between water balance and growth in maize plants with and without irrigation.] - Rev. Invest. agropec., INTA (Buenos Aires) Ser. 2, Biol. Prod. veg. *6* (3): 35-64, 1969. [Growth analysis, Chl.]

9323 - BERNER, E. jr.: Studies in the nitrogen metabolism of barley leaves. Part

I. The assimilation of nitrate. - Physiol. Plant. (Suppl. VI): 5-45, 1971. [Chl.]

9324 - BERNER, E. Jr.: Studies in the nitrogen metabolism of barley leaves. Part II. The effect of nitrate and ammonium on respiration and photosynthesis. - Physiol. Plant. (Suppl. VI): 46-56, 1971.

9325 - BERRY, J.A.: The β-carboxylation pathway of photosynthesis. - Diss. Abstr. int. B 31: 6451-B, 1971.

9326 - BERRY, J.A.: The compartmentation of reactions in β-carboxylation photosynthesis. - Carnegie Inst. Year Book 69: 649-655, 1971.

9327 - BERRY, J.A.: The effect of oxygen on CO_2 fixation by carboxydismutase *in vitro* and an examination of a possible reaction of ribulose diphosphate with oxygen. - Carnegie Inst. Year Book 70: 526-530, 1971.

9328 - BERSHTEĬN, B.I., IVANISHCHEVA, S.Yu.: Aktivnost' ATF-az v tkanyakh sakharnoĬ svekly pri razlichnom soderzhanii v nikh kaliya. [Activity of ATPases in sugar beet tissue containing various amounts of potassium.] - Fiziol. Biokhim. kul't. Rast. 3: 145-150, 1971. [In R, ab: E.]

9329 - BERSHTEĬN, B.I., IVANISHCHEVA, Yu.S., IL'YASHCHUK, E.M., BELOUS, I.I., PSHENICHNAYA, A.K., OKANENKO, A.S.: Fotosintez, dykhanie i fosfornyĬ obmen u rasteniĬ v svyazi s raspredeleniem kaliya pri ego defitsite. [Photosynthesis, respiration and phosphorus metabolism in plants in relation to potassium distribution during its deficit.] - Fiziol. Rast. 18: 518-525, 1971. [In R, ab: E.]

9330 - BERSHTEĬN, B.I., OKANENKO, A.S.: KaliĬ u rasteniĬ kak element mekhanizma usvoeniya solnechnoĬ energii. [Potassium in plants as a factor in the mechanism of solar energy utilization.] - In: ZALENSKIĬ, O.V. /ed./: Fotosintez i Ispol'zovanie SolnechnoĬ Energii. Pp. 213-217. Nauka, Leningrad 1971. [In R, ab: E.]

9331 - BERSHTEIN, B.I., VOLKOVA, N.V., VOLOVIK, O.I., ZAITSEVA, N.A., OKANENKO, A.S., OSTROVSKAYA, L.K., PETRENKO, S.G., POLISHCHUK, A.I., SEMENYUK, I.I., YASNIKOV, A.A.: O mekhanizme "protonnogo nasosa" i funktsiyakh enolfosfatov i epoksikarotinoidov pri fotofosforilirovanii v khloroplastakh. ["Proton pump" mechanism and functions of enol-phosphates and epoxycarotenoids during photophosphorylation in chloroplasts.] - In: Biokhimiya i Biofizika Fotosinteza. Pp. 21-27. Irkutsk 1971. [In R.]

9332 - BERTSCH, W., LURIE, S.: Delayed-light studies on photosynthetic energy conversion - IV. Effect of Tris poisoning, and of its reversal by hydrogen donors, on the millisecond emission from chloroplasts. - Photochem. Photobiol. 14: 251-260, 1971.

9333 - BERTSCH, W., WEST, J., HILL, R.: Delayed-light studies on photosynthetic energy conversion - III. Effect of 3-(3,4-dichlorophenyl)-1,1-dimethyl urea on the millisecond emission from chloroplasts performing photoreduction of ferricyanide. - Photochem. Photobiol. 14: 241-250, 1971.

*9334 - BETLACH, J., FRYDRYCH, J.: Intenzita fotosyntézy heterosních kombinací papriky. [Photosynthetic rate of two heterosis combinations of sweet pepper.] - Bull. výzk. Ústavu zelin. Olomouci 10: 99-112, 1966. [In Czech, ab: E, G.]

9335 - BEUERLEIN, J.E., PENDLETON, J.W.: Photosynthetic rates and light saturation curves of individual soybean leaves under field conditions. - Crop Sci. 11: 217-219, 1971.

9336 - BEUERLEIN, J.E., PENDLETON, J.W., BAUER, M.E., GHORASHY, S.R.: Effect of branch removal and plant populations at equidistant spacings on yield and light use efficiency of soybean canopies. - Agron. J. 63: 317-319, 1971.

*9337 - BEZINGER, E.N., MOLCHANOV, M.I., SISAKYAN, N.M.: Lamelly khloroplastov i ikh svyaz' s biosintezom belka. [Chloroplast lamellae and their relation to protein synthesis.] - Zh. evolyuts. Biokhim. Fiziol. 2: 145-151, 1966. [In R, ab: E.]

9338 - BEZUGLOV, V.K.: Izmenenie okislitel'no-vosstanovitel'nogo rezhima suspenziĭ khloroplastov i gomogenatov list'ev pri preryvistom osveshchenii pod deĭstviem povyshennoĭtemperatury. [Changes in the redox potential of chloroplast suspension and leaf homogenates under flash irradiation and increased temperature.] - In: ZALENSKII, O.V. (ed.): Fotosintez i Ispol'zovanie Solnechnoĭ Energii. Pp. 199-205. Nauka, Leningrad 1971. [In R, ab: E.]

9339 - BHANDARI, M.C., SEN, D.N.: Effect of cytokinins on seedling growth, excised cotyledons, chlorophylls and protein contents of *Citrullus lanatus* (THUNB.)MANSF. - Biochem. Physiol. Pflanzen *162*: 481-488, 1971.

9340 - BHOSALE, L.J., JOSHI, G.V.: Effect of IAA and IAA + aspartic acid on the growth in *Terminalia catappa* LINN. - Botanique /Nagpur/ *2*: 103-106, 1971. [Chl.]

*9341 - BIDWELL, R.G.S.: Photorespiration. - Science *161*: 79-80, 1968. [Conference report.]

9342 - BIDWELL, R.G.S.: Is the level of soluble protein dependent on the rate of CO_2 fixation ? - Acta agron. Acad. Sci. hung. *20*: 443-445, 1971.

*9343 - BIDWELL, R.G.S., LEVIN, W.B. /TURNER/, TAMAS, I.A.: The effects of auxin on photosynthesis and respiration. - In: WIGHTMAN, F., SETTERFIELD, G. (ed.): Biochemistry and Physiology of Plant Growth Substances. Pp. 361-376. Runge Press Ltd., Ottawa 1968.

9344 - BIDZILYA, N.I., SHAKHOV, A.A.: Obrazovanie nesparennykh elektronov v rasteniyakh pod vliyaniem vysokoenergeticheskogo izlucheniya. [Formation of impaired electrons in plants under the effect of high-energy radiation.] - In: Svetoimpul'snaya Stimulyatsiya RasteniĭT. Pp. 45-76. - Nauka, Moskva 1971. [In R.]

9345 - BIEBL, R., McROY, C.P.: Plasmatic resistance and rate of respiration and photosynthesis of *Zostera marina* at different salinities and temperatures. - Mar. Biol. *8*: 48-56, 1971.

9346 - BIEDERMANN, M.: Über die Untereinheiten der Thylakoide von *Rhodospirillum rubrum* nach Einwirkung des Detergentiums Triton X-100. - Hoppe Seyler's Z. physiol. Chem. *352*: 567-574, 1971.

9347 - BIGGINS, J.: Protoplasts of algal cells. - In: COLOWICK, S.P., KAPLAN, N.O. (ed.): Methods in Enzymology. Vol. 23. Pp. 209-211. Academic Press, New York - London 1971.

9348 - BILLINGS, W.D., GODFREY, P.J., CHABOT, B.F., BOURQUE, D.P.: Metabolic acclimation to temperature in arctic and alpine ecotypes of *Oxyria digyna*. - Arctic alpine Res. *3*: 277-289, 1971. [Ps, Chl.]

9349 - BINET, A., VOLFIN, P.: A cytoplasmic molecule active on membranar Mg^{2+} movements. II - Its role as a factor of the membrane integrity in mitochondria, chloroplasts and bacteria. - FEBS Lett. *17*: 197-202, 1971.

9350 - BÍNOVÁ, J., ONDOK, J.P.: Hodnocení fotosyntetické produktivity mladých rostlin slunečnice. [Evaluation of photosynthetic productivity of young sunflower plants.] - Rostl. Výr. (Praha) *17*: 483-492, 1971. [In Czech, ab: E, G, R.]

*9351 - BIRECKA, H.: Contribution of carbon assimilated before and after heading to the accumulation of organic compounds in the grain of some cereals. - Qual. Plant. Mat. veg. *13*: 381-383, 1966.

9352 - BISALPUTRA, T. The fine structure of the chloroplast envelope by certain algae. - Amer. J. Bot. *58*: 483, 1971.

9353 - BISALPUTRA, T.: The nature of the thylakoids of brown algae. - Amer. J. Bot. *58*: 483, 1971.

9354 - BISHOP, D.G., ANDERSEN, K.S., SMILLIE, R.M.: Incomplete membrane bound-photosynthetic electron transfer pathway in agranal chloroplasts. - Biochem. biophys. Res. Commun. *42*: 74-81, 1971.

9355 - BISHOP, D.G., ANDERSEN, K.S., SMILLIE, R.M.: Lamellar structure and composition in relation to photochemical activity. - In: HATCH, M.D., OSMOND,

C.B., SLATYER, R.O. (ed.): Photosynthesis and Photorespiration. Pp. 372-381 .Wiley-Interscience, New York - London - Sydney - Toronto 1971.

9356 - BISHOP, D.G., ANDERSEN, K.S., SMILLIE, R.M.: The distribution of galacto-lipids in mesophyll and bundle sheath chloroplasts of maize and sorghum. - Biochim. biophys. Acta *231*: 412-414, 1971.

9357 - BISHOP, D.G., SMILLIE, R.M., ANDERSEN, K.S.: The relationship between lamellar structure and photosynthetic electron transfer. - Proc. aust. biochem. Soc. *4*: 23, 1971.

9358 - BISHOP, N.I.: Photosynthesis: the electron transport system of green plants. - Annu. Rev. Biochem. *40*: 197-226, 1971.

9359 - BISHOP, N.I.: Preparation and properties of mutants: *Scenedesmus*. - In: COLOWICK, S.P., KAPLAN, N.O. /ed./: Methods in Enzymology, Vol. 23. Pp. 130-143. Academic Press, New York - London 1971.

9360 - BISHOP, N.I. SENGER, H.: Preparation and photosynthetic properties of synchronous cultures of *Scenedesmus*. - In: COLOWICK, S.P., KAPLAN, N.O. /ed./: Methods in Enzymology. Vol. 23. Pp. 53-66. Academic Press, New York - London 1971.

9361 - BISHOP, N.I., WONG, J.: Observations on Photosystem II mutants of *Scenedesmus*: Pigments and proteinaceous components of the chloroplast. - Biochim. biophys. Acta *234*: 433-445, 1971.

9362 - BISKUPSKY, V.: Productivité primaire des forêts tempérées caducifoliées en Europe orientale (Tchécoslovaquie, Bulgarie, Hongrie). - In: DUVIGNEAUD, P. (ed.): Productivity of Forest Ecosystems. Pp. 289-292. Unesco, Paris 1971.

9363 - BJÖRKMAN, O.: Comparative photosynthetic CO_2 exchange in higher plants. - In: HATCH, M.D., OSMOND, C.B., SLATYER, R.O. /ed./: Photosynthesis and Photorespiration. Pp. 18-32. Wiley-Interscience, New York - London - Sydney - Toronto 1971.

9363 - BJÖRKMAN, O.: Interaction between the effects of oxygen and CO_2 concentration on quantum yield and light-saturated rate of photosynthesis in leaves of *Atriplex patula* ssp. *spicata*. - Carnegie Inst. Year Book *70*: 520-526, 1971.

9365 - BJÖRKMAN, O., NOBS, M.A., BERRY, J.A.: Further studies no hybrids between C_3 and C_4 species of *Atriplex*. - Carnegie Inst. Year Book *70*: 507-511, 1971.

9366 - BJÖRKMAN, O., NOBS, M., PEARCY, R., BOYNTON, J., BERRY, J.: Characteristics of hybrids between C_3 and C_4 species of *Atriplex*. - In: HATCH, M.D., OSMOND, C.B., SLATYER, R.O. (ed.): Photosynthesis and Photorespiration, Pp. 105-119. Wiley-Interscience, New York - London - Sydney - Toronto 1971.

9367 - BJÖRKMAN, O., PEARCY, R.W.: Effect of growth temperature on the temperature dependence of photosynthesis *in vivo* and on CO_2 fixation by carboxydismutase *in vitro* in C_3 and C_4 species. - Carnegie Inst. Year Book *70*: 511-520, 1971.

9368 - BJÖRKMAN, O., PEARCY, R.W., NOBS, M.A.: Photosynthetic characteristics. - Carnegie Inst. Year Book *69*: 640-648, 1971.

9369 - BJÖRN, L.O.: Far-red induced, long-lived afterglow from photosynthetic cells. Size of afterglow unit and paths of energy accumulation and dissipation. - Photochem. Photobiol. *13*: 5-20, 1971.

9370 - BJÖRN, L.O.: Simple methods for the calibration of light measuring equipment. - Physiol. Plant. *25*: 300-307, 1971.

9371 - BJÖRN, L.O.: Effects of some chemical compounds on the slow, far-red induced afterglow from leaves of *Elodea* and the lack of effect of N-methyl phenazonium methosulfate on far-red induced glucose uptake in *Chlorella*. - Physiol. Plant. *25*: 316-323, 1971. [Chl.]

9372 - BJÖRN, L.O., SIGFRIDSSON, B.: Luminescence induced in intact leaves and isolated chloroplasts by alcohols and other substances. - Physiol. Plant. *25*: 308-315, 1971.

9373 - BLACK, C.C.: Ecological implications of dividing plants into groups with distinct photosynthetic production capacities. - In: CRAGG, J.B. (ed.): Advances in Ecological Research. Vol. 7. Pp. 87-114. Academic Press, New York - London 1971.

9374 - BLACK, C.C. Jr.: Chloroplasts (and grana): heptane treated. - In: COLOWICK, S.P., KAPLAN, N.O. (ed.): Methods in Enzymology. Vol. 23. Pp. 253-256. Academic Press, New York - London 1971.

9375 - BLACK, C.C. Jr.: Phosphodoxin. - In: COLOWICK, S.P., KAPLAN, N.O. (ed.): Methods in Enzymology. Vol. 23. Pp. 582-586. Academic Press, New York - London 1971.

9376 - BLACK, C.C. Jr., MOLLENHAUER, H.H.: Structure and distribution of chloroplasts and other organelles in leaves with various rates of photosynthesis. - Plant Physiol. 47: 15-23, 1971.

*9377 - BLAGONRAVOVA, L.N.: K voprosu o vliyanii nekotorykh khlororganicheskikh pestitsidov na zhiznedeyatel'nost' rastitel'nogo organizma. [Effect of some chlororganic pesticides on the viability of plant organisms.] - In: O Merakh, Predotvrashchayushchikh Nakoplenie Pestitsidov v Pishchevykh Produktakh. Pp. 95-98. "Krym", Simferopol' 1967. [Chl, in R.]

*9378 - BLAGONRAVOVA, L.N.: Fitotoksicheskoe deĭstvie khlororganicheskikh pestitsidov na vechnozelenye khvoĭnye i listvennye rateniya. [Phototoxic influence of chlororganic pesticides on evergreen coniferous and deciduous plants.] - Tr. gos. nikit. bot. Sada 46: 123-126, 1970. [In R, ab: E.]

*9379 - BLAIN, J.A., PATTERSON, J.D.E., PEARCE, M.: Carotene bleaching activity of tomato extracts. - J. Sci. Food Agr. 19: 713-717, 1968.

9380 - BLOOMQUIST, R.V., KUST, C.A.: Translocation pattern of soybeans as affected by growth substances and maturity. - Crop Sci. 11: 390-393, 1971. [Ps.]

*9381 - BLYUMENFEL'D, L.A., KAFALIEVA, D.N., LIVSHITS, V.A., SOLOV'EV, I.S., CHETVERIKOV, A.G.: Spektry deĭstviya fotoprovodimosti zelenykh list'ev i khloroplastov na chastote 10^{10} gts. [Action spectra of photoconductivity of green leaves and chloroplasts at 10^{10} Hz.] - Dokl. Akad. Nauk SSSR 193: 700-702, 1970. [In R.]

9382 - BOARDMAN, N.K.: Subchloroplast fragments: digitonin method. - In: COLOWICK, S.P., KAPLAN, N.O. (ed.): Methods in Enzymology. Vol. 23. Pp. 268-276. Academic Press, New York - London 1971.

9383 - BOARDMAN, N.K.: The photochemical systems in C_3 and C_4 plants. - In: HATCH, M.D., OSMOND, C.B., SLATYER, R.O. (ed.): Photosynthesis and Photorespiration. Pp. 309-322. Wiley-Interscience, New York - London - Sydney - Toronto 1971.

9384 - BOARDMAN, N.K., ANDERSON, J.M., HILLER, R.G. Photooxidation of cytochromes in leaves and chloroplasts at liquid-nitrogen temperature. - Proc. aust. biochem. Soc. 4: 24, 1971.

9385 - BOARDMAN, N.K., ANDERSON, J.M. HILLER, R.G.: Photooxidation of cytochrome in leaves and chloroplasts at liquid-nitrogen temperature. - Biochim. biophys. Acta 234: 126-136, 1971.

9386 - BOARDMAN, N.K., ANDERSON, J.M., KAHN, A., THORNE, S.W., TREFFRY, T.E.: Formation of photosynthetic membranes during chloroplast development. - In: BOARDMAN, N.K., LINNANE, A.W., SMILLIE, R.M. (ed.): Autonomy and Biogenesis of Mitochondria and Chloroplasts. Pp. 70-84. North-Holland Publ. Comp., Amsterdam - London 1971.

9387 - BOARDMAN, N.K., THORNE, S.W.: Sensitive fluorescence method for the determination of chlorophyll a/ chlorophyll b ratios. - Biochim. biophys. Acta 253: 222-231, 1971.

9388 - BOBODZHANOV, V.A., SIDOROVA, K.K. Izuchenie geneticheskoĭ prirody i fiziologicheskikh osobennosteĭ khlorofil'nykh mutantov gorokha. [Genetic nature and physiological peculiarities of chlorophyll mutants of pea.] - In: NASYROV, Yu. S. (ed.): Geneticheskie Aspekty Fotosinteza. Pp. 159-170. Donish, Dushanbe 1971. [In R, ab: E.]

*9389 - BOBROV, Yu. A.: Zavisimost' skorosti fotosinteza na raznykh glubinakh ot temperatury i sveta. [The dependence of the rate of photosynthesis at various

depths on temperature and light.] - Vestn. mosk. Univ., Ser. 6 - Biol. Pochvoved. 25 (5): 91-93, 1970. [In R.]

9390 - BOBST, A.M.: Light-induced electron-transfer reactions between chlorophyll a
and hydrogenated pteridine derivatives in solution. - Proc. nat. Acad. Sci.
USA 68: 541-543, 1971.

9391 - BOCHMANN, R., BLUMENFELD, L.A., KUKUSCHKIN, A.K., RUUGE, E.K.: Spektrale
und kinetische Untersuchungen des lichtinduzierten EPR-Signal I in grünen
Blättern höherer Pflanzen. - Stud. biophys. 28: 165-169, 1971.

*9392 - BODEA, C., TAMAS, V., NEAMTU, G.: Semisynthesis of dehydrocarotenoids. -
Tidsskr. Kjemi, Bergv., Metallurgi 26 (6-7): 132, 1966.

9393 - BODROGKÖZY, Gy., HORVÁTH, I.: Production tests in plant communities of
meadow-land with solonetz soil. II. The effect of climatic and soil factors on the dry-substance, carbohydrate and nitrogen concentration in the
species of draining sodic marshland. - Acta biol. (Szeged) 17: 35-48, 1971.

9394 - BODROV, V.P.: Vliyanie gibberellina i geteroauksina na fiziologo-biokhimicheskie protsessy v rasteniyakh podsolnechnika. [Effect of gibberellin and
heteroauxin on physiological and biochemical processes in sunflower plants.]
- Khim. sel'. Khoz. 9: 607-610, 1971. [Ps; In R.]

9395 - BÖGER, P.: Algal preparations with photophosphorylation activity. - In: CO-
LOWICK, S.P., KAPLAN, N.O. (ed.): Methods in Enzymology. Vol. 23. Pp. 242-
248. Academic Press, New York - London 1971.

9396 - BÖGER, P.: Einfluss von Ferredoxin auf Ferredoxin-NADP-Reduktase. - Planta
99: 319-338, 1971.

9397 - BÖGER, P.: Zusammenhang zwischen Transhydrogenase- und Diaphorase-Aktivität
der Ferredoxin-NADP-Reduktase und der photosynthetischen NADP-Reduktion. -
Z. Naturforsch. 26 b: 807-815, 1971.

9398 - BOGORAD, L., FALK, R.H., FORGER, J.M. III, LOCKSHIN, A.: Properties of etio-
plast membranes and membrane development in maize. - In: BOARDMAN, N.K.,
LINNANE, A.W., SMILLIE, R.M. (ed.): Autonomy and Biogenesis of Mitochondria
and Chloroplasts. Pp. 85-91. North-Holland Publ. Comp., Amsterdam-London
1971.

9399 - BOGORAD, L., WOODCOCK, L.F.: Rifamycins: the inhibition of plastid RNA synthesis in vivo and in vitro and variable effects on chlorophyll formation in
maize leaves. - In: BOARDMAN, N.K., LINNANE, A.W., SMILLIE, R.M. (ed.): Autonomy and Biogenesis of Mitochondria and Chloroplasts. Pp. 92-97. North-
Holland Publ. Comp., Amsterdam - London 1971.

9400 - BÖHME, H., CRAMER, W.A.: PLastoquinone mediates electron transport between
cytochrome b-559 and cytochrome f in spinach chloroplasts. - FEBS Lett. 15:
349-351, 1971.

9401 - BÖHME, H., REIMER, S., TREBST, A.: The effect of dibromothymoquinone, an
antagonist of plastoquinone, on non cyclic and cyclic electron flow systems
in isolated chloroplasts. - Z. Naturforsch. 26 b: 341 -352, 1971.

9402 - BOKHMAN, R., BLYUMENFEL'D, L.A., KUKUSHKIN, A.K., RUUGE, E.K.: Vozniknovenie
paramagnitnykh tsentrov v zelenykh list'yakh pod deĭstviem sveta s dlinoĭ
volny bol'she 725 mµ. [Formation of paramagnetic centres in green leaves
under the action of radiation with greater than 725 nm wavelengths.] - Dokl.
Akad. Nauk SSSR 201: 1138-1139, 1971. [In R.]

*9403 - BOLL, M.: Enzyme der Electronentransportpartikel aus Rhodospirillum rubrum:
Eigenschaften von NADH- und Succinat-Cytochrom c-Reduktase. - Arch. Mikrobiol. 64: 85-102, 1968.

*9404 - BOLLI, M., FRENGUELLI, G.: Periodo di adattamento alla luce relativo al
fenomeno fotosintetico. [Photosynthetic adaption to light.] - Ann. Fac.
Agrar. Univ. Studi Perugia 24: 93-101, 1969. [In Ital.]

9405 - BONAVENTURA, G., KINDERGAN, M.: Kinetic and spectral resolution of two components of delayed emission from Chlorella pyrenoidosa. - Biochim. biophys.
Acta 234: 249-265, 1971.

*9406 - BONAVENTURA, C., MYERS, J.: Involvement of oxygen in photosynthesis. - Plant
         Physiol. 42 (Suppl.): S-33, 1967.

9407 - BONDARENKO, V.I., VETROGONOVA, K.Z.: MorozostoĬkost' i produktivnost' ozimoĬ
        pshenitsy. [Cold resistance and productivity of winter wheat.] - Fiziol.
        Biokhim. kul't. Rast. 3: 637-641, 1971. [In R, ab: E.]

*9408 - BONHOMME, R.: Contribution à l'étude de la composition spectrale des rayonne-
         ments d'origine solaire à l'air libre et sous une serre. - Ann. agron. 20:
         183-200, 1969. [Radiation measurement.]

*9409 - BONHOMME, R.: Composition spectrale des rayonnements solaire et celeste et
         énergie disponible pour les végétaux. - Cahiers Ass. fr. Biometeorol. 3(3):
         23-31, 1970.

9410 - BONHOMME, R., CHARTIER, P., VARLET-GRANCHER, C.: Assimilation nette, uti-
        lisation de l'eau et microclimatologie d'un champ de maĬs. II. - Éclairement
        d'un plan horizontal au-dessus et aux différents niveaux de la culture. -
        Ann. agron. 22: 383-396, 1971.

9411 - BONITATI, J., ELLIOTT, W.B., MILES, P.G.: Errors in volumetric respirometry:
        Description of an improved microrespirometer. - Anal. Biochem. 43: 481-491,
        1971.

*9412 - BOOTS, M.R., BOOTS, S.G., MORELAND, D.E.: Conformational aspects of ureas in
         the inhibition of the Hill reaction. - J. med. Chem. 1: 144, 1970.

9413 - BORDERIE, R., SAUVEZON, R.: Étude théorique de l'évaluation des échanges ga-
        zeux à partir d'enceintes climatisées. Conception et utilisation d'une régu-
        lation automatique de la teneur en CO_2 de l'air. - Oecol. Plant. 6: 387-406,
        1971.

*9414 - BORISOV, A.Yu.: Stabilizatsiya energii vo vremeni v fotoindutsirovannykh
         okislitel'no-vosstanovitel'nykh protsessakh. [Time-course of energy stabi-
         lization in photo-induced redox processes.] - Nauch. Dokl. vyssh. Shkoly,
         biol. Nauki 13 (3): 136, 1970. [In R.]

9415 - BORISOV, A.Yu., FETISOVA, Z.G.: Issledovaniya rezonansnoĬ migratsii ener-
        gii v geterogennom pigmentnom komplekse. I. Geterogennost' kak faktor, usk-
        oryayushchiĬ lokalizatsiyu elektronnogo vozbuzhdeniya na lovushkakh. [Reso-
        nance energy migration in heterogeneous pigment complexes. I. Heterogeneity
        as a factor enhancing localization of electron excitation in traps.] - Mol.
        Biol. (Moskva) 5: 509-517, 1971. [In R, ab: E.]

9416 - BORISOV, A.Yu., IL'INA, M.D.: Kolebaniya fluorestsentsii u lista gorokha i
        vydelennykh iz nego organell. [Fluorescence oscillations in pea leaf and
        its isolated organelles.] - Biofizika 16: 157-160, 1971. [In R, ab: E.]

9417 - BORISOV, A.Yu., IL'INA, M.D.: Vremya zhizni i vykhod fluorestsentsii dvukh
        fotokhimicheskikh sistem vysshikh rasteniĬ. [Fluorescence lifetimes and
        quantum yields of two photochemical systems of higher plants.] - Biokhimiya
        36: 822-825, 1971. [In R, ab: E.]

*9418 - BORISOV, A.Yu., IVANOVSKIĬ, R.N.: Teoreticheskoe i eksperimental'noe iss-
         ledovanie tsitokhromnykh uchastkov tsepi perenosa elektronov u purpurnykh
         bakteriĬ. [Theoretical and experimental study of cytochrome components of
         electron transfer chain in purple bacteria.] - Nauch. Dokl. vyssh. Shkoly,
         biol. Nauki 13 (3): 138, 1970. [In R.]

*9419 - BORISOV, A.Yu., IVANOVSKIĬ, R.N., KONDRAT'EVA, E.N.: Fotoindutsirovannoe
         okislenie tsitokhromov Rhodopseudomonas sp. [Light-induced oxidation of
         cytochromes in Rhodopseudomonas sp.] - Mol. Biol. (Moskva) 2: 489-497,
         1968. [In R, ab: E.]

9420 - BOROVIKOVA, A.M.: Intensivnost' fotosinteza derev'ev raznykh klassov rosta
        v odnovozrastnoĬ sosnovoĬ kul'ture. [Photosynthetic rate of trees of various
        growth classes in an even-aged pine culture.] - Fiziol. Rast. 18: 427-429,
        1971. [In R.]

*9421 - BORZENKOVA, R.A.: Peredvizhenie C^{14}-assimilyatov iz list'ev raznykh yarusov
         u kartofelya. [Transport of ^{14}C-photosynthates from leaves of various leaf
         insertion levels of potato.] - Uch. Zap. ural'. gos. Univ. 113, Ser. biol.
         8 (Voprosy Regulyatsii Fotosinteza, Sb. 2): 147-152, 1970. [In R.]

9422 - BORZENKOVA, R.A.: Raspredelenie assimilyatov u kartofelya pri izmenenii
       aktivnosti potreblyayushchikh organov. [The distribution of potato photo-
       synthates with a change in activity of consuming organs.] - Sel'sko-khoz.
       Biol. 6: 371-375, 1971. [In R, ab: E.]

*9423 - BORZENKOVA, R.A., SUVOROVA, L.I.: Zavisimost' pogloshcheniya i peredvizheniya
       P^{32} ot fotosinteticheskoi aktivnosti list'ev prorostkov zlakov. [Dependence
       of ^{32}P absorption and translocation of the photosynthetic activity of leaves
       of cereal seedlings.] - Uch. Zap. ural'. gos. Univ. 113, Ser. biol. 8 (Vopro-
       sy Regulyatsii Fotosinteza, Sb.2): 153-162, 1970. [In R.]

9424 - BOSCHETTI, A.: Plastidiale DNS und extrachromosomale Vererbung. - Chimia
       25: 47-54, 1971. [Chloroplast.]

9425 - BOSCHETTI, A., GROB, E.C., STRASSER-SCHULLER, C.: Wachstum, Feinstruktur
       und Pigmentierung chromosomaler und nicht-chromosomaler Mutanten von
       Chlamydomonas reinhardi unter Streptomycineinfluss. - Verhandl. schweiz.
       naturforsch. Ges. 1971: 102-104, 1971.

*9426 - BOTHE, H.: Photosynthetische Stikstoffixierung mit einem zellfreien Extrakt
       aus der Blaualge Anabaena cylindrica. - Ber. deut. bot. Ges. 83: 421-432,
       1970.

9427 - BOTKIN, D.B., SMITH, W.H., CARLSON, R.W.: Ozone supression of white pine
       net photosynthesis. - J. Air Pollut. Control Ass. 21: 778-780, 1971.

9428 - BOUGES, B.: Action de faibles concentrations d'hydroxylamine sur l'émission
       d'oxygène des algues Chlorella et des chloroplastes d'Épinards. - Biochim.
       biophys. Acta 234: 103-112, 1971.

*9429 - LE BOURHIS, J., WAUTHY, B.: Quelques aspects de la distribution de la pro-
       duction primaire le long du méridien 170° E entre 20°S et 5°N. - Cah. O.R.
       S.T.O.M., Sér. Océanogr. 7 (4): 83-93, 1969. [Chl.]

9430 - BOURQUE, D.P., NAYLOR, A.W.: Large effects of small water deficits on chlo-
       rophyll accumulation and ribonucleic acid synthesis in etiolated leaves of
       Jack Bean (Canavalia ensiformis [L.] DC.). - Plant Physiol. 47: 591-594,
       1971.

9431 - BOWES, G., OGREN, W.L.: Properties of RuDP and PEP carboxylases from soy-
       bean and corn. - Plant. Physiol. 47 (Suppl.): 10, 1971.

9432 - BOWES, G., OGREN, W.L., HAGEMAN, R.H.: Phosphoglycolate production cataly-
       zed by ribulose diphosphate carboxylase. - Biochem. biophys. Res. Commun.
       45: 716-722, 1971.

9433 - BOWES, G.W., GEE, R.W.: Inhibition of photosynthetic electron transport by
       DDT and DDE. - J. Bioenerg. 2: 47-60, 1971.

9434 - BOWMAN, G.E.: Measurement of light and solar radiation. - Chemistry Indus-
       try 1971 (30): 840-843, 1971.

9435 - BOYD, C.E.: Further studies on productivity, nutrient and pigment relation-
       ships in Typha latifolia populations. - Bull. Torrey bot. Club 98: 144-150,
       1971.

9436 - BOYD, C.E.: The dynamics of dry matter and chemical substances in a Juncus
       effesus population. - Amer. Midland Nat. 86: 28-45, 1971. [Chl.]

9437 - BOYER, J.S.: Recovery of photosynthesis in sunflower after a period of low
       leaf water potential. - Plant Physiol. 47: 816-820, 1971.

9438 - BOYER, J.S.: Nonstomatal inhibition of photosynthesis at low leaf water
       potentials in sunflower. - Plant Physiol. 47 (Suppl.): 39, 1971.

9439 - BOYER, J.S.: Nonstomatal inhibition of photosynthesis in sunflower at low
       leaf water potentials and high light intensities. - Plant Physiol. 48: 532-
       536, 1971.

9440 - BOZHUKOVA, E.E., FEDOROV, V.D., SHESTAKOV, S.V.: Izuchenie mutageneza u
       sinezelenoi vodorosli Synechocystis aquatilis SANV. [Mutagenesis in the
       blue-green alga, Synechocystis aquatilis SANV.] - Nauch. Dokl. vyssh.
       Shkoly, biol. Nauki 14 (10): 100-104, 1971. [Chl, Car; in R.]

9441 - BRACH, E.J., MASON, W.J.: A digital radiant energy integrator. - Lab. Prac-
       tice 20: 320-323. 330. 1971.

9442 - BRADBEER, J.W.: Plastid development in primary leaves of *Phaseolus vulgaris*. The effects of short blue, red, far-red, and white light treatments on dark-grown plants. - J. exp. Bot. *22*: 382-390, 1971.

9443 - BRADLEY, S., CARR, N.G.: The absence of a functional photosystem II in heterocysts of *Anabena cylindrica*. - J. gen. Microbiol. *68* (3): XIII-XIV, 1971.

9444 - BRADY, C.J., PATTERSON, B.D., HENG FONG TUNG, SMILLIE, R.M.: Protein and RNA synthesis during ageing of chloroplasts in wheat leaves. - In: BOARDMAN, N.K., LINNANE, A.W., SMILLIE, R.M. (ed.): Autonomy and Biogenesis of Mitochondria and Chloroplasts. Pp. 453-465. North-Holland Publ. Comp., Amsterdam - London 1971.

9445 - BRAND, J., BASZYNSKI, T., CRANE, F.L., KROGMANN, D.W.: Photosystem I inhibition by polycations. - Biochem. biophys. Res. Commun. *45*: 538-543, 1971.

9446 - BRAND, J., KROGMANN, D.W., CRANE, F.L.: A lipid requirement for photosystem I activity in heptane-extracted spinach chloroplasts. - Plant Physiol. *47*: 135-138, 1971.

*9447 - BRANDL, Z., BRANDLOVÁ, J., POŠTOLKOVÁ, M.: The influence of submerged vegetation on the photosynthesis of phytoplankton in ponds. - In: STRAŠKRABA, M., PIECZYŃSKA, E., BRANDL, Z., BRANDLOVÁ, J., POŠTOLKOVÁ, M., DVOŘÁK, J., LIŠKOVÁ, E.: Relations of Aquatic Macroflora of Phytoplankton, Periphyton and Macrofauna. Rozpravy ČSAV, Ser. MPV (Praha) *80* (6): 33-62, 1970.

9448 - BRÄNDLE, E., ZETSCHE, K.: Die Wirkung von Rifampicin auf die RNA- und Proteinsynthese sowie die Morphogenese und den Chlorophyllgehalt kernhaltiger und kernloser *Acetabularia*-Zellen. - Planta *99*: 46-55, 1971.

9449 - BRÄNDLE, R., MARTI, J.: Einbau von Schwefelwasserstoff in Verbindungen des Photosynthesestoffwechsels. - Experientia *27*: 513-514, 1971.

9450 - BRANDON, P.C.: Inhibition of photophosphorylation by β-bromo-β-nitrostyrene. - FEBS Lett. *14*: 153-156, 1971.

9451 - BRANGEON, J., KRIVITZKY, M., BOURDU, R.: Le fractionnement des deux formes chloroplastiques des feuilles de *Zea mays*. - Photosynthetica *5*: 384-388, 1971.

9452 - BRAUNE, W.: Zur Ermittlung der potentiellen Produktivität von Flusswasserproben im Algentest. - Int. Rev. ges. Hydrobiol. *56*: 795-810, 1971.

9453 - BRAVDO, B.: Carbon dioxide compensation points of leaves and stems and their relation to net photosynthesis. - Plant Physiol. *48*: 607-612, 1971.

9454 - BRENCHLEY, R.G., APPLEBY, A.P.: Effect of magnesium and photoperiod on atrazine toxicity to tomatoes. - Weed Sci. *19*: 524-525, 1971. [Chl.]

9455 - BRETON, J., ROUX, E.: Chlorophylls and carotenoids states *in vivo*. I. A linear dichroism study of pigments orientation in spinach chloroplasts. - Biochem. biophys. Res. Commun. *45*: 557-563, 1971.

9456 - BRETT, W.J., SINGER, A.C.: Long term chlorophyll fluctuation in an evergreen - *Juniperus virginiana*. - Proc. Indiana Acad. Sci. *80*: 95, 1971.

9457 - BRIN, G.P., ALIEV, Z.Sh., KRASNOVSKIĬ, A.A.: Primenenie khemilyuminestsentnogo metoda dlya izmereniya kisloroda pri reaktsii Khilla. [Oxygen determination during Hill reaction by the chemoluminescence technique.] - Fiziol. Rast. *18*: 455-459, 1971. [In R, ab: E.]

9458 - BRIN, G.P., KRASNOVSKIĬ, A.A., KOMAROVA, L.F.: Obratimoe narushenie reaktsii Khilla pri deĭstvii dimetilsul'foksida i metanola na khloroplasty. [Reversible inhibitor of Hill reaction induced by dimethyl sulfoxide and methanol treatments of chloroplasts.] - Dokl. Akad. Nauk SSSR *197*: 713-716, 1971. [In R.]

9459 - BRINCKMANN, E., LÜTTGE, U., FISCHER, K.: Die Wirkung von SO_2 und Bisulfitverbindungen auf Photosynthese, Ionentransport und Spaltöffnungsregulation bei Blättern von höheren Pflanzen. - Ber. deut. bot. Ges. *84*: 523-524, 1971.

9460 - BRITTON, G.: General aspects of carotenoid biosynthesis. - In: GOODWIN, T.

W. (ed.): Aspects of Terpenoid Chemistry and Biochemistry. Pp. 255-289. Academic Press, London - New York 1971.

9461 - BRITTON, G., GOODWIN, T.W.: Biosynthesis of carotenoids. - In: COLOWICK, S.P., KAPLAN, N.O. (ed.): Methods in Enzymology. Vol. 18 (Pt.C). Pp. 654-701. Academic Press, New York - London 1971.

9462 - BRIX, H.: Effects of nitrogen fertilization on photosynthesis and respiration in Douglas-fir. - Forest Sci. *17*: 407-414, 1971.

9463 - BROCKMANN, H.Jr.: Zur absoluten Konfiguration der Chlorophylle, V. Die absolute Konfiguration der Chlorophylle *a* und *b*. - J. Liebigs Ann. Chem. *754*: 139-148, 1971.

9464 - BROCKMANN, H.Jr., KNOBLOCH, G., PLIENINGER, H., EHL, K., RUPPERT, J., MOSCOWITZ, A., WATSON, C.J.: The absolute configuration of natural /-/-stercobilin and other urobilinoid compounds. - Proc. nat. Acad. Sci. USA *68*: 2141-2144, 1971. [Phycobilins.]

9465 - BRODY, M.: The influence of two naturally occurring factors, from the leaves of *Ricinus communis*, on the fluorescence induction of higher plant chloroplasts and intact algae. - Biophys. J. *11*: 189-203, 1971.

9466 - BRODY, M., BRODY, S.S.: Internal fluorescence standards for light scattering biological material. - Photochem. Photobiol. *13*: 293-297, 1971.

9467 - BRODY, S.S.: Interactions of chlorophyll monolayers with electron donors and acceptors. - Z. Naturforsch. *26 b*: 134-141, 1971.

9468 - BRODY, S.S.: Interactions between ferredoxin and chlorophyll in a monolayer system. - Z. Naturforsch. *26 b*: 922-929, 1971.

9469 - BROOKS, K.N., THORUD, D.B.: Antitranspirant effects on the transpiration and physiology of tamarisk. - Water Resour. Res. *7*: 499-510, 1971. [Ps, Chl.]

*9470 - BROUARDEL, J., SERRUYA, C.: Détermination de la production de matière organique dans le lac Léman à l'aide du ^{14}C. - Compt. rend. Acad. Sci. Paris, Sér. D *264*: 2733-2736, 1967.

*9471 - BROVCHENKO, M.I.: Gidroliz sakharozy v svobodnom prostranstve tkaneĭ lista i lokalizatsiya invertazy. [Sucrose hydrolysis in the free space of leaf tissues and localization of invertase.] - Fiziol. Rast. *17*: 31-39, 1970. [Ps, in R, ab: E.]

9472 - BROWN, J., FRENCH, C.S., CHAPMAN, G., LAWRENCE, M.: Action spectra for two partial reactions of photosynthesis. - Plant Physiol. *47* (Suppl.): 31, 1971.

*9473 - BROWN, J.M.: The photosynthetic regime of some Southern Arizona ponderosa pine. - Diss. Abstr. B *29*: 3217-B, 1969.

9474 - BROWN, J.S.: Biological forms of chlorophyll *a*. - In: COLOWICK, S.P., KAPLAN, N.O. (ed.): Methods in Enzymology. Vol. 23. Pp. 477-487. Academic Press, New York - London 1971.

9475 - BROWN, J.S.: Photosystem 1 activity and the long-wavelength forms of chlorophyll. - Carnegie Inst. Year Book *69*: 678-682, 1971.

9476 - BROWN, J.S.: Photochemical activity of chloroplast particle fractions. - Carnegie Inst. Year Book *70*: 499-504, 1971.

9477 - BROWN, J.S., GRIFFITH, O.M.: Separation of spinach chloroplast fractions with the zonal rotor. - Carnegie Inst. Year Book *69*: 704-706, 1971.

9478 - BROWN, K.W., ROSENBERG, N.J.: Energy and CO_2 balance of an irrigated sugar beet *(Beta vulgaris)* field in the Great Plains. - Agron. J. *63*: 207-213, 1971.

9479 - BROWN, R.D., HASELKORN, R.: Chloroplast RNA populations in dark-grown, light-grown, and greening *Euglena gracilis*. - Proc. nat. Acad. Sci. USA *68*: 2536-2639, 1971.

9480 - BRZOSKA, W.: Energiegehalte verschiedener Organe von nivalen Sprosspflanzen im Laufe einer Vegetationsperiode. - Photosynthetica *5*: 183-189, 1971.

9481 - BUCHANAN, B.B., ARNON, D.I.: Ferredoxins from photosynthetic bacteria, al-
       gae, and higher plants. - In: COLOWICK, S.P., KAPLAN, N.O. (ed.): Methods
       in Enzymology. Vol. 23. Pp. 413-440. - Academic Press, New York - London
       1971.

9482 - BUCHANAN, B.B., SCHÜRMANN, P., KALBERER, P.P.: Ferredoxin-activated fructose
       diphosphatase of spinach chloroplasts. Resolution of the system, properties
       of the alkaline fructose diphosphatase component, and physiological signifi-
       cance of the ferredoxin-linked activation. - J. biol. Chem. 246: 5952-5959,
       1971.

9483 - BUCHECKER, R., HAMM, P., EUGSTER, C.H.: Absolute Konfiguration von Xantho-
       phyll (Lutein). - Chimia 25: 192-194, 1971.

9484 - BUCHWALD, H.-E., WOLFF, C.: Further evidence for carotenoids engaged in a
       metastable state in photosynthesis. - Z. Naturforsch. 26 b: 51-53, 1971.

9485 - BUCKE, C., LONG, S.P.: Release of carboxylating enzymes from maize and
       sugar cane leaf tissue during progressive grinding. - Planta 99: 199-210,
       1971.

9486 - BUDAGOVSKIĬ, A.I.: Energeticheskiĭ balans lista. [Energy balance of the leaf.]
       - In: ZALENSKIĬ, O.V. (ed.): Fotosintez i ispol'zovanie solnechnoĭ energii.
       Pp. 87-97. Nauka, Leningrad 1971. [In R, ab: E.]

*9487 - BUDZIKIEWICZ, H., v.d. HAAR, F., INHOFFEN, H.H.: Zur weiteren Kenntnis des
       Chlorophylls und des Hämins, X. Zum Aussagewert metastabiler Ionen bei Struk-
       turermittlungen mit Hilfe von Massenspektren: Kombinierte Zerfallsreaktionen
       bei Chlorinen. - J. Liebigs Ann. Chem. 701: 23-27, 1967.

9488 - BUDZIKIEWICZ, H., TARAZ, K.: Chlorophyll c. - Tetrahedron 27: 1447-1460,
       1971.

9489 - BULL, T.A.: The C_4 pathway related to growth rates in sugarcane. - In:
       HATCH, M.D., OSMOND, C.B., SLATYER, R.O. (ed.): Photosynthesis and Photo-
       respiration. Pp. 68-75. Wiley-Interscience, New York - London - Sydney -
       Toronto 1971.

9490 - BULLEY, N.R., TREGUNNA, E.B.: Photorespiration and the postillumination
       CO_2 burst. - Can. J. Bot. 49: 1277-1284, 1971.

9491 - BULYCHEV, A.A., ANDRIANOV, V.K., KURELLA, G.A., LITVIN, F.F.: Transmembrannyĭ
       potentsial khloroplasta i ego fotoindutsirovannye izmeneniya. [Chloroplast
       transmembrane potential and its photoinduced changes.] - Dokl. Akad. Nauk
       SSSR 197: 473-476, 1971. [In R.]

9492 - BULYCHEV, A.A., ANDRIANOV, V.K., KURELLA, G.A., LITVIN, F.F.: Transmembran-
       nyĭ potentsial kletki i khloroplasta vysshego nazemnogo rasteniya. [Trans-
       membrane potential of higher plant cells and chloroplasts.] - Fiziol. Rast.
       18: 248-256, 1971. [In R, ab: E.]

*9493 - BUNT, J.S.: Microalgae of the Antarctic pack ice zone. - Symp. antarctic
       Oceanogr. 1966: 198-218, 1966. [Chl.]

9494 - BUNT, J.S.: Levels of dissolved oxygen and carbon fixation by marine micro-
       algae. - Limnol. Oceanogr. 16: 564-566, 1971.

9495 - BUNT, J.S., HEEB, M.A.: Consumption of O_2 in the light by Chlorella pyre-
       noidosa and Chlamydomonas reinhardtii. - Biochim. biophys. Acta 226: 354-
       359, 1971.

*9496 - BUNT, J.S., LEE, C.C.: Seasonal primary production in Antarctic Sea ice at
       McMurdo Sound in 1967. - J. mar. Res. 28: 304-320, 1970, [Ps, Chl.]

9497 - BURNS, E.R., BUCHANAN, G.A., CARTER, M.C.: Inhibition of carotenoid synthe-
       sis as a mechanism of action of amitrole, dichlormate, and pyriclor. - Plant
       Physiol. 47: 144-148, 1971.

*9498 - BURNS, J.C., GROSS, H.D., WOODHOUSE, W.W. Jr., NELSON, L.A.: Seasonal dry
       matter distribution and annual yields of a cool-season sward as altered by
       frequency and rate of nitrogen application. - Agron. J. 62: 453-458, 1970.

*9499 - BURR, G.O.: Photosynthesis via the PGA and malic acid pathways. - Taiwan
       Sugar exp. Sta., Res. Rep. 1970 (6): 1-23, 1970.

9500 - BURRAGE, S.W.: The micro-climate at the leaf surface. - In: PREECE, T.F., DICKINSON, C.H. (ed.): Ecology of Leaf Surface Micro-organisms. Pp. 91-101. Academic Press, London - New York 1971. [Growth analysis.]

9501 - BURTON, H., BISALPUTRA, T.: Origin of proplastids in the red alga *Antithamnion subulatum*. - Amer. J. Bot. *58*: 484, 1971.

9502 - BUSH, K.J., SWEENEY, B.M.: Ribulose diphosphate carboxylase from *Gonyaulax*. - Plant Physiol. *47* (Suppl.): 10, 1971.

9503 - BUSH, P.B., GRUNWALD, C., DAVIS, D.L.: Changes in sterol composition during greening of etiolated barley shoots. - Plant Physiol. *47*: 745-749, 1971.

9504 - BUTLER, W.L.: The relationship between C-550 and delayed fluorescence. - FEBS Lett. *19*: 125-127, 1971.

9505 - BUTLER, W.L., EPEL, B., NINNEMANN, H.: Inhibition of respiration by light. - In: QUAGLIARIELLO, E., PAPA, S., ROSSI, C.C. (ed.): Energy Transduction in Respiration and Photosynthesis. P. 609. Adriatica Editrice, Bari 1971.

9506 - BUTLER, W.L., OKAYAMA, S.: The photoreduction of C-550 in chloroplasts and its inhibition by lipase. - Biochim. biophys. Acta *245*: 237-239, 1971.

9507 - BUTTERFASS, T.: Fourfold exact duplication of chloroplasts in cells of *Sphagnum*. - Naturwissenschaften *58*: 420, 1971.

*9508 - BUTTERY, B.R.: Analysis of the growth of soybeans as affected by plant population and fertilizer. - Can. J. Plant Sci. *49*: 675-684, 1969.

9509 - BÜTTNER, R.: Untersuchungen zur Ökologie und Physiologie des Gasstoffwechsels bei einigen Strauchflechten. - Flora *160*: 72-99, 1971.

9510 - BUTTROSE, M.S., HALE, C.R.: Effects of temperature on accumulation of starch or lipid in chloroplasts of grapevine. - Planta *101*: 166-170, 1971.

9511 - BYKOV, O.D., BYKOVA, M.A.: Analiz izmeneniĬ udel'noĬ radioaktivnosti uglekislogo gaza v zakrytoĬ sisteme pri fotosinteze. [Analysis of changes in carbon dioxide specific activity in a closed system during photosynthesis.] - In: ZALENSKIĬ, O.V. (ed.): Fotosintez i Ispol'zovanie SolnechnoĬ Energii. Pp. 108-113. Nauka, Leningrad 1971. [In R, ab: E.]

*9512 - BYKOV, O.D., IVANOV, O.V.: InfrakrasnyĬ gazovyĬ analiz v issledovaniyakh po fotosintezu. [Infra-red gas analysis in studies in photosynthesis.] - In: Tezisy Dokladov Vsesoyuznogo Soveshchaniya po Unifikatsii Metodov i Priborov dlya Massovykh IzmereniĬ Intensivnosti Fotosinteza. Pp. 16-20. Nauch. issled. Inst. Rastenievod. N. I. Vavilova, Leningrad - Pushkin 1970. [In R.]

*9513 - BYKOV, O.D., KOSHKIN, V.A., IVANOV, O.V.: Eksperimental'nye metody massovykh izmereniĬ fotosinteza s pomoshch'yu $C^{14}O_2$. [Experimental methods of routine measurements of photosynthesis with $^{14}CO_2$.] - In: Tezisy Dokladov Vsesoyuznogo Soveshchaniya po Unifikatsii Metodov i Priborov dlya Massovykh IzmereniĬ Intensivnosti Fotosinteza. Pp. 11-15. Nauch. issled. Inst. Rastenievod. N.I. Vavilova, Leningrad - Pushkin 1970. [In R.]

9514 - BYSTROVA, M.I., KRASNOVSKIĬ, A.A.: Fotokhimicheskie svoĬstva raznykh tipov agregirovannykh form khlorofilla *a* i bakterioviridina. [Photochemical properties of different aggregated forms of chlorophyll *a* and bakterioviridin.] - Mol. Biol. (Moskva) *5*: 291-301, 1971. [In R, ab: E.]

9515 - BYSTRZEJEWSKA, G., MALESZEWSKI, S., POSKUTA, J.: Photosynthesis, ^{14}C-photosynthetic products and transpiration of bean leaves as influenced by gibberellic acid. - Bull. Acad. pol. Sci., Sér. Sci. biol. *19*: 533-536, 1971.

9516 - CALABRESE, G.: Resistenza di *Pterocladia capillacea* (GMEL.) BORN et THUR. a variazione du salinità. [Resistance of *Pterocladia capillacea* (GMEL.) BORN et THUR. to salinity variations.] - G. bot. ital. *105*: 125-130, 1971. [Ps, Chl, Car, in Ital., ab: E.]

9517 - CALÈ, M.T., FIGLIOLIA, A.: Indagini sulla reazione di Hill, sull'attività fosforilante e catalasica dei cloroplasti di *Vicia faba* nel corso del ciclo vegetativo. [Hill reaction, photosynthetic phosphorylation and catalase activity of isolated *Vicia faba* chloroplasts during vegetative period.] - Ann. Ist. sper. Nutr. Piante (Roma) *2*: 3-14, 1971. [In Ital., ab: E.]

9518 - **CAMMACK, R., HALL, D., RAO, K.**: Ferredoxins: are they living fossils ? - New Scientist *51*: 696-698, 1971.

9519 - **CAMMACK, R., NEUMANN, J., NELSON, N., HALL, D.O.**: Circular dichroism studies of the complex between ferredoxin and ferredoxin-NADP reductase. - Biochem. biophys. Res. Commun. *42*: 292-297, 1971.

9520 - **CAMMACK, R., RAO, K.K., HALL, D.O.**: Effects of chaotropic agents on the spectroscopic properties of spinach ferredoxins. - Biochem. biophys. Res. Commun. *44*: 8-14, 1971.

9521 - **CAMMACK, R., RAO, K.K., JOHNSON, C.E., HALL, D.O.**: The iron-sulphur proteins: Physico-chemical properties and mechanism of electron transfer. - In: BRODA, E., LOCKER, A., SPRINGER-LEDERER, H. (ed.): First European Biophysics Congress. Vol. 4. Pp. 259-263. Verlag Wiener med. Akad., Wien 1971.

*9522 - **CAMPBELL, C.A., READ, D.W.L.** Influence of air temperature, light intensity and soil moisture on the growth, yield and some growth analysis characteristics of Chinook wheat grown in the growth chamber. - Can. J. Plant Sci. *48*: 299-311, 1968.

9523 - **CANNELL, M.G.R.**: Production and distribution of dry matter in trees of *Coffea arabica* L. in Kenya as affected by seasonal climatic differences and the presence of fruits. - Ann. appl. Biol. *67*: 99-120, 1971.

9524 - **CANNY, M.J.**: Translocation: Mechanisms and kinetics. - Annu. Rev. Plant Physiol. *22*: 237-260, 1971. [Ps.]

9525 - **CANVIN, D.T.**: Is there any basis to recommend the use of protein content as a base on which to express photosynthetic rate? - Acta agron. Acad. Sci. hung. *20*: 438-439, 1971.

9526 - **CARLSON, G.E., PEARCE, R.B., LEE, D.R., HART, R.H.**: Photosynthesis and photorespiration in two clones of orchardgrass. - Crop Sci. *11*: 35-37, 1971.

9527 - **CARLSON, R.E., YARGER, D.N.**: An evaluation of two methods for obtaining leaf transmissivity from leaf reflectivity measurements. - Agron. J. *63*: 78-81, 1971.

9528 - **CARLSON, R.E., YARGER, D.N., SHAW, R.H.**: Factors affecting the spectral properties of leaves with special emphasis on leaf water status. - Agron. J. *63*: 486-489, 1971.

9529 - **CARPENTER, S.B.**: Developmental changes in assimilation and translocation of photosynthate in black walnut (*Juglans nigra* L.) and honeylocust (*Gleditsia triacanthos* L.) seedlings. - Diss. Abstr. int. B *32*: 3096-B, 1971.

9530 - **CARR, L.J., HIEBERT, R.D., CURRIE, W.D., GREGG, C.T.**: A stable, sensitive, and inexpensive amplifier for oxygen electrode studies. - Anal. Biochem. *41*: 492-502, 1917.

9531 - **CARR, L.J., LARKINS, J.H., GREGG, C.T.**: A multichannel recording oxygen electrode amplifier for biochemical studies. - Anal. Biochem. *41*: 503-509, 1971.

*9532 - **CARROLL, T.W., KOSUGE, T.**: Changes in structure of chloroplasts accompanying necrosis of tobacco leaves systematically infected with tobacco mosaic virus. Phytopathology *59*: 953-962, 1969.

9533 - **CASE, G.D., PARSON, W.W.**: Thermodynamics of the primary and secondary photochemical reactions in *Chromatium*. - Biochim. biophys. Acta *253*: 187-202, 1971.

9534 - **ČATSKÝ, J., LAKE, J.V., BEGG, J.E., VOZNESENSKII, V.L.**: Physico-chemical measurement of pCO_2 and chemical determination of carbon dioxide. - In: ŠESTÁK, Z., ČATSKÝ, J., JARVIS, P.G. (ed.): Plant Photosynthetic Production. Manual of Methods. Pp. 198-237. - Dr. W. Junk N.V. Publ., The Hague 1971.

*9535 - **ČATSKÝ, J., METZNER, H., ŠESTÁK, Z.**: Bibliography. - Photosynthetica *1*: 137-147, 304-315, 1967; Photosynthetica 2: 50-56, 127-134, 220-226, 307-313, 1968. [Ps.]

9536 - **CAUBERGS, R., DE GREEF, J.**: Pigment studies in the genus *Vaucheria*. - Bot. Mag. (Tokyo) *84*: 222-230, 1971. [Chl.]

9537 - CERNUSCA, A.: Ökophysik: Neue Wege zur quantitativen Okologie. - Umschau
Wiss. Tech. *1971* (18): 663-668, 1971. [Models of Ps.]

9538 - CERVIGNI, T., TEOFANI, F., BASSANELLI, C.: Effect of CO_2 on carbonic anhydrase in *Avena sativa* and *Zea mays*. - Phytochemistry *10*: 2991-2994, 1971.

9539 - CHAÏKA, M.T., SAVCHENKO, G.E.: Metabolizm khlorofilla v svyazi s nekotorymi
voprosami razvitiya khloroplastov. [Chlorophyll metabolism in relation to
some aspects of chloroplast development.] - In: Problemy Biosinteza Khlorofillov. Pp. 138-172. Nauka i Tekhnika, Minsk 1971. [In R.]

9540 - CHAKRAVARTI, A., THANASSI, J.W.: Liquid scintillation counting of C^{14} and
H^3 samples on filter paper. - Anal. Biochem. *40*: 484-487, 1971.

*9541 - CHALUPA, V.: Počátek, trvání a ukončení vegetační činnosti u lesních
dřevin. [Start, duration and end of growing activity of forest trees.]
- Práce výzk. Úst. les. Hosp., Mysl. Zbraslav-Strnady *37*: 43-68, 1969.
[Ps, Chl; in Czech, ab: E, R.]

9542 - CHAMPIGNY, M.L., MIGINIAC-MASLOW, M.: Relations entre l'assimilation photosynthétique de CO_2 et la photophosphorylation des chloroplastes isolés. I.
Stimulation de la fixation de CO_2 par l'antimycine A, antagoniste de son
inhibition par le phosphate. - Biochim. biophys. Acta *234*: 335-343, 1971.

*9543 - CHAMURA, S., SUZUKI, R., TAKIZAWA, T.: [Effects of nitrogen on the chloro-
phyll amount of the leaves of rice plants.] - Niigata agr. Sci. *19*: 7-11,
1967. [In Jap.]

9544 - CHANDLER, M.T., VIDAVER, W.: Stationary platinum electrode for measurement
of O_2 exchange by biological systems under hydrostatic pressure. - Rev. sci.
Instrum. *42*: 143-146, 1971.

9545 - CHANT, S.R., BATEMAN, J.G., BATES, D.C.: Effect of cassava mosaic virus
infection on the metabolism of cassava leaves. - Trop. Agr. (Trinidad) *48*:
263-270, 1971. [Ps.]

9546 - CHAPMAN, G.S., BROWN, J.S., LAWRENCE, M.C., FRENCH, C.S.: Continuous action
spectra of photosynthetic partial reactions. - Carnegie Inst. Year Book *70*:
498-499, 1971.

9547 - CHARDARD, R.: Action de l'obscurité, suivie de réillumination, sur la structure des plastes de l'Algue verte: *Cosmarium lundellii* (DELP.). - Compt.
rend. Acad. Sci. Paris, Sér. D *272*: 2872-2874, 1971.

9548 - CHARDARD, R.: Ultrastructure des plastes d'une desmidiale (*Closterium ace-
rosum* EHRENB.) cultivée en lumière atténuée. - Compt. rend. Acad. Sci. Paris,
Sér. D *273*: 569-571, 1971.

9549 - CHARLES-EDWARDS, D.A.: A simple kinetic model for leaf photosynthesis and
respiration. - Planta *101*: 43-50, 1971.

9550 - CHARLES-EDWARDS, D.A.: Photosynthesis and photorespiration in *Lolium multiflorum* and *Lolium perenne*. - J. exp. Bot. *22*: 663-669, 1971.

9551 - CHARLES-EDWARDS, D.A., CHARLES-EDWARDS, J., COOPER, J.P.: The influence
of temperature on photosynthesis and transpiration in ten temperate grass
varieties grown in four different environments. - J. exp. Bot. *22*: 650-662,
1971.

9552 - CHARTIER, M., CHARTIER, P.: Design of an air-conditioned assimilation
chamber for detached leaves. - Photosynthetica *5*: 74-75, 1971.

9553 - CHARTIER, P., PERRIER, A., VERBRUGGHE, M.: Assimilation nette, utilisation
de l'eau et microclimatologie d'un champ de maïs. I. - Objectifs et pré-
sentation du dispositif expérimental de la Minière. - Ann. agron. *22*: 367
-381, 1971.

9554 - CHEN, L.H., MEDERSKI, H.J., CURRY, R.B.: Water stress effects on photosynthesis and stem diameter in soybean plants. - Crop Sci. *11*: 428-431, 1971.

*9555 - CHEN, S.: Carbohydrate metabolism in the *Narcissus* leaf. - J. exp. Bot. *20*:
302-316, 1969. [Ps.]

9556 - CHEN, T.M., BROWN, R.H., BLACK, C.C. Jr.: Photosynthetic $^{14}CO_2$ fixation

products and activities of enzymes related to photosynthesis in bermudagrass and other plants. - Plant Physiol. *47*: 199-203, 1971.

9557 - CHENIAE, G.M., MARTIN, I.F.: Effects of hydroxylamine on photosystem II. I.
Factors affecting the decay of O_2 evolution. - Plant Physiol. *47*: 568-575,
1971.

9558 - CHENIAE, G.M., MARTIN, I.F.: Photoactivation of the manganese catalyst of
O_2 evolution. I. Biochemical and kinetic aspects. - Biochim. biophys. Acta
253: 167-181, 1971.

9559 - CHERNAVINA, I.A., VORONINA, E.I.: Fotosinteticheskoe fosforilirovanie u
salata pri razlichnykh usloviyakh pitaniya molibdenom. [Photophosphorylation
in lettuce under various molybdenum nutrition.] - Nauch. Dokl. vyssh. Shkoly,
biol. Nauki *14* (8): 78-82, 1971. [In R.]

*9560 - CHERNOMORSKIĬ, S.A., FRAGINA, A.I., KURNYGINA, V.T., SOLODKIĬ, F.T.: K pre-
parativnomu polucheniyu feofitina. [Preparation of pheophytin.] - Rast. Resursy *4*: 395-396, 1968. [In R.]

9561 - CHERNOV, I.A., KHLYUSTOVA, T.M., SHAKIROVA, D.Z., GRISHINA, L.I.: Posledeĭstvie temperatury na fotosintez izolirovannykh khloroplastov. [After-effect of
temperature on photosynthesis of isolated chloroplasts.] - In: ZALENSKIĬ, O.V.
(ed.): Fotosintez i Ispol'zovanie Solnechnoĭ Energii. Pp. 205-209. Nauka, Leningrad 1971. [In R, ab: E.]

9562 - CHERNOV, I.A., KRAĬNOVA, N.N.: Mnogokomponentnaya sreda dlya polucheniya preparatov khloroplastov, sposobnykh k intensivnoĭ fiksatsii $C^{14}O_2$. [Multicompo-
nent medium for the isolation of chloroplast preparations for intensive $^{14}CO_2$
fixation.] - Fiziol. Rast. *18*: 450-454, 1971. [In R, ab: E.]

9563 - CHERRY, R.J., HSU, K., CHAPMAN, D.: Absorption spectroscopy of chlorophyll
in bimolecular lipid membranes. - Biochem. biophys. Res. Commun. *43*: 351-358,
1971.

*9564 - CHESNOKOV, V.A.: Nekotorye zadachi v sovremennykh issledovaniyakh po fotosin-
tezu. [Problems of recent studies of photosynthesis.] - In: Tezisy Dokladov
Vsesoyuznogo Soveshchaniya po Unifikatsii Metodov i Priborov dlya Massovykh
Izmereniĭ Intensivnosti Fotosinteza. Pp. 115-116. Nauch. issled. Inst. Rastenievod. N.I. Vavilova, Leningrad - Pushkin 1970. [In R.]

9565 - CHI, E.Y.: Brown algal pyrenoids. - Protoplasma 72: 101-104, 1971.

9566 - CHICHEV, P., ĬORDANOV, I.: Vliyanie tetratsiklina na soderzhanie plastidnykh
pigmentov, vklyuchenie i raspredelenie C^{14} v yuvenil'nykh list'yakh intaktnykh i dekapitirovannykh rastenii fasoli. [Effect of tetracycline on the content of plastid pigments, incorporation and distribution of ^{14}C in juvenile
leaves of intact and decapitated bean plants.] - Fiziol. Rast. *18*: 12-17, 1971.
[In R, ab: E.]

9567 - CHICHEV, P., ĬORDANOV, I., POPOV, K.: Intenzivnost i produkti na fotosintezata
pri rasteniya s razlichno fiziologichno s"stoyanie na fona na s"zdalen fosfo-
ren defitsit. [Rate and products of photosynthesis in plants of different
physiological state affected by phosphorus deficiency.] - Izv. Inst. Fiziol.
Rast. M. Popov b"lg. Akad. Nauk *17*: 231-243, 1971. [In Bulg., ab: E, R.]

*9568 - CHIGIREV, V.S., MOLCHANOV, M.I., BEZINGER, E.N.: Izuchenie biosinteza lipoa-
minokislotnykh soedinenii v khloroplastakh *in vivo*. [Biosynthesis of lipoami-
no-acid compounds in chloroplasts *in vivo*.] - Dokl. Akad. Nauk SSSR *195*: 743
-745, 1970. [In R.]

9569 - CHIMIKLIS, P.E., KARLANDER, E.P.: Effects of NaCl and light intensity on photosynthetic O_2 evolution and CO_2 fixation in *Chlorella sorokiniana*. - Plant
Physiol. *47* (Suppl.): 30, 1971.

9570 - CHINNADURAI, G., CHANDRASEKHRAN NAIR, M. Effect of Dolichos yellow mosaic on
chlorophyll and leaf proteins. - Curr. Sci. *40*: 18-19, 1971.

9571 - CHIRKOV, V.N., MANSUROV, Ya.V.: Nekotorye osobennosti fotosinteticheskoĭ deyatel'nosti kukuruza v zavisimosti ot gustoty stoyanina i fona pitaniya. [Some
characteristics of photosynthetic activity of maize in dependence on canopy
density and level of nutrition.] - Uzbek. biol. Zh. *15* (4): 17-19, 1971. [In
R.]

*9572 - CHMIELEWSKI, T., BERGER, S.: Genetic aspects of some carotenoid synthesis in tomatoes. - Qual. Plant. Mat. veg. *13*: 219-227, 1966.

9573 - CHMORA, S.N., POYARKOVA, N.M., VOSKRESENSKAYA, N.P.: Uglekislotnyĭ gazoobmen kukuruzy na sinem i krasnom svetu. [Carbon dioxide exchange of maize under the effect of blue and red light.] - Fiziol. Rast. *18*: 1120-1126, 1971. [In R, ab: E.]

9574 - CHOLLET, R., OGREN, W.L., Cytoplasmic, photorespiratory, and photosynthetic enzyme activities during greening of a virescent corn mutant. - Plant Physiol. *47* (Suppl.): 44, 1971.

9575 - CHUA, N.-H.: The methyl viologen-catalyzed Mehler reaction and catalase activity in blue-green algae and *Chlamydomonas reinhardi*. - Biochim. biophys. Acta *245*: 277-287, 1971.

9576 - CHUDNOVSKIĬ, A.F., GAPRINDASHVILI, I.S.: Kolichestvennaya teoriya vliyaniya dinamiki razvitiya rasteniya na protsessy teplo- i vlagoobmena na sel' skokhozyaĭstvennom pole. [Quantitative theory of the effect of dynamics of plant development on the heat- and water-exchange processes in a crop field.] - In: Sbornik Trudov po Agronomicheskoĭ Fizike *30* - Matematicheskie Modeli v Agrofizike i Biologii. Pp. 33-43. Gidrometeorol. Izdat., Leningrad 1971. [In R.]

*9577 - CHULANOVSKAYA, M.V.: Rezul'taty izmereniĭ fotosinteticheskogo koeffitsienta u khlorelly pri raznoĭ temperature. [Measuring the photosynthetic quotient at different temperature in *Chlorella* cells.] - In: Materialy V. Rabochego Soveshchaniya po Voprosu Krugovorota Veshchestv v Zamknutoĭ Sisteme na Osnove Zhiznedeyatel'nosti Nizshikh Organizmov. Pp. 176-178. Naukova Dumka, Kiev 1968. [In R.]

9578 - CHULANOVSKAYA, M.V., ZALENSKIĬ, O.V.: O sopryazhennosti gazoobmena O_2 i CO_2 pri fotosinteze khlorelly. [Interdependence of O_2 and CO_2 exchange during photosynthesis in *Chlorella*.] - In: ZALENSKIĬ, O.V. (ed.): Fotosintez i Ispol'zovanie Solnechnoĭ Energii. Pp. 174-177. Nauka, Leningrad 1971. [In R, ab: E.]

9579 - CLARKE, G.C.S., GREENE, S.W., GREENE, D.M.: Productivity of bryophytes in polar regions. - Ann. Bot. *35*: 99-108, 1971.

9580 - CLAYTON, R.K.: Photochemical reaction centers and photosynthetic membranes. - Advances chem. Phys. *19*: 353-378, 1971.

9581 - CLAYTON, R.K., WANG, R.T.: Photochemical reaction centers from *Rhodopseudomonas spheroides*. - In: COLOWICK, S.P., KAPLAN, N.O. (ed.): Methods in Enzymology. Vol. 23. Pp. 696-704. Academic Press, New York - London 1971.

9582 - CLEMENT-METRAL, J.D., LEFORT-TRAN, M.: Fluorescence transfer in glutaraldehyde fixed particles of the red alga *Porphyridium cruentum* (N.). - FEBS Lett. *12*: 225-228, 1971.

*9583 - CODD, G.A., MERRETT , M.J. Enzymes of the glycollate pathway in relation to greening in *Euglena gracilis*. - Planta *95*: 127-132, 1970.

9584 - CODD, G.A., MERRETT, M.J.: Phosphopyruvate carboxylase activity and carbon dioxide fixation via C_4 acids over the division cycle in synchronized *Euglena* cultures. - Planta *100*: 124-130, 1971.

9585 - CODD, G.A., MERRETT, M.J.: Photosynthetic products of division synchronized cultures of *Euglena*. - Plant Physiol. *74*: 635-639, 1971.

9586 - CODD, G.A., MERRETT, M.J.: The regulation of glycolate metabolism in division synchronized cultures of *Euglena*. - Plant Physiol. *47*: 640-643, 1971.

9587 - COHEN, W.S., JAGENDORF, A.T.: Reversible inhibition of energy-linked chloroplast reactions by polygalacturonate. - Plant Physiol. *47* (Suppl.): 32, 1971.

9588 - COHEN, W.S., SHERMAN, L.A.: Proton translocation and energy dependent quenching of chlorophyll *a* fluorescence. - FEBS Lett. *16*: 319-323, 1971.

9589 - COHEN-BAZIRE, G.: The photosynthetic apparatus of procaryotic organisms. - In: HARRIS, P.J. (ed.): Biological Ultrastructure: The Origin of Cell Organelles. Pp. 65-90. Oregon State Univ. Press, Corvallis 1971.

B9590 – COLOWICK, S.P., KAPLAN, N.O. (ed.): Methods in Enzymology. Volume 23. Photosynthesis. Part A. – Academic Press, New York – London 1971.

9591 – CONNELLY, P.R., BARTHOLOMEW, D.P.: Effects of temperature on carbon fixation by pineapple. – Plant Physiol. 47 (Suppl.): 30, 1971.

9592 – CONNOR, D.J.: Photosynthesis of Rhodes grass and wheat communities. – Pac. Sci. Congr. Proc. 1: 54, 1971.

9593 – CONNOR, D.J., CARTLEDGE, O.: Structure and photosynthesis of wheat communities. – J. appl. Ecol. 8: 469-475, 1971.

9594 – CONNOR, D.J., TUNSTALL, B.R., van den DRIESSCHE, R.: An analysis of photosynthetic response in a brigalow forest. – Photosynthetica 5: 218-225, 1971. [Model.]

9595 – COOK, J.R.: Synchronous cultures: Euglena. – In: COLOWICK, S.P., KAPLAN, N. O. (ed.): Methods in Enzymology. Vol. 23. Pp. 74-78. Academic Press, New York – London 1971. [Ps, Chl.]

9596 – COOMBS, J.: The potential of higher plants with the phosphopyruvic acid cycle. – Proc. roy. Soc. London, Ser. B 179: 221-235, 1971.

9597 – COOMBS, J., BALDRY, C.W.: Effects of intermediates of the photosynthetic carbon reduction cycle on carbon metabolism in spinach chloroplasts illuminated on filter paper discs. – Plant Physiol. 48: 379-381, 1971.

*9598 – COOPER, J.P., TAINTON, N.M.: Light and temperature requirements for the growth of tropical and temperate grasses. – Herbage Abstr. 38: 167-176, 1968. [Ps.]

9599 – COOPER, T.G., WOOD, H.G.: The carboxylation of phosphoenolpyruvate and pyruvate. II. The active species of "CO_2" utilized by phosphoenolpyruvate carboxylase and pyruvate carboxylase. – J. biol. Chem. 246: 5488-5490, 1971.

9600 – CORNIC, G., SCHMITT, A.: Action de l'ombrage sur le dégagement de CO_2 à la lumière de Quercus sessiliflora SALISB. – Physiol. vég. 9: 453-460, 1971.

9601 – COSTES, C., BAZIER, R., BRANGEON, J., BOURDU, R.: Lipides et pigments dans les deux formes des chloroplastes de la feuille de Maïs. – Compt. rend. Acad. Sci. Paris, Sér. C 272: 1597-1600, 1971.

*9602 – COUTINHO, L.M.: Novas observações sôbre a ocorrência do "efeito de de Saussure" e suas relações com a suculência, a temperatura folhear e os movimentos estomáticos. [New observations on "de Saussure effect" and its relations to succulence, leaf temperature and stomatal movements.] – Bol. 331, Fac. Filosof., Ciênc. Letras Univ. São Paulo – Bôt. 24: 77-102, 1969. [In Portugal., ab: E.]

*9603 – COUTINHO, L.M., SCHRAGE, C.A.F.: Sôbre o efeito da temperatura na ocorrência de fixação noturna de CO_2 em orquídeas e bromélias. [Temperature effect on night CO_2 fixation in Orchidaceae and Bromeliaceae.] – An. Acad. brasil. Cienc. 42: 843-849, 1970. [In Portugal., ab: E.]

*9604 – COVI, A., NAVARI-IZZO, F.: La presenza dell'ALA-deidratasi in alcune piante leguminose e chenopodiacee. [Presence of δ-aminolaevulinic acid dehydratase in some leguminous plants and Chenopodiaceae.] – Agrochimica 14: 4-9, 1969. [In Ital., ab: E, G, F, Span.]

9605 – COWAN, I.R.: Light in plant stand with horizontal foliage. – J. appl. Ecol. 8: 579-580, 1971.

9606 – COWAN, I.R., TROUGHTON, J.H.: The relative role of stomata in transpiration and assimilation. – Planta 97: 325-336, 1971.

*9607 – COX, R.E., MAXWELL, J.R., EGLINTON, G., PILLINGER, C.T.: The geological fate of chlorophyll: the absolute stereochemistries of a series of acyclic isoprenoid acids in a 50 million year old lacustrine sediment. – Chem. Commun. 1970: 1639-1641, 1970.

9608 – CRAIG, I.: Chloroplasts: little green slaves? – New Sci. Sci. J. 51: 313-314, 1971.

9609 – CRAIG, I.W., GIBOR, A.: Chloroplast development in Charophyceae. – J. Cell Biol. 49: 950-953, 1971.

9610 - CRAIGIE, J.S., LEIGH, C., CHEN, L.C.-M., McLACHLAN, J.: Pigments, polysaccharides, and photosynthetic products of *Phaeosaccion collinsii*. - Can.
J. Bot. *49*: 1067-1074, 1971.

9611 - CRAMER, W.A., FAN, H.N., BÖHME, H.: High and low potential states of the
chloroplast cytochrome *b*-559 and thermodynamic control of non-cyclic electron transport. - J. Bioenergetics *2*: 289-303, 1971.

*9612 - CRANE, F.L.: Quinones in electron transport. - In: SINGER, T.P. (ed.): Biol-
ogical Oxidations. Pp. 533-580. Intersci. Publ., New York - London - Sydney
1968. [Ps.]

9613 - CRANE, F.L., ARNTZEN, C.J., HALL, J.D., RUZICKA, F.J., DILLEY, R.A.: Binary
membranes in mitochondria and chloroplasts. - In: BOARDMAN, N.K., LINNANE,
A.W., SMILLIE, R.M. (ed.): Autonomy and Biogenesis of Mitochondria and Chloroplasts. Pp. 53-69. North-Holland Publ. Comp., Amsterdam - London 1971.

9614 - CRAWFORD, C.G., JENSEN, R.G.: Isolation and partial characterization of ferredoxin from *Zea mays*. - Plant Physiol. *47*: 447-449, 1971.

9615 - CREED, D., WERBIN, H., STROM, E.T.: Photochemistry of electron-transport
quinones. II. Model studies with plastoquinone-1[2,3-dimethyl-5-(3-methyl-
but-2-enyl)-1,4-benzoquinone]. - J. amer. chem. Soc. *93*: 502-511, 1971.

9616 - CRISWELL, J.G.: Physiological basis for variation of net photosynthesis in
oat (*Avena* spp.) leaves as affected by genotype and sink-source ratios. -
Diss. Abstr. Int. B *31*: 5116-B, 1971.

9617 - CRISWELL, J.G., SHIBLES, R.M.: Physiological basis for genotypic variation
in net photosynthesis of oat leaves. - Crop Sci. *11*: 550-553, 1971.

9618 - CROFTS, A.R.: The potential of bacterial photosynthesis in recycling of hu-
man wastes. - Proc. roy. Soc. London, Ser. B *179*: 209-219, 1971.

9619 - CROFTS, A.R., COGDELL, R.J., JACKSON, J.B.: The mechanism of H^+ uptake in
Rhodopseudomonas spheroides. - In: QUAGLIARIELLO, E., PAPA, S., ROSSI, C.S.
(ed.): Energy Transduction in Respiration and Photosynthesis. Pp. 883-901.
Adriatica Editrice, Bari 1971.

9620 - CROFTS, A.R., WRAIGHT, C.A., FLEISCHMANN, D.E.: Energy conservation in the
photochemical reactions of photosynthesis and its relation to delayed fluorescence. - FEBS Lett. *15*: 89-100, 1971.

9621 - CUMMINS, W.R., KENDE, H., RASCHKE, K.: Specificity and reversibility of the
rapid stomatal response to abscisic acid. - Planta *99*: 347-351, 1971. [Ps.]

*9622 - ĆUPINA, T.: Značaj lisne površine pri radu na selekciji biljaka. [Importance
of leaf area characteristics for plant selection.] - Savrem. Poljoprivreda
(Novi Sad) *16*: 805-814, 1968. [Ps, in Croat.]

*9623 - ĆUPINA, T., BOROJEVIĆ, S.: Proučavanje dinamike sadržaja i metabolizma karo-
tinoida u različitim listovima i drugim zelenim organima u nekih genotipova
pšenice. [Dynamics of content and metabolism of carotenoids in different
leaves and other green organs in some wheat genotypes.] - Agron. Glasnik
(Zagreb) *31*: 349-362, 1969. [In Croat., ab: E.]

9624 - CURRY, R.B.: Dynamic simulation of plant growth - Part I. Development of a
model. - Trans. ASAE *14*: 946-949, 959, 1971. [Ps.]

9625 - CURRY, R.B., CHEN, L.H.: Dynamic simulation of plant growth - Part II. Incorporation of actual daily weather and partitioning of net photosynthate.
- Trans. ASAE *14*: 1170-1174, 1971.

9626 - CUSANOVICH, M.A.: Subchromatophore fragments: *Rhodopseudomonas spheroides*.
- In: COLOWICK, S.P., KAPLAN, N.O. (ed.): Methods in Enzymology. Vol. 23.
Pp. 321-324. Academic Press, New York - London 1971.

9627 - CUSANOVICH, M.A., EDMONDSON, D.E.: The isolation and characterization of
Rhodospirillum rubrum flavodoxin. - Biochem. biophys. Res. Commun. *45*: 327-
336, 1971.

9628 - CZOPEK, M.: Metody pomiaru promieniowania fotosyntetycznie czynnego. [Methods
for measurement of photosynthetically active radiation.] - Wiadomości ekol.
17: 30-52, 1971. [In Pol., ab: E.]

9629 - CZOPEK, M.: Produktywność fotosyntetycznych systemów w Międzynarodowym programie biologicznym. [Productivity of photosynthetic systems in the International Biological Programme.] - Wiadomości bot. 15: 13-28, 1971. [In Pol.]

9630 - CZUCHAJOWSKA, Z., WOJTASZEK, T.: Drogi i mechanizmy oddziaływania herbicydów pochodnych mocznika na fizjologiczne i biochemiczne procesy organizmów roślinnych. [Pathways and mechanisms of action of herbicidal urea derivatives on physiological and biochemical processes in plant organisms.] - Postępy Nauk roln. 18: 37-70, 1971. [Ps, in Pol.]

9631 - CZYGAN, F.-C., EICHENBERGER, W.: Die Fettsäuren der Sekundär-Carotinoid-Ester von Ankistrodesmus braunii (NAEGELI) COLLINS (Chlorophyta, Chlorococcales). - Z. Naturforsch. 26 b: 264-267, 1971.

9632 - DĂBALĂ, I., MAGHIARU, M.: Variante cantitative de măsurare a intensităţii fotosintezei, respiraţiei si transpiraţiei. [Quantitative variants in the measurement of rates of photosynthesis, respiration and transpiration.] - Natura (Bucureşti) 22: 53-56, 1971. [In Roum.]

9633 - DABES, J.N.: The behavior of Chlorella pyrenoidosa in steady state continuous culture. - Diss. Abstr. Int. B 32: 902-B, 1971. [Ps, Chl.]

9634 - DĄBROWSKA, J.: Korelacja między liczbą chloroplastów w komórkach szparkowych a poziomem poliploidalności czternastu taksonów Achillea L. [Correlation of chloroplast amount in stomatal cells and polyploidity in fourteen taxons of Achillea L.] - Herba pol. 17: 200-208, 1971. [In Pol., ab: E, R.]

*9635 - DALETSKAYA, I.A., ZALENSKIǏ, O.V.: Temperaturnaya zavisimost' fotosinteza razlichnykh shtammov khlorelly. [Temperature dependence of photosynthesis in different strains of Chlorella.] - In: Fotosintez i Ispol'zovanie Solnechnoǐ Radiatsii. Tezisy Dokladov. Pp. 48-49. Dushanbe 1967. [In R.]

9636 - DALETSKAYA, I.A., ZALENSKIǏ, O.V.: Deǐstvie temperatury na fotosintez razlichnykh shtammov khlorelly. [Effect of temperature on photosynthesis of different Chlorella strains.] - In: ZALENSKIǏ, O.V. (ed.): Fotosintez i Ispol'zovanie Solnechnoǐ Energii. Pp. 184-189. Nauka, Leningrad 1971. [In R, ab: E.]

*9637 - DANCS, K., CSORBA, Z., FERENCZ, V., POZSÁR, B.: Fungicidek alkalmazásának radiobiológiai értékelése az almafa varasodás elleni védekezésben. [Radiobiological evaluation of the application of fungicides on apple scab.] - Agrártudományi Közlem. 26: 313-325, 1967. [Ps, Chl; in Hung., ab: G, R.]

9638 - DARASELIYA, Z.G.: Kharakteristika otdel'nykh chasteǐ chaǐnoǐ fleshi po nekotorym fiziko-mekhanicheskim i khimicheskim pokazatelyam. [Characteristics of separate parts of tea shoots based on some physicomechanical and chemical indexes.] - Subtrop. Kul't. 1971 (3): 49-52, 1971. [Chl, Car; in R.]

9639 - DAS, B.K., SEN, S.P.: Effect of mineral deficiency on photosynthetic $^{14}CO_2$ fixation by crop plants. - In: Proc. Dept. At. Energy Symp. Rad. and Radioisotop. Soil Stud. and Plant Nutrition Bangalore 1970. Pp. 207-215. Bombay 1971.

*9640 - DAS, C.R.: Thermal adaptation of photosynthesis in two algal strains. - Current Sci. 38: 396-397, 1969.

9641 - DAUNICHT, H.-J.: Ein Verfahren zur exakten automatischen Photosynthese-Kompensation. - Ber. deut. bot. Ges. 83: 499, 1971.

9642 - DAVEY, M.R., FOWLER, M.W., STREET, H.E.: Cell clones contrasted in growth, morphology and pigmentation isolated from a callus culture of Atropa belladonna var. lutea. - Phytochemistry 10: 2559-2575, 1971. [Chl, Car.]

9643 - DAVID, K.A.V., THOMAS, J.: Nitrogenase activity and its relation to photosystems I and II of photosynthesis in heterocysts of a blue-green alga. - In: Proc. Dept. At. Energy Symp. Rad. and Radioisotop. Soil Stud. and Plant Nutrition Bangalore 1970. Pp. 435-444. Bombay 1971.

*9644 - DAVIES, B.H.: Some aspects of carotenoid analysis. - Tidskr. Kjemi, Bergv., Metallurgi 26: 132, 1966.

9645 - DAVIS, J.T., SPARKS, D.: Assimilation of $^{14}CO_2$ by catkins of Carya illinoensis and apparent translocation to the pollen. - Amer. J. Bot. 58: 932-938, 1971.

*9646 - DAVTYAN, V.A.: O razlichiyakh v funktsional'noi aktivnosti list'ev i kornei sukhovershinnogo i normal'nogo rastushchego topolya. [Differences in the functional activity of leaves and roots in dry-topped and normally growing poplars.] - In: KAZARYAN, V.O. (ed.): Ontogenez Vysshikh Rastenii. Pp. 280-287. Izdat. Akad. Nauk arm. SSR, Erevan 1970. [Ps, Chl; in R.]

9647 - DAYNARD, T.B.: Characterization of corn (*Zea mays* L.) canopies from measurements of individual plants. - Agron. J. *63*: 133-135, 1971. [Computer prediction of canopy characteristics.]

9648 - DECLEIRE, M., BASTIN, R.: Etudes de la force de liaison chlorophylle-protéine et de la solubilisation par la soude diluée des pigments et protéines du tissu foliaire en fonction de divers trempages en eau chaude. - Rev. Ferment. Ind. alim. *26*: 215-222, 1971.

9649 - DEDONDER, A., VAN SUMERE, C.F.: Effect of some phenolics and related compounds on the cyclic and noncyclic photophosphorylation in isolated *Spinacia* chloroplasts. - Z. Pflanzenphysiol. *65*: 176-182, 1971.

*9650 - DEHADRAI, P.V.: Observations on certain environmental features at the Dona Paula point in Marmugao Bay, Goa. - Proc. indian Acad. Sci., Sect. B *72*: 56-67, 1970. [Ps.]

*9651 - DEHADRAI, P.V.: Changes in the environmental features of the Zuari and Mandovi estuaries in relation to tides. - Proc. indian Acad. Sci., Sect. B *72*: 68-80, 1970. [Ps, Chl.]

9652 - DEKKER, R.F.H., RICHARDS, G.N.: Determination of starch in plant material. - J. Sci. Food Agr. *22*: 441-444, 1971.

9653 - DEKOCK, P.C., RUTHERFORD, M., CHESHIRE, M.V.: The fine structure of leaf cells of copper-deficient oats. - Ann. Bot. *35*: 193-199, 1971. [Chloroplast.]

9654 - DELANEY, M.E., OWEN, W.J., ROGERS, L.J.: Inhibition of photosynthetic electron transport by 1,1,1-trichloro-2, 2-bis-(p-chlorophenyl)ethane (DDT) at a site before photosystem 2. - Biochem. J. *124* (2) - Proc. biochem. Soc.: 24 P, 1971.

9655 - DELAPORTE, N., LAVAL-MARTIN, D.: Analyse spectrophotométrique des chlorophylles et des phéophytines a et b en milieu hydroacetonique. I. Détermination des extinctions molaires. - Anal. chim. Acta *55*: 415-424, 1971.

9656 - DELAPORTE, N., LAVAL-MARTIN, D.: Analyse spectrophotométrique des chlorophylles et des phéophytines a et b en milieu hydroacetonique. II. Méthode cinétique de dosage. - Anal. chim. Acta *55*: 425-435, 1971.

9557 - DELOSME, R.: Variations du rendement de fluorescence de la chlorophylle *in vivo* sous l'action d'éclairs de forte intensité. - Compt. rend acad. Sci. Paris, Sér. D *272*: 2828-2831, 1971.

9658 - DELRIEU, M.-J., DE KOUCHKOVSKY, Y.: Relationships between the photon distribution between the two photosystems, the concentration of System II reaction centers and the intersystem equilibrium constant in *Chlorella pyrenoidosa*. - Biochim. biophys. Acta *226*: 409-421, 1971.

9659 - DELVAUX, J.: Des tables de production aux bilans énergétiques. - In: DUVIGNEAUD, P. (ed.): Productivity of Forest Ecosystem. Pp. 177-184. Unesco, Paris 1971.

9660 - DeMAGGIO, A.E., STETLER, D.A.: Polyploidy and gene dosage effects on chloroplasts of fern gametophytes. - Exp. Cell Res. *67*: 287-294, 1971. [Chl.]

9661 - DEMETRADZE, T.Ya., BAKANIDZE, M.Sh.: Vliyanie shpalernoi podrezki na stroenie, fiziologicheskuyu akvitnost' i biokhimicheskii sostav lista chainogo rasteniya. [Effect of espalier pruning on the structure, physiological activity, and biochemical composition of a tea plant leaf.] - Subtrop. Kul't. *1971* (3): 23-29, 1971. [Chloroplast; in R.]

9662 - DEMETRADZE, T.Ya., SIKHARULIDZE, A.M.: Vliyanie povrezhdeniya tsitrusovoi belokrylkoi na nekotorye fiziologicheskie i biokhimicheskie protsessy v list'yakh subtropicheskoi khurmy. [Effect of damage caused by the citrus whitefly on physiological and biochemical processes in the leaves of the subtropical persimmon.]- Subtrop. Kul't *1971* (4): 129-130, 1971. [Ps; in R.]

*9663 - DENMEAD, O.T.: A strip net radiometer. - Aust. J. Instr. Control *23*: 61, 1967.

9664 - DENMEAD, O.T., McILROY, I.C.: Measurement of carbon dioxide exchange in the field. - In: ŠESTÁK, Z., ČATSKÝ, J., JARVIS, P.G. (ed.): Plant Photosynthetic Production. Manual of Methods. Pp. 467-516. Dr. W. Junk N.V. Publ., The Hague 1971.

9665 - DENNE, M.P., SMITH, C.J.: Daylength effects on growth, tracheid development, and photosynthesis in seedlings of *Picea sitchensis* and *Pinus sylvestris*. - J. exp. Bot. *22*: 347-361, 1971.

*9666 - DENNIS, D.T., STUBBS, M., COULTATE, T.P.: The inhibition of Brussels sprout leaf senescence by kinins. - Can. J. Bot. *45*: 1019-1024, 1967. [Chl.]

9667 - DEROCHE, M.-E.: Amélioration de la séparation chromatographique des pigments liposolubles en carotènes, xanthophylles et chlorophylles. - Chim. anal. *53*: 704-709, 1971.

9668 - DESCOMPS, S.: Action de l'éclairement continu sur le développement des pleuridies de *Chara vulgaris* et sur l'infrastructure de leurs chloroplastes. - Compt. rend. Acad. Sci. Paris. Sér. D *273*: 2249-2252, 1971.

*9669 - Determination of photosynthetic pigments. - In: Determination of Photosynthetic Pigments in Sea-water. Pp. 9-18. Unesco, Paris 1966. [Filtration, extraction, spectrophotometry,]

9670 - DEVAULT, D.: Energy transduction in electron transport. - Biochim. biophys. Acta *226*: 193-199, 1971.

9671 - DÉVAY, M.: Does the amount of soluble proteins in the plastids or plasm depend on the photosynthetic CO_2 fixation? - Acta agron. Acad. Sci. hung. *20*: 451-452, 1971.

B9672 - DEVLIN, R.M., BARKER, A.V.: Photosynthesis. - Van Nostrand Reinhold Comp., New York - Cincinnati - Toronto - London - Melbourne 1971.

9673 - DICKMANN, D.I.: Chlorophyll, ribulose-1, 5-diphosphate carboxylase, and Hill reaction activity in developing leaves of *Populus deltoides*. - Plant Physiol. *48*: 143-145, 1971.

9674 - DICKMANN, D.I.: Photosynthesis and respiration by developing leaves of cottonwood (*Populus deltoides* BARTR.). - Bot. Gaz. *132*: 253-259, 1971.

9675 - DICKSON, M.H., SHANNON, S.: Genetic and maturity effects on chlorophylls and carotenoids in wax beans. - J. amer. Soc. hort. Sci. *96*: 510-513, 1971.

9676 - DIERS, L.: Übertragung von Plastiden durch den Pollen bei *Antirrhinum majus*. II. Der Einfluss verschiedener Temperaturen auf die Zahl der Schecken. - Mol. gen. Genet. *113*: 150-153, 1971.

9677 - DIETRICH, W.E. Jr., THORNBER, J.P.: The P700-chlorophyll *a*-protein of a blue-green alga. - Biochim. biophys. Acta *245*: 482-493, 1971.

9678 - DILLEY, R.A.: Coupling of ion and electron transport in chloroplasts. - In: SANADI, D.R. (ed.): Current Topics in Bioenergetics. Vol. 4. Pp. 237-271. - Academic Press, New York - London 1971.

9679 - DILMY, A.: The primary productivity of equatorial tropical forests in Indonesia. - In: DUVIGNEAUD, P. (ed.): Productivity of Forest Ecosystems. Pp. 333-337. Unesco, Paris 1971.

9680 - DILOVA, S.A., ZEINALOV, H.: Influence of *o*-dinitrophenol on the oxidation of zeaxanthin. - Dokl. bolg. Akad. Nauk *24*: 667-670, 1971.

9681 - DINER, B.A., MAUZERALL, D.C.: 3-(3,4-dichlorophenyl)-1, 1-dimethylurea-insensitive oxygen production in a cell-free preparation from *Phormidium luridum* that shows redox potential dependent coupling to one-electron oxidants. - Biochim. biophys. Acta *226*: 492-497, 1971.

9682 - DJAVANCHIR, A., ČATSKÝ, J.: Diffusion resistances of the adaxial and abaxial epidermes of kale leaves in a controlled environment. - Photosynthetica *5*: 267-268, 1971.

*9683 - DJENDOV, C.: Despre efectul sărurilor cu fier asupra plantelor de floarea-soarelui. [Effect of iron salts on sunflower plants.] - Stud. Cercet. Biol., Ser. bot. *21*: 429-435, 1969. [Chl; in Roum., ab: F.]

9684 - DODD, W.A., BIDWELL, R.G.S.: Photosynthesis and gas exchange of *Acetabularia* chloroplasts in an artificial leaf. - Nature *234*: 45-47, 1971.

9685 - DODD, W.A., BIDWELL, R.G.S.: The effect of Ph on the products of photosynthesis in $^{14}CO_2$ by chloroplast preparations from *Acetabularia mediterranea*. - Plant Physiol. *47*: 779-783, 1971.

9686 - DODGE, A.D., ALEXANDER, D.J., BLACKWOOD, G.C.: The contribution of photosynthesis to chlorophyll formation in etiolated mung bean leaves. - Physiol. Plant. *25*: 71-74, 1971.

9687 - DOEMEL, W.N., BROCK, T.D.: The physiological ecology of *Cyanidium caldarium*. - J. gen. Microbiol. *67*: 17-32, 1971. [Ps, Chl.]

9688 - DÖHLER, G., BRAUN, F.: Untersuchung der Beziehung zwischen extracellulärer Glykolsäure-Ausscheidung und der photosynthetischen CO_2-Aufnahme bei der Blaualge *Anacystis nidulans*. - Planta *98*: 357-361, 1971.

9689 - DOMAN, N.G., ROMANOVA, A.K.: Primenenie izotopov dlya issledovaniya assimilyatsii ugleroda pri foto- i khemosinteze. [Use of isotopes to study the assimilation of carbon during photo- and chemosynthesis.] - In: Primenenie Izotopov i Yadernykh Izluchenii v Sel'skom Khozyaĭstve. Pp. 124-130. Atomizdat, Moskva 1971. [In R.]

9690 - DOMES, W.: Unterschiedliche CO_2-Abhängigkeit des Gasaustausches beider Blattseiten von *Zea mays*. - Planta *98*: 186-189, 1971.

9691 - DONAHER, D.J., PARTANEN, C.R.: The role of light in the interrelated processes of morphogenesis and photosynthesis in the fern gametophyte. - Physiol. Plant. *25*: 461-468, 1971. [Ps.]

9692 - DÖRFLING, P., DUMMLER, W., MÜCKE, D.: Bedeutung der Citronensäure und anderer Säuren des Tricarbonsäurecyklus für das Wachstum und die Pigmentbildung der einzelligen Alge *Poteriochromonas stipitata*. - Z. allg. Mikrobiol. *11*: 163-167, 1971.

9693 - DÖRFLING, P., DUMMLER, W., MÜCKE, D.: Über die Chinone mit isoprenoider Seitenkette in *Poteriochromonas stipitata*. - Biochem. Physiol. Pflanzen *162*: 159-164, 1971. [Plastoquinones, ubiquinones.]

9694 - DÖRFLING, P., SEIFERT, P., DUMMLER, W., MÜCKE, D.: Der Einfluss von 4,6-Dinitro-o-kresol auf Atmung sowie Wachstum und Pigmentbildung von *Poteriochromonas stipitata*. - Z. allg. Mikrobiol. *11*: 169-172, 1971.

9695 - DORNHOFF, G.M.: Net photosynthesis of soybean leaves as influenced by anatomy, respiration, and variety. - Diss. Abstr. Int. B *32*: 1956-B, 1971.

*9696 - DORODNEVA, V.I.: Identifikatsiya karotinoidov list'ev gretskogo orekha *Juglans regia* L. metodami spektrofotometrii i tonkosloĭnoĭ khromatografii. [Identification of carotenoids of *Juglans regia* L. leaves by spectrophotometry and thin layer chromatography.] - Rast. Resur. *3*: 266-268, 1967. [In R.]

9697 - DOROKHOV, B.L., ZHAKOTE, A.G.: Opticheskie svoĭstva, intensivnost' i KPD fotosinteza list'ev fasoli pri usilenii mineral'nogo pitaniya. [Optical properties, photosynthetic rate and coefficient of use of radiant energy for bean leaves under high levels of mineral nutrition.] - In: Biokhimiya i Biofizika Fotosinteza. Pp. 223-228. Irkutsk 1971. [In R.]

9698 - DOWDELL, R.J., DODGE, A.D.: Chlorophyll formation and the development of photosynthesis in illuminated etiolated pea leaves. - Planta *98*: 11-19, 1971.

9699 - DOWNES, R.W.: Adaptation of sorghum plants to light intensity: its effect on gas exchange in response to changes in light, temperature, and CO_2. - In: HATCH, M.D., OSMOND, C.B., SLATYER, R.O. (ed.): Photosynthesis and Pho-

torespiration. Pp. 57-62. Wiley-Interscience, New York - London - Sydney -
Toronto 1971.

9700 - DOWNES, R.W.: Relationship between evolutionary adaptation and gas exchange
characteristics of diverse sorghum taxa. - Aust. J. biol. Sci. *24*: 843-852,
1971.

9701 - DOWNEY, L.A.: Effect of gypsum and drought stress on maize (*Zea mays* L.). I.
Growth, light absorption and yield. - Agron. J. *63*: 569-572, 1971.

9702 - DOWNTON, J., SLATYER, R.O.: Variation in levels of some leaf enzymes. -
Planta *96*: 1-12, 1971. [Chl.]

9703 - DOWNTON, W.J.S.: Adaptive and evolutionary aspects of C_4 photosynthesis. - In:
HATCH, M.D., OSMOND, C.B., SLATYER, R.O. (ed.): Photosynthesis and Photorespiration. Pp. 3-17. Wiley-Interscience, New York - London - Sydney - Toronto
1971.

9704 - DOWNTON, W.J.S.: The chloroplasts and mitochondria of bundle sheath cells in
relation to C_4 photosynthesis. - In: HATCH, M.D., OSMOND, C.B., SLATYER, R.O.
(ed.): Photosynthesis and Photorespiration. Pp. 419-425. Wiley-Interscience,
New York - London - Sydney - Toronto 1971.

9705 - DOWNTON, W.J.S.: Check list of C_4 species. - In: HATCH, M.D., OSMOND, C.B.,
SLATYER, R.O. (ed.): Photosynthesis and Photorespiration. Pp. 554-558. Wiley-
Interscience, New York - London - Sydney - Toronto 1971.

9706 - DOWNTON, W.J.S.: Further evidence for two modes of carboxyl transfer in plants
with C_4-photosynthesis. - Can. J. Bot. *49*: 1439-1442, 1971.

9707 - DOWNTON, W.J.S., PYLIOTIS, N.A.: Loss of photosystem II during ontogeny of
sorghum bundle sheath chloroplasts. - Can. J. Bot. *49*: 179-180, 1971.

9708 - DOZIER, W.A. Jr.: The influence of 2-chloroethylphosphonic acid (ethephon)
on growth, net photosynthesis, respiration rate, chlorophyll content, and
leaf abscission of the apple *Malus sylvestris* MILL. - Diss. Abstr. int. B
32: 650-B-651-B, 1971.

9709 - DOZIER, W.A. Jr., BARDEN, J.A.: Net photosynthesis and respiration of apple
leaves influenced by (2-chloroethyl) phosphonic acid. - J. amer. Soc.hort.
Sci. *96*: 789-790, 1971.

*9710 - DRABKOVA, V.G., LETANSKAYA, G.I.: Pervichnaya produktsiya fitoplanktona i
rol' bakteriĭ v destruktsii organicheskogo veshchestva v oz. Byalozero (Kol'-
skiĭ poluostrov). [Primary production of phytoplankton and role of bacteria
in destruction of organic matter in Lake Byalozero (The Kola Peninsula).] -
Gidrobiol. Zh. *6* (3): 37-43, 1970. [In R, ab: E.]

9711 - DRAPER, S.R., SIMON, E.W.: Changes in free fatty-acid content and respiratory activity during the senescence of cotyledons of cucumber. - J. exp. Bot.
22: 481-486, 1971. [Chl.]

9712 - DREWS, G., WITZEMANN, V.: Zur Taxonomie von *Rhodopseudomonas palustris*. -
Arch. Mikrobiol. *78*: 322-329, 1971. [Chl.]

9713 - van den DRIESSCHE, R.: Growth of one-year-old Douglas fir plants at four
spacings. - Ann. Bot. *35*: 117-126, 1971. [Growth analysis.]

9714 - van den DRIESSCHE, R., CONNOR, D.J., TUNSTALL, B.R.: Photosynthetic response
of brigalow to irradiance, temperature and water potential. - Photosynthetica
5: 210-217, 1971.

9715 - DRING, M.J.: Light quality and the photomorphogenesis of algae in marine
environments. - In: CRISP, D.J. (ed.): Fourth European Marine Biology Symposium. Pp. 375-392. Cambridge Univ. Press, London 1971. [Ps.]

9716 - DROKOVA, I.G.: Stereoizomery β-karotynu vodorosti *Dunaliella salina* TEOD.
[Stereoisomers of β-carotene of *Dunaliella salina* TEOD.] - Ukr. bot. Zh.
28: 670-673, 1971. [In Ukr., ab: E.]

9717 - DROKOVA, I.G., POPOVA, R.Ts.: Pigmentnyĭ sklad deyakykh karotynonosnykh
shtamiv vodorosti *Dunaliella salina* TEOD. [Pigment composition of some ca-
rotenoid-containing strains of the alga *Dunaliella salina* TEOD.] - Ukr. bot.
Zh. *28* (2): 153-155, 1971. [In Ukr.]

9718 - DRUMMOND, A.J., ÅNGSTRÖM, A.K.: Derivation of the photometric flux of day-
       light from filtered measurements of global (sun and sky) radiant energy. -
       Appl. Optics *10*: 2024-2030, 1971.

*9719 - DUBOVA, K.P., VIVAL'KO, I.G.: Osoblivosti vidtoku michenykh po C^{14} asymilya-
       tiv u roslyn kartopli. [Peculiarities of outflow of ^{14}C-labelled assimilates
       in potato plants.] - Dopovidi Akad. Nauk ukr. RSR, Ser. B (Geol., Geofiz.,
       Khim. Biol.) *32*: 462-465, 480, 1970. [In Ukr., ab: E, R.]

9720 - DUNCAN, W.G.: Leaf angles, leaf area, and canopy photosynthesis. - Crop Sci.
       11: 482-485, 1971.

9721 - DUNCAN, W.G., BARFIELD, B.J.: Description of photosynthesis within plant
       canopies. - Trans. ASAE *14*: 523-526, 1971.

9722 - DUNHAM, W.R., BEARDEN, A.J., SALMEEN, I.T., PALMER, G., SANDS, R.H., ORME-
       JOHNSON, W.H., BEINERT, H.: The two-iron ferredoxins in spinach, parsley,
       pig adrenal cortex, *Azotobacter vinelandii*, and *Clostridium pasteurianum*:
       Studies by magnetic field Mössbauer spectroscopy. - Biochim. biophys. Acta
       253: 134-152, 1971.

9723 - DUNHAM, W.R., PALMER, G., SANDS, R.H., BEARDEN, A.J.: On the structure of
       the iron-sulfer complex in the two-iron ferredoxins. - Biochim. biophys.
       Acta *253*: 373-384, 1971.

9724 - DUNIWAY, J.M., SLATYER, R.O.: Gas exchange studies on the transpiration and
       photosynthesis of tomato leaves affected by *Fusarium oxysporum* f. sp. *lyco-
       persici*. - Phytopathology *61*: 1377-1381, 1971.

9725 - DUNN, E.L.: Seasonal patterns of carbon dioxide metabolism in evergreen
       sclerophylls in California and Chile. - Diss. Abstr. int. B *31*: 5831-B-
       5832-B, 1971.

9726 - DUPAIGNE, F., LOUGUET, P.: Étude de la transmission des mouvements stomati-
       ques d'une partie à l'autre de la feuille de *Zea mays* au moyen d'une métho-
       de porométrique utilisant la diffusion de l'hydrogène à travers la feuille.
       - Physiol. vég. *9*: 129-136, 1971.

9727 - DUS, K., TEDRO, S., BARTSCH, R.G., KAMEN, D.M.: The primary structure of
       Chromatium high-potential iron-sulfur protein. - Biochem. biophys. Res.
       Commun. *43*: 1239-1245, 1971.

*9728 - DUTHIE, H.C., KIRTON, W.L. Jr.: Primary productivity and standing crops of
       phytoplankton in Belwood Reservoir. - Can. J. Bot. *48*: 665-670, 1970.

9729 - DUTTON, P.L.: Oxidation-reduction potential dependence of the interaction
       of cytochromes, bacteriochlorophyll and carotenoids at 77 °K in chromato-
       phores of *Chromatium* D and *Rhodopseudomonas gelatinosa*. - Biochim. biophys.
       Acta *226*: 63-80, 9171.

9730 - DUTTON, P.L., KIHARA, T., McCRAY, J.A., THORNBER. J.P.: Cytochrome C_{553} and
       bacteriochlorophyll interaction at 77 °K in chromatophores and a subchroma-
       tophore preparation from *Chromatium* D. - Biochim. biophys. Acta *226*: 81-87,
       1971.

9731 - DUVIGNEAUD, P.: Concepts sur la productivité primaire des écosystèmes fores-
       tiers. - In: DUVIGNEAUD, P. (ed.): Productivity of Forest Ecosystems. Pp.111-
       140. Unesco, Paris 1971.

9732 - DUVIGNEAUD, P., KESTEMONT, P., AMBROES, P.: Productivité primaire des forêts
       tempérées d'essences feuillues caducifoliées en Europe occidentale. - In:
       DUVIGNEAUD, P. (ed.): Productivity of Forest Ecosystems. Pp. 259-270. Unesco,
       Paris 1971.

9733 - DUYSENS, L.N.M.: Mechanisms of photosynthesis. - J. gen. Microbiol. *69* (3):
       V-VI, 1971.

9734 - DUYSENS, L.N.M.: Transfer of electronic excitation energy from pigments to
       photosynthetic reaction centers and *vice versa*. - In: BRODA, E., LOCKER, A.,
       SPRINGER-LEDERER, H. (ed.): Proceedings of the First European Biophysics Con-
       gress. Vol. IV. Pp. 13-17. Verlag Wiener med. Akad., Wien 1971.

9735 - DVOŘÁK, J., NÁTR, L.: Carbon dioxide compensation points of *Triticum* and
       Aegilops species. - Photosynthetica *5*: 1-5, 1971.

9736 - DWYER, M.R., SMILLIE, R.M.: β-1,3-glucan: a source of carbon and energy for chloroplast development in *Euglena gracilis*. - Aust. J. biol. Sci. *24*: 15-22, 1971.

*9737 - DYBING, C.D.: Maturity and yield of seedflax in controlled environments: effects of root environment. - Crop Sci. *9*: 572-575, 1969. [Chl.]

9738 - DYER, T.A., OSBORNE, D.J.: Leaf nucleic acids. II. Metabolism during senescence and the effect of kinetin. - J. exp. Bot. *22*: 552-560, 1971. [Chl.]

9739 - DYKYJOVÁ, D.: Production, vertical structure and light profiles in littoral stands of reed-bed species. - Hidrobiologia (Bucureşti) *12*: 361-376, 1971.

9740 - DYKYJOVÁ, D.: Productivity and solar energy conversion of reedswamp stands in comparison with outdoor mass cultures of algae in the temperate climate of Central Europe. - Photosynthetica *5*: 329-340, 1971.

9741 - DYKYJOVÁ, D., VÉBER, K., PŘIBÁŇ, K.: Productivity and root/shoot ratio of reedswamp species growing in outdoor hydroponic cultures. - Folia geobot. phytotax. *6*: 233-254, 1971.

9742 - DYLIS, N.: Primary production of mixed forests. - In: DUVIGNEAUD, P. (ed.): Productivity of Forest Ecosystems. Pp. 227-232. Unesco, Paris 1971.

9743 - DYSON, P.W., WATSON, D.J.: An analysis of the effects of nutrient supply on the growth of potato crops. - Ann. appl. Biol. *69*: 47-63, 1971. [Growth analysis.]

9744 - DZHAGAROV, B.M., SAGUN, E.I., GURINOVICH, G.P.: Dikhroizm triplet-tripletnogo pogloshcheniya v molekulakh tetrapirrol'nykh pigmentov. [Dichroism of triplet-triplet absorption in molecules of tetrapyrrole pigments.] - Zh. prikl. Spektroskop. *15*: 476-480, 1971. [In R.]

9745 - DZHANUMOV, D.A., SHCHERBAKOV, A.A.: Issledovanie ustoĭchivosti rasteniĭ k neblagopriyatnym faktoram khemilyuminestsentnym metodom. [Study of plant resistance to unfavourable conditions by a chemiluminescence method.] - Nauch. Dokl. vyssh. Shkoly, biol. Nauki *14* (10): 51-53, 1971. [Chl; in R.]

9746 - DZHOKHADZE, G.K.: [Effect of microelements and strains of *Rhizobium* on soybean photosynthesis.] - Soobshch. Akad. Nauk gruz. SSR *64*: 445-447, 1971. [In Georg., ab: E, R.]

*9747 - DZIĘCIOŁ, U., ANTOSZEWSKI, R.: Distribution of assimilates in strawberry plant. - Annu. Rep. Isotope Lab. Fruit Biochem., Res. Inst. Pomol. Skierniewice *IV*: 1-8, 1967.

9748 - EAGLES, C.F.: Changes in net assimilation rate and leaf-area ratio with time in *Dactylis glomerata* L. - Ann. Bot. *35*: 63-74, 1971.

9749 - EAGLES, C.F., ØSTGÅRD, O.: Variation in growth and development in natural populations of *Dactylis glomerata* from Norway and Portugal. I. Growth analysis. - J. appl. Ecol. *8*: 367-381, 1971.

9750 - EBATA, T., FUJITA, Y.: Changes in photosynthetic activity of the diatom *Phaeodactylum tricornutum* in a culture of limited volume. - Plant Cell Physiol. *12*: 533-541, 1971.

9751 - EBREY, T.G.: Anomalous energy transfer behaviour of light absorbed by bacteriochlorophyll in several photosynthetic bacteria. - Biochim. biophys. Acta *253*: 385-395, 1971.

*9752 - EBRINGER, L., MEGO, J.L., JURÁŠEK, A.: Mitomycins and the bleaching of *Euglena gracilis*. - Arch. Mikrobiol. *64*: 229-234, 1969. [Chl.]

9753 - ECKARDT, F.E., HEIM, G., METHY, M., SAUGIER, M., SAUVEZON, R.: Fonctionnement d'un écosystème au niveau de la production primaire. Mesures effectuées dans une culture d'*Helianthus annuus*. - Oecol. Plant. *6*: 51-100, 1971.

9754 - EDELMAN, J., HANSON, A.D.: Sucrose suppression of chlorophyll synthesis in carrot callus cultures. - Planta *98*: 150-156, 1971.

9755 - EDELMAN, J., HANSON, A.D.: Sucrose suppression of chlorophyll synthesis in carrot tissue cultures: the role of invertase. - Planta *101*: 122-132, 1971.

9756 - EDELMAN, J., SCHOOLAR, A.I.: The effects of iodoacetate and analogous compounds upon sucrose secretion by sugar-cane and other leaf tissue. - J. exp. Bot. *22*: 118-124, 1971. [Ps.]

9757 - EDELMAN, J., SCHOOLAR, A.I., BONNOR, W.B.: Permeability of sugar-cane chloroplasts to sucrose. - J. exp. Bot. *22*: 534-545, 1971.

*9758 - EDREVA, A., BAJLOV, D., NIKOLOV, S.: Contribution to the biochemical characteristics of certain wild species of *Nicotiana* with a view to their use in interspecific hybridization. - Dokl. Akad. Sel'skokhoz. Nauk Bolg. *3*: 55-62, 1970. [Chl, Car.]

9759 - EDWARDS, D.I., SCHOOLAR, A.I.: Inhibition of sugar synthesis and potentiation of chlorophyll degradation in sugar cane leaf tissue by metronidazole. - Z. Pflanzenphysiol. *64*: 73-76, 1971.

9760 - EDWARDS, G.E., BLACK, C.C.: Photosynthesis in mesophyll cells and bundle sheath cells isolated from *Digitaria sanguinalis* (L.) SCOP. leaves. - In: HATCH, M.D., OSMOND, C.B., SLATYER, R.O. (ed.): Photosynthesis and Photorespiration. Pp. 153-168. Wiley-Interscience, New York - London - Sydney - Toronto 1971.

9761 - EDWARDS, G.E., BLACK, C.C. Jr.: Insolation of mesophyll cells and bundle sheath cells from *Digitaria sanguinalis* /L./ SCOP. leaves and a scanning microscopy study of the internal leaf cell morphology. - Plant Physiol. *47*: 149-156, 1971. [Chl.]

9762 - EDWARDS, G.E., KANAI, R., BLACK, C.C.: Phosphoenolpyruvate carboxykinase in leaves of certain plants which fix CO_2 by the C_4-dicarboxylic acid cycle of photosynthesis. - Biochem. biophys. Res. Commun. *45*: 278-285, 1971.

9763 - EDWARDS, M.R., GANTT, E.: Phycobilisomes of the thermophilic blue-green alga *Synochococcus lividus*. - J. Cell Biol. *50*: 896-900, 1971.

*9764 - EFIMOV, M.V.: Osobennosti vodnogo rezhima i fotosinteza sel'skokhozyaĭstvennykh rasteniĭ v usloviyakh ZabaĭkaI'ya. [Peculiarities of water regime and photosynthesis of agricultural crops in the Trans-Baikal conditions.] - In: KLIMASHEVSKIĬ, E.L. (ed.): 14-e Nauchnye Chteniya Pamyati Mikhaila Grigor'evicha Popova. Pp. 3-12. Sibir. Inst. Fiziol. Biokhim. Rast. sibir. Otd. Akad. Nauk SSSR, Irkutsk 1970. [In R.]

*9765 - EFIMOV, M.V., KASHIN, V.K.: Deĭstvie Ĭoda i nikelya na fotosinteticheskuyu deyatel'nost' rasteniĭ kukuruzy v usloviyakh ZabaĭkaI'ya. [Effect of iodine and nickel on the photosynthetic activity of maize plants in the Trans-Baikal conditions.] - In: Mikroelementy v Sel'skom KhozyaĭstvE i Meditsine. Pp. 468-473. Ulan-Ude 1968. [In R.]

*9766 - EFIMOV, M.V., KHARLOVA (BUINOVA), M.G.: Osobennosti pigmentnoĭ sistemy rasteniĭ v usloviyakh ZabaĭkaI'ya. [Pigment system of plants in the Transbaikal region.] - Tr. buryat. Inst. estestv. Nauk, buryat. Filial sib. Otd. Akad. Nauk SSSR *4*: 45-48, 1969. [In R.]

*9767 - EFIMOV, M.V., KHARLOVA, M.G.: Sootnoshenie mezhdu khlorofillami *"a"* i *"b"* v rasteniyakh Buryatii. [Relationships between the chlorophylls *a* and *b* in the plants of Buryatia.] - Inform. Byull. (Irkutsk) *8*: 54-55, 1970. [In R.]

9768 - EFIMOVA, N.A., KUPCHENKO, G.S.: Zavisimost' meteorologicheskogo rezhima ot biometricheskikh kharakteristik poseva. [Dependence of meteorological regime on biometrical characteristics of a canopy.] - In: ZALENSKIĬ, O.V. (ed.): Fotosintez i Ispol'zovanie Solnechnoĭ Energii. Pp. 25-30. Nauka, Leningrad 1971. [In R, ab: E.]

*9769 - EFREMOV, V.V.: Izuchenie vliyaniya pozdnikh azotnykh podkormok na fiziologicheskie i biokhimicheskie protsessy, urozhaĭ i kachestvo zerna ozimoĭ pshenitsy. [Effect of late nitrogen fertilizing on physiological and biochemical processes, yield and grain quality in winter wheat.] - Tr. khar'kov. sel'skokhoz. Inst. *90*(Issledovaniya Fiziologii Biokhimii Rasteniĭ): 37-42, 1970. [Ps, Chl; in R.]

*9770 - EFREMOV, V.V.: Vliyanie razlichnykh doz mocheviny pri vnekornevoĭ podkormke v faze kolosheniya na kachestvo zerna ozimoĭ pshenitsy. [Effect of various urea doses in extra-root fertilizing during the heading phase on the grain quality in winter wheat.] - Tr. khar'kov. sel'-skokhoz. Inst. *90* (Issledo-vaniya Fiziologii Biokhimii Rastenii): 43-47, 1970. [In R.]

9771 - EGED, Š., KOLEK, J., DUDA, M.: Sezónne zmeny rastu a hodnoty čistej asimi-lácie u slnečnice a fazule. [Seasonal changes in growth and net assimilation rates in sunflower and bean.] - Biológia (Bratislava) *26*: 41-48, 1971. [In Slovak, ab: E.]

9772 - EGUNJOBI, J.K.: Ecosystem processes in a stand of *Ulex europaeus* L. I. Dry matter production, litter fall and efficiency of solar energy utilization. - J. Ecol. *59*: 31-38, 1971.

*9773 - EHEART, M.S.: Effect of storage and other variables on composition of frozen broccoli. - Food Technol. *24*: 69-71, 1970. [Chl.]

*9774 - EĬDEL'MAN, Z.M.: Poznanie Taĭny Zelenogo Rasteniya. (K 200-letiyu Ucheniya o Fotosinteze.) [Recognizing Green Plant's Mystery. (On the Occasion of the 200th Anniversary of Photosynthesis Science.)] - Znanie, Moskva 1970. [In R.]

9775 - EĬDEL'MAN, Z.M., SYCHEVA, S.P.: Vliyanie insektitsidov na fotofosforili-rovaniya. [Effects of insecticides on photophosphorylation.] - In: ZALENSKIĬ, O.V. (ed.): Fotosintez i Ispol'zovanie Solnechnoĭ Energii. Pp. 209-213. Nau-ka, Leningrad 1971. [In R, ab: E.]

9776 - EILAM, Y., BUTLER, R.D., SIMON, E.W.: Ribosomes and polysomes in cucumber leaves during growth and senescence. - Plant Physiol. *47*: 317-323, 1971. [Chl.]

9777 - EĬNOR, L.O., KORDYUM, V.A., BUKH, I.G.: Izuchenie vliyaniya nekotorykh ros-taktiviruyushchikh veshchestv na fotosintez i rost avtotrofnoĭ termofil'noĭ khlorelly. [Effect of some growth activating substances on photosynthesis and growth of autotrophic thermophilic *Chlorella*.] - Gidrobiol. Zh. *7* (2): 69-73, 1971. [In R, ab: E.]

9778 - EĬNOR, L.O., L'VOVS'KA, N.R.: Zastosuvannya metodu Louri dlya vyznachennya bilka v khloroplastakh. [Application of the Lowry method for protein con-tent determination in chloroplasts.] - Ukr. bot. Zh. *28*: 241-245, 1971. [In Ukr., ab: E, R.]

9779 - EISA, H.M., DOBRENZ, A.K.: Morphological and anatomical aspects of oedema in eggplants (*Solanum melongena* L.). - J. amer. Soc. hort. Sci. *96*: 766-769, 1971. [Chl.]

9780 - ELEY, D.D., PETHIG, R.: Microwave Hall mobility measurements on rat liver mitochondria and spinach chloroplasts. - J. Bioenerg. *2*: 39-45, 1971.

9781 - EL-FOULY, M.M., EL-HAMAWI, H.A., FAWZI, A.F.A.: Different effects of chlor-mequat on chlorophyll, protein and yield in cotton grown under varying water regimes. - Plant Soil *35*: 183-186, 1971.

9782 - ELLIS, R.J.: Lincomycin as a chloroplast probe. - Biochem. J. *124*: 52P-53P, 1971.

9783 - ELLSWORTH, R.K.: Simple method for separation of milligram quantities of protochlorophyll *a* from seed oils in extracts of whole pumpkin seeds. - Anal. Biochem. *39*: 540-542, 1971.

9784 - ELLSWORTH, R.K.: Studies on chlorophyllase I. Hydrolytic and esterification activities of chlorophyllase from wheat seedlings. - Photosynthetica *5*: 226-232, 1971.

9785 - EL-MANSY, H.I., SALISBURY, F.B.: Biochemical responses of *Xanthium* leaves to ultraviolet radiation. - Rad. Bot. *11*: 325-328, 1971. [Chl.]

9786 - EMEL'YANOV, N.A.: Izmerenie vydeleniya ili pogloshcheniya gazov volyumetri-cheskim metodom s pomoshch'yu apparata Varburga. [Volumetric method for measuring the gas consumption or evolution with the Warburg apparatus.] - Ukr. biokhim. Zh. *43*: 390-392, 1971. [In R, ab: E.]

9787 - ENGELBRECHT, L.: Cytokinins in buds and leaves during growth, maturity and aging (with a comparison of two bioassays). - Biochem. Physiol. Pflanzen *162*: 547-558, 1971. [Chl.]

9788 - ENGLISH, L.S., LEES, D.N., ROGERS, L.J.: Aggregated states of chromatophore membrane protein from *Rhodopseudomonas spheroides*. - Biochem. J. *125*: 42P-43P, 1971.

9789 - EPEL, B.L., LEVINE, R.P.: Mutant strains of *Chlamydomonas reinhardi* with lesions on the oxidizing side of photosystem II. - Biochim. biophys. Acta *226*: 154-160, 1971.

9790 - ERDELSKÝ, K., REGULA, Š.: Influence of red and blue lights on the growth and pigment formations in tissue cultures of *Nicotiana tabacum* L. - Acta Fac. Rerum nat. Univ. Comenianae - Physiol. Plant. *III - 1971*: 11-51, 1971.

9791 - ERGASHEV, A., ABDURAKHMANOVA, Z.N.: Deĭstvie vysokogornoĭ UF-radiatsii na biosintez aminokislot pri fotosinteze. [Action of high-altitude ultraviolet radiation on amino acid biosynthesis during photosynthesis.] - Dokl. Akad. Nauk tadzh. SSR *14* (12): 53-57, 1971. [In R, ab: Tadzh.]

9792 - ERGASHEV, A., ABDURAKHMANOVA, Z.N., KICHITOV, V.K., NASYROV, Yu.S.: Vliyanie estestvennoĭ vysokogornoĭ UF-radiatsii na fotosinteticheskuyu assimilyatsiyu ugleroda. [Action of natural UV-radiation on photosynthesis.] - In: ZALENSKIĬ, O.V. (ed.): Fotosintez i Ispol'zovanie Solnechnoĭ Energii. Pp. 226-231. Nauka, Leningrad 1971. [In R, ab: E.]

9793 - ERIXON, K., BUTLER, W.L.: Light-induced absorbance changes in chloroplasts at -196 ºC. - Photochem. Photobiol. *14*: 427-433, 1971.

9794 - ERIXON, K., BUTLER, W.L.: The relationship between Q, C-550 and cytochrome b 559 in photoreactions at - 196º in chloroplasts. - Biochim. biophys. Acta *234*: 381-389, 1971.

9795 - ERIXON, K., BUTLER, W.L.: Destruction of C-550 by ultraviolet radiation. - Biochim. biophys. Acta *253*: 483-486, 1971.

9796 - EROKHINA, L.G., KRASNOVSKIĬ, A.A.: Vliyanie denaturiruyushchikh vozdeĭstviĭ na spektral'nye svoĭstva fikotsianina. [The effect of denaturing agents on the absorption and fluorescence spectra of phycocyanin.] - Mol. Biol. (Moskva) *5*: 399-408, 1971. [In R, ab: E.]

*9797 - ESHEL, Y.: Tolerance of cotton to diuron, fluometuron, norea, and prometryne. - Weed Sci. *17*: 492-496, 1969. [Ps.]

9798 - EVANS, L.T.: Evolutionary, adaptive, and environmental aspects of the photosynthetic pathway: assessment. - In: HATCH, M.D., OSMOND, C.B., SLATYER, R.O. (ed.): Photosynthesis and Photorespriation. Pp. 130-136. Wiley-Interscience, New York - London - Sydney - Toronto 1971.

*9799 - EVANS, M.C.W., HALL, D.O., JOHNSON, C.E.: Hyperfine structure of [^{57}Fe] iron in the Mössbauer spectrum of the high-potential iron protein from *Chromatium*. - Biochem. J. *119*: 289-291, 1970.

9800 - EVANS, W.R.: Adenosine diphosphoribose phosphorylase from *Euglena gracilis*. - In: COLOWICK, S.P., KAPLAN, N.O. (ed.): Methods in Enzymology. Vol. 23. Pp. 566-570. Academic Press, New York - London 1971.

9801 - EVANS, W.R., SMILLIE, R.M.: Inhibition of light-dependent development of chloloplasts by 5-fluorouracil. - J. exp. Bot. *22*: 371-381, 1971.

9802 - EVERSON, R.G.: Carbonic anhydrase in photosynthesis. - In: HATCH, M.D., OSMOND, C.B., SLATYER, R.O. (ed.): Photosynthesis and Photorespiration. Pp. 275-281. Wiley-Interscience, New York - London - Sydney - Toronto 1971.

9803 - EVERSON, R.G., GRAHAM, D.: Addendum: Effects of an inhibitor of carbonic anhydrase on light-induced pH changes in pea chloroplasts. - In: HATCH, M.D., OSMOND, C.B., SLATYER, R.O. (ed.): Photosynthesis and Photorespiration. Pp. 281-282. Wiley-Interscience, New York - London - Sydney - Toronto 1971.

9804 - EVERT, R.F., DESHPANDE, B.P.: Plastids in sieve elements and companion cells of *Tilia americana*. - Planta *96*: 97-100, 1971.

9805 - EVSTIGNEEV, V.B., GAVRILOVA, V.A.: O vosstanovlenii khlorofilla askorbinovoĭ kislotoĭ v dioksane. [Reduction of chlorophyll by ascorbic acid in dioxane.] - Dokl. Akad. Nauk SSSR *200*: 725-728, 1971. [In R.]

9806 - EVSTIGNEEV, V.B., SADOVNIKOVA, N.A., KOSTIKOV, A.P., GRIBOVA, Z.P., KAYUSHIN, L.P.: Zavisimost' signala elektronnogo paramagnitnogo rezonansa pri fotookis-

lenii khlorofilla khinonom ot kislotnosti sredy. [Effect of acidity of the
medium on the electron spin resonance signal during quinone photooxidation
of chlorophyll.] - Biofizika *16*: 431-436, 1971. [In R, ab: E.]

9807 - EVSTIGNEEV, V.B., SHVEDOVA, T.A.: O fotopotentsiale pri obratimykh fotokhimicheskikh reaktsiyakh khlorofilla s okislitelyami. [Photopotential in reversible photochemical reactions of chlorophyll with oxidizing agents.] -
Dokl. Akad. Nauk SSSR *198*: 706-709, 1971. [In R.]

9808 - EVSTIGNEEV, V.B., SHVEDOVA, T.A.: O vliyanii kislotnosti sredy na fotovos-
stanovlenie i fotosensibiliziruyushchuyu sposobnost' khlorofilla. [Effect
of medium acidity on photoreduction and photosensitizing properties of chlorophyll.] - Biofizika *16*: 25-31, 1971. [In R, ab: E.]

9809 - FABIAN, I.: Der Einfluss von Phosphor und Kalium im Licht und im Dunkel auf
den Carotinoid-Gehalt der Sonnenblumenblätter. - Rev. roum. Biol. - Sér. Bot.
16: 347-356, 1971.

9810 - FABIAN, I.: Pigmenţii cloroplastelor. [Pigments of chloroplasts.] - Natura
(Bucureşti) *23* (3): 9-20, 1971. [In Roum.]

9811 - FABIAN-GALAN, G.: Produşii fotosintezei. [Products of photosynthesis.] -
Natura (Bucureşti) *23* (5): 23-32, 1971. [In Roum.]

9812 - FABIAN-GALAN, G.: The transport of assimilates in P- and K-deficient plants.
- Rev. roum. Biol. - Sér. Bot. *16*: 33-40, 1971.

9813 - FABIAN-GALAN, G.: P and K influence on photosynthesis and on its products in
sunflower leaves. - Rev. roum. Biol. - Sér. Bot. *16*: 357-363, 1971.

9814 - FADRUS, H., MALÝ, J.: Photometric determination of small amounts of oxygen
in water with 3,3'-dimethylnaphthidine. - Analyst *96*: 591-597, 1971.

9815 - FAGADE, S.O., de DATTA, S.K.: Leaf area index, tillering capacity, and
grain yield of tropical rice as affected by plant density and nitrogen level. - Agron. J. *63*: 503-506, 1971.

9816 - FANICA-GAIGNIER, M., CLEMENT-METRAL, J., KAMEN, M.D.: Adenine nucleotide
levels and photopigment synthesis in a growing photosynthetic bacterium. -
Biochim. biophys. Acta *226*: 135-143, 1971.

9817 - FARINEAU, J.: A comparative study of the activities of photosynthetic carboxylation in a C_4 and a Calvin-type plant (the sites of CO_2 fixation in C_4
plants). - In: HATCH, M.D., OSMOND, C.B., SLATYER, R.O. (ed.): Photosynthesis
and Photorespiration. Pp. 202-210. Wiley-Interscience, New York - London -
Sydney - Toronto 1971.

9818 - FARRELL, G.M.: Localization of photosynthetic products in potato leaves infected by *Phytophtora infestans*. - Physiol. Plant Pathol. *1*: 457-467, 1971.

*9819 - FASULO, M.P., DALL'OLIO G.: Alcuni effetti dei ritardanti di crescita "cy-
cocel" e "AMO-1618" sul metabolismo di *Euglena gracilis* KLEBS. [Some effects
of growth retardants CCC and AMO-1618 on the metabolism of *Euglena gracilis*
KLEBS.] - Atti Accad. Sci. Ferrara *45-46*: 1-12, 1968/69. [Ps; in Ital., ab:
E, F.]

9820 - FEDELI, E., CAMURATI, F., JACINI, G.: Structure of monohydroperoxides formed
by chlorophyll photo-sensitized oxidation of methyl linoleate. - J. amer.
Oil Chem. Soc. *48*: 787-789, 1971.

*9821 - FEDENKO, E.P., KONDRAT'EVA, E.N., KRASNOVSKIĬ, A.A.: Protokhlorofill'nye mu-
tanty *Rhodopseudomonas palustris*. [Protochlorophyll mutants of *Rhodopseudomonas palustris*.] - Nauch. Dokl. vyssh. Shkoly, biol. Nauki *12* (8): 102-111,
1969. [In R.]

*9822 - FEDENKO, E.P., LANG, F.: O predshestvennikakh bakteriokhlorofilla *a* u pigmentnykh mutantov *Rhodopseudomonas palustris*. [Precursors of bacteriochlorophyll *a* in pigment mutants of *Rhodopseudomonas palustris*.] - Nauch. Dokl.
vyssh. Shkoly, biol. Nauki *13* (3): 134, 1970. [In R.]

9823 - FEDERER, C.A.: Solar radiation absorption by leafless hardwood forests. -
Agr. Meteorol. *9*: 3-20, 1971.

*9824 - FEDIN, P.E., SHIROBOKOVA, E.S.: Sravnitel'naya.otsenka priborov dlya opre-

deleniya Intensivnosti fotosinteza u zernobobovykh kul'tur. [Comparative
evaluation of apparatuses for determining photosynthetic rate in leguminous
crops.] - In: Tezisy Dokladov Vsesoyuznogo Soveshchaniya po Unifikatsii
Metodov I Priborov dlya Massovykh Izmerenii Intensivnosti Fotosinteza. Pp.
109-111. Nauch. -issled. Inst. Rastenievod. N.I. Vavilova, Leningrad - Pushkin 1970. [In R.]

9825 - FEDOROV, V.D., KHROMOV, V.M.: Deistvie sveta na fotosinteticheskuyu aktivnost' fitoplanktona v zavisimosti ot ego obespechennosti mineral'nym pitaniem. [Effect of light on the photosynthetic activity of phytoplankton
depending on the supply of mineral nutrition.] - Izv. Akad. Nauk SSSR, Ser.
biol. *1971*: 503-517, 1971. [In R, ab: E.]

9826 - FEDOROVA, A.I.: Vliyanie udobrenii na soderzhanie osnovnykh makroelementov
i khlorofilla v khvoe listvennitsy sibirskoi. [Effect of mineral nutrition
on the content of basic macroelements and chlorophyll in needles of *Larix
sibirica*.] - In: Fiziologo-biokhimicheskie Osobennosti Drevesnykh Rastenii
Sibiri. Pp. 47-54. Nauka, Moskva 1971. [In R.]

9827 - FEE, J.A., MAYHEW, S.G., PALMER, G.: The oxidation-reduction potentials of
parsley ferredoxin and its selenium-containing homolog. - Biochim. biophys.
Acta *245*: 196-200, 1971.

9828 - FEE, J.A., PALMER, G.: The properties of parsley ferredoxin and its selenium-
containing homolog. - Biochim. biophys. Acta *245*: 175-195, 1971.

9829 - FEHER, G.: Some chemical and physical properties of a bacterial reaction center particle and its primary photochemical reactants. - Photochem. Photobiol.
14: 373-387, 1971.

9830 - FEHR, W.R., THORNE, J.C., HAMMOND, E.G.: Relationship of fatty acid formation
and chlorophyll content in soybean seed. - Crop Sci. *11*: 211-213, 1971.

*9831 - FEIGE, B.: Stoffwechselphysiologische Untersuchungen an der tropischen Basi-
diolichene *Cora pavonia* (SW.) FR. - Flora A *160*: 169-180, 1969. [Ps.]

*9832 - FEIGE, B.: Untersuchungen zur Stoffwechselphysiologie der Flechten unter Ver-
wendung radioaktiver Isotope. - Vortr. Gesamtgeb. Bot. *1970* (4): 35-44, 1970.
[Ps.]

*9833 - FEIGE, B.: Zur Verwertung uniform ^{14}C-markierter Glukose und uniform ^{14}C-mar-
kierten Glycerins durch die Flechte *Cladonia convoluta* (LAM.) P. COUT. - Z.
Pflanzenphysiologie *63*: 211-213, 1970. [Ps.]

*9834 - De FEKETE, M.A.R.: Die Rolle der Phosphorylase im Stoffwechsel der Stärke in
den Plastiden. - Planta *79*: 208-221, 1968.

9835 - FEL'DMAN, N.L., KAMENTSEVA, I.E. Teploustoichivost' kletok i vnutrikletochnykh
belkov vesennego i letnego belotsvetnikov. [Thermostability of cells and of
intracellular proteins in spring and summer snowflakes.] - Tsitologiya *13*:
479-483, 1971. [Ps; in R; ab: E.]

9836 - FENCHEL, T., STRAARUP, B.J.: Vertical distribution of photosynthetic pigments
and the penetration of light in marine sediments. - Oikos *22*: 172-182, 1971.

9837 - FERREE, M.E., BARDEN, J.A.: The influence of strains and rootstocks on photosynthesis, respiration, and morphology of "Delicious" apple trees. - J. amer.
Soc. hort. Sci. *96*: 453-457, 1971.

9838 - FERRON, F.: Observation de la régulation stomatique dans la feuille de Blé
par une méthode isotopique. - Physiol. vég. *9*: 281-301, 1971.

9839 - FIALA, V., TETTER, M.: Změny intenzity fotosyntézy, obsah chlorofylu a cel-
kového dusíku v listech srhy laločnaté během vegetace a vlivem hnojení.
[Changes in photosynthetic rate, content of chlorophyll and total nitrogen
in leaves of cocksfoot in the course of vegetation as influenced by fertilizing.] - Rostlinná Výroba (Praha) *17*: 511-518, 1971. [In Czech, ab: E, R.]

9840 - FILIP, L., BANU, M., SCUTARU-UNGUREAN, C.: Metodă de calcul expeditiv pentru
stabilirea producţiei medii evaluate la hectar. [A rapid calculation method
for establishing the estimated average yield per hectare.] - Probl. agr. *23*
(6): 69-73, 1971. [In Roum.]

*9841 - FILIPPOVA, L.A.: Vydelenie khloroplastov v organicheskoĭ srede metodom diffe-
rentsial'nogo tsentrifugirovaniya. [Isolation of chloroplasts in organic
medium by the differential centrifugation technique.] - In: KIRICHEŇKO, E.B.
(ed.): Metody Vydeleniya Khloroplastov. Pp. 116-125. Pushchino-na-Oke 1970.
[In R, ab: E.]

*9842 - FILIPPOVA, L.A., ZALENSKIĬ, O.V.: O vnutrikletochnoĭ lokalizatsii assimilya-
tov i ikh ispol'zovanii v protsesse dykhaniya. [Intracellular localisation
of assimilates and their utilization in the process of respiration.] - In:
Fotosintez i Ispol'zovanie Solnechnoĭ Radiatsii. Tezisy Dokladov. Pp. 69-70.
Dushanbe 1967. [In R.]

9843 - FILIPPOVA, L.A., ZALENSKIĬ, O.V.: O vnutrikletochnom transporte assimilyatov
pri ingibirovanii okislitel'nogo fosforilirovaniya. [Intracellular transport
of assimilates during the inhibition of oxidative phosphorylation.]- In:
Biokhimiya i Biofizika Fotosinteza. Pp. 183-186. Sibir. Inst. Fiziol. Biokhim.
Rast., Irkutsk 1971. [In R.]

9844 - FILIPPOVA, L.A., ZALENSKIĬ, O.V.: Vliyanie nekotorykh faktorov na vnutri-
kletochnyĭ transport assimilyatov. [Effect of some factors on the intracellular transport of photosynthates.] - In: ZALENSKIĬ, O.V. (ed.): Fotosintez
i Ispol'zovanie Solnechnoĭ Energii. Pp. 181-184. Nauka, Leningrad 1971. [In
R, ab: E.]

9845 - FILIPPOVICH, I.I., ALIEV, K.A., SVETAĬLO, E.N., SATAROVA, N.A.: Sravnitel'noe
izuchenie beloksinteziruyushcheĭ sistemy khloroplastov i tsitoplazmy prorostkov gorokha. [Comparative study of protein-synthesizing system of chloroplasts
and cytoplasm in pea seedlings.] - In: ZALENSKIĬ, O.V. (ed.): Fotosintez i
Ispol'zovanie Solnechnoĭ Energii. Pp. 169-173. Nauka, Leningrad 1971. [In R,
ab: E.]

*9846 - FIRGER, V.V., KARPOVA, T.B.: Vliyanie mineral'nogo pitaniya na metabolizm
soedineniĭ ugleroda-14 i gazoustoĭchivost' gazonnykh trav. [Effect of mineral
nutrition on the metabolism of ^{14}C compounds and gas resistance of lawn grasses.] - Uch. Zap. perm. Ordena trud. kras. Znameni gos. Univ. A.M. Gor'kogo
222: 85-97, 1969. [Ps; in R.]

9847 - FISCHER, K.S., WILSON, G.L.: Measurement of distribution of photosynthesis
in plant canopies. - Nature - new Biol. *229*: 30, 1971.

*9848 - FIŠERE, Dž.: Dažas vienkāršas lauka metodes fotosintēzes un elpošanas notei-
kšanai. [Some simple field methods of determining photosynthetic and respiration rates.]- In: Fotosintēzes Pētīšana Sējumos. Pp. 63-81. Zinātne, Rīgā
1970. [In Latvian, ab: E.]

9849 - FISHER, R.R., GUILLORY, R.J.: Resolution of enzymes catalyzing energy-linked
transhydrogenation. II. Interaction of transhydrogenase factor with the *Rhodospirillum rubrum* chromatophore membrane. - J. biol. Chem. *246*: 4679-4686,
1971.

9850 - FISHER, R.R., GUILLORY, R.J.: Resolution of enzymes catalyzing energy-linked
transhydrogenation. III. Preparation and properties of *Rhodospirillum rubrum*
transhydrogenase factor.- - J. biol. Chem. *246*: 4687-4693, 1971.

9851 - FLAHIVE, W.J.: The effect of blue, green, red and white light of varying intensities on the pigment content and photosynthetic capabilities of a red
alga *Porphyra umbilicalis*. - Diss. Abstr. Int. B *31*: 5204-B, 1971.

9852 - FLECK- GERNDT G.: Untersuchungen uber die Photosynthese alternder Blätter. I.
Beziehungen zwischen Chlorophyllgehalt, Stickstoffhaushalt und Assimilations-
intensität. - Biol. Zentralbl. *90*: 479-506, 1971.

9853 - FLECK-GERNDT, G.: Untersuchungen über die Photosynthese alternder Blätter. II.
Der Einbau von Radiokohlenstoff in die Assimilate. - Biol. Zentralbl. *90*: 723-
743, 1971.

9854 - FLEISCHMANN, D.E.: Glow curves from photosynthetic bacteria. - Photochem.
Photobiol. *14*: 65-70, 1971.

9855 - FLEISCHMANN, D.E.: Luminescence in photosynthetic bacteria. - Photochem.
Photobiol. *14*: 277-286, 1971.

9856 - FLEISCHMAN, D.E., COOKE, J.A.: Electron transport in *Rhodopseudomonas viridis* at low temperatures. - Photochem. Photobiol. *14*: 71-83, 1971.

9857 - FLETCHER, R.A., McCULLAGH, D.: Cytokinin-induces chlorophyll formation in cucumber cotyledons. - Planta *101*: 88-90, 1971.

9858 - FLOHRS, H., HAUPT, W.: Tagesperiodische Empfindlichkeitsschwankungen der lichtinduzierten Chloroplastenbewegung von *Mougeotia*. - Z. Pflanzenphysiol. *65*: 65-69, 1971.

9859 - FLOYD, R.A., CHANCE, B., DEVAULT, D.: Low temperature photo-induced reactions in green leaves and chloroplasts. - Biochim. biophys. Acta *226*: 103-112, 1971.

9860 - FLOYD, R.A., KEYHANI, E., CHANCE, B.: Membrane structure and function. II. Alterations in the photo-induced absorption changes after treatment of isolated chloroplasts with large pulses of the ruby laser. - Arch. Biochem. Biophys. *146*: 627-634, 1971. [Chl.]

9861 - FOCK, H., CANVIN, D.T., GRANT, B.R.: Effects of oxygen and carbon dioxide on photosynthetic O_2 evolution and CO_2 uptake in sunflower and *Chlorella*. - Photosynthetica *5*: 389-394, 1971.

9862 - FOCKE, R.: Assimilationsmessungen an einigen Kulturpflanzen unter winterlichen Temperaturbedingungen. - Biol. Zbl. *90*: 161-173, 1971.

*B9863 - FOGG, G.E.: Photosynthesis. - English Univ. Press, London 1968; 2nd Ed. 1969.

9864 - FOGG, G.E.: Recycling through algae. - Proc. roy. Soc. London, Ser. B *179*: 201-207, 1971. [Ps.]

9865 - FOOS, K.: Untersuchungen zur Feinstruktur von *Mougeotia* spec. und zum Bewegungsmechanismus der Chloroplasten. - Z. Pflanzenphysiol. *64*: 369-386, 1971.

9866 - FOPPEN, F.H.: Tables for the identification of carotenoid pigments. - Chromatogr. Rev. *14*: 133-298, 1971.

9867 - FORBUSH, B., KOK, B., McGLOIN, M.P.: Cooperation of charges in photosynthetic O_2 evolution - II. Damping of flash yield oscillation, deactivation. - Photochem. Photobiol. *14*: 307-321, 1971.

9868 - FORD, E.D.: The potential production of forest crops. - In: WAREING, P.F., COOPER, J.P. (ed.): Potential Crop Production. A Case Study. Pp. 172-186. Heinemann Educational Books, London 1971.

9869 - FORD, E.D., NEWBOULD, P.J.: The leaf canopy of a coppiced deciduous weodland. I. Development and structure. - J. Ecol. *29*: 843-862, 1971. [Chl.]

9870 - FORGER, J.M. III., BOGORAD, L.: Acid to base phosphorylation and membrane integrity in plastids of greening maize. - Biochim. biophys. Acta *226*: 383-392, 1971.

9871 - FORK, D.C.: Studies on the oxidation-reduction reactions of grana particles prepared from spinach chloroplasts by a nondetergent method. - Carnegie Inst. Year Book *70*: 472-482, 1971.

9872 - FORK, D.C., JACOBI, G.: Photochemical reactions of P700 and cytochrome *f* and the action of plastocyanin in a grana stack preparation from spinach. - Carnegie Inst. Year Book *69*: 690-695, 1971.

9873 - FORK, D.C., MURATA, N.: Oxidation-reduction reactions of P700 and cytochrome *f* in fraction 1 particles prepared from spinach chloroplasts by French press treatment. - Photochem. Photobiol. *13*: 33-44, 1971.

9874 - FORK, D.C., MURATA, N.: P700 and cytochrome *f* oxidation-reduction reactions in fraction 1 particles from spinach. - Carnegie Inst. Year Book *69*: 682-690, 1971.

9875 - FORTI, G.: NADPH-cytochrome *f* reductase from spinach. - In: COLOWICK, S.P., KAPLAN, N.O. (ed.): Methods in Enzymology. Vol. 23. Pp. 447-451. Academic Press, New York - London 1971.

9876 - FORTI, G., MEYER, E.M.: Effect of pyrophosphate on photosynthesis and photophosphorylation. - In: QUAGLIARIELLO, E., PAPA, S., ROSSI, C.S. (ed.): Energy Transduction in Respiration and Photosynthesis. Pp. 559-564. Adriatica Editrice, Bari 1971.

9877 - FORTI, G., ROSA, L.: On the pathway of electron transport in cyclic photophosphorylation. - FEBS Lett. *18*: 55-58, 1971.

B9878 - Fotosintez i Ispol'zovanie SolnechnoT Energii. [Photosynthesis and Solar
Energy Utilization.] - Nauka, Leningrad 1971. [In R, ab: E.]

B9879 - Fotosintez, Rost i UstoTchivost' RasteniT. [Photosynthesis, Growth and
Resistance of Plants.] - Naukova Dumka, Kiev 1971. [In R.]

9880 - FOWLER, C.F.: Light-induced electron transport in photosynthetic bacteria.
- Diss. Abstr. Int. B *31*: 7128-B, 1971.

9881 - FOWLER, C.F., NUGENT, N.A., FULLER, R.C.: The isolation and characterization of a photochemically active complex from *Chloropseudomonas ethylica*.
- Proc. nat. Acad. Sci. U.S.A. *68*: 2278-2282, 1971.

9882 - FØYN, E., HANNEBORG, S.: Determination of ^{14}C labelled carbonate in solution. - Mar. Biol. *8*: 57-59, 1971.

9883 - FRACKOWIAK, B.: Struktura i funkcja błon chloroplastów. [Structure and function of chloroplast membranes.] - Postępy Biochem. *17*: 235-248, 1971. [In
Pol.]

9884 - FRACKOWIAK, B., KANIUGA, Z.: Studies on the enzyme systems involved in electron and energy transfer in isolated chloroplasts. I. Effect of endogenous
phosphate on the photophosphorylation coupled with noncyclic electron transport in intact chloroplasts. - Biochim. biophys. Acta *226*: 360-365, 1971.

9885 - FRĄCKOWIAK, D., GRABOWSKI, J.: Low temperature spectra of biliproteins. -
Photosynthetica *5*: 146-152, 1971.

9886 - FRAKER, P.J., KAPLAN, S.: Isolation and fractionation of the photosynthetic
membranous organelles from *Rhodopseudomonas spheroides*. - J. Bacteriol. *108*:
465-473, 1917.

9887 - FREDERICK, S.E., NEWCOMB, E.H.: Ultrastructure and distribution of microbodies in leaves of grasses with and without CO_2-photorespiration. - Planta
96: 152-174, 1971.

9888 - FREDRICKS, W.W., GEHL, J.M.: Stimulation of the transhydrogenase activity
of spinach ferredoxin-nicotinamide adenine dinucleotide phosphate reductase
by ferredoxin. - J. biol. Chem. *246*: 1201-1205, 1971.

9889 - FRENCH, C.S.: The distribution and action in photosynthesis of several forms
of chlorophyll. - Proc. nat. Acad. Sci. U.S.A. *68*: 2893-2897, 1971.

9890 - FRENCH, C.S., BERRY, J.A.: Curve analysis of low-temperature spectra of mesophyll and bundle sheath chloroplasts of *Sorghum sudanense* in comparison
to naturally and artificially separated pigment system of higher plants. -
Carnegie Inst. Year Book *70*: 495-498, 1971.

9891 - FRENCH, C.S., BROWN, J.S., LAWRENCE, M.C.: The forms of chlorophyll in chloroplast fractions of various algae. - Carnegie Inst. Year Book *70*: 487-495,
1971.

9892 - FRENCH, C.S., BROWN, J.S., WIESSNER, W., LAWRENCE, M.C.: Four common forms
of chlorophyll *A*. - Carnegie Inst. Year Book *69*: 662-670, 1971.

9893 - FRENKEL, A.W., NELSON, R.A.: Bacterial chromatophores. - In: COLOWICK, S.P.,
KAPLAN, N.O. (ed.): Methods in Enzymology. Vol. 23. Pp. 256-268. Academic
Press, New York - London 1971.

9894 - FRIDVALSZKY, L.: Are the changes caused by virus infections or experimental
treatments in the photosynthetic activity connected with molecular and ultrastructural changes? - Acta agron. Acad. Sci. hung. *20*: 449, 1971.

9895 - FRIEDBERG, I., GOLDBERG, I., OHAD, I.: A prolamellar body-like structure in
Chlamydomonas reinhardi. - J. Cell Biol. *50*: 268-275, 1971.

9896 - FRIIS-NIELSEN, B.: A basic system for evaluating optimum utilization of the
transpiration and photosynthetic capacity of plants. - In: SAMISH, R.M.
(ed.): Recent Advances in Plant Nutrition. Vol. 2. Pp. 657-668. Gordon
and Breach Sci. Publ., New York - London - Paris 1971.

9897 - FRITZ, J., ANDERSON, R., FEE, J., PALMER, G., SANDS, R.H., TSIBRIS, J.C.M.,

GUNSALUS, I.C., ORME-JOHNSON, W.H., BEINERT, H.: The iron electron-nuclear double resonance (ENDOR) of two-iron ferredoxins from spinach, parsley, pig adrenal cortex and *Pseudomonas putida*. - Biochim. biophys. Acta *253*: 110-133, 1971.

9898 - FRÖHLICH, J., SCHÖN, W.J.: Einige Photoreaktionen isolierter Chloroplasten der Wildformen und zweier Mutanten von *Melilotus albus* und *Arabidopsis thaliana*; eine vergleichende Untersuchung. - Angew. Bot. *45*: 249-283, 1971.

9899 - FROMENT, A., TANGHE, M., DUVIGNEAUD, P., CALOUX, A., DENAEYER-De SMET, S., SCHNOCK, G., GRULOIS, J., MOMMAERTS-BILLIET, F., VANSÉVEREN, J.P.: La chênaie mélangée calcicole de Virelles-Blaimont, en haute Belgique. - In: DUVIGNEAUD, P. (ed.): Productivity of Forest Ecosystems. Pp. 635-665. Unesco, Paris 1971. [Production.]

*9900 - FRYDRYCH, J.: Studium fotosyntetické aktivity rané vývojové fáze nízkých odrůd rajčat ve vztahu k výnosu plodu. [Photosynthetic activity in the early phase of development of some dwarf cultivars of tomato with regard to the fruit field.] - Bull. výzk. Ust. zelin. v Olomouci *12/13*: 33-4ᶠ, 1969. [In Czech, ab: E, R.]

*9901 - FRYDRYCH, J.: Příspěvek k růstové a fotosyntetické charakteristice dvou odrůd papriky (*Capsicum annuum* L.). [Growth and photosynthetic characteristics of two cultivars of sweet pepper (*Capsicum annuum* L.).] - Zahradnictví (Praha) *1*: 37-46, 1970. [In Czech, ab: E, G, R.]

9902 - FRYDRYCH, J.: Některé růstové a fotosyntetické charakteristiky ve vztahu k tvorbě výnosu u brukve (*Brassica oleracea* var. *gongylodes*). [Some growth and photosynthetic characteristics in relation to yield in kohlrabi (*Brassica oleracea* var. *gongylodes*).] - Rostlinná Výroba (Praha) *17*: 475-482, 1971. [In Czech, ab: E, G, R.]

9903 - FRYDRYCH, J.: Photosynthetische Aktivität diploider und tetraploider Formen von *Brassica oleracea* var. *gongylodes*. - Photosynthetica *5*: 38-43, 1971.

9904 - FUHRHOP, J.-H., MAUZERALL, D.: The photooxygenation of magnesium-octaethyl-porphin. - Photochem. Photobiol. *13*: 453-458, 1971.

9905 - FUJITA, Y., EBATA, T., SHIMURA, S.: Light-dependent oxygen uptake by non-living, chlorophyll-containing particles obtained from oceanic environments. - Plant Cell Physiol. *12*: 543-550, 1917.

9906 - FUJITA, Y., MYERS, J.: Cytochrome reducing substance. - In: COLOWICK, S.P., KAPLAN, N.O. (ed.): Methods in Enzymology. Vol. 23. Pp. 613-618. Academic Press, New York - London 1971.

9907 - FUJITA, Y., SUZUKI, R.: Studies on the Hill reaction of membrane fragments of blue-green algae. I. Stabilizing effect of various media on the 2,6-dichlorphenol indophenol-Hill activity of membrane fragments obtained from *Anabaena cylindrica* and *Anabaena variabilis*. - Plant Cell Physiol. *12*: 641-651, 1971.

9908 - FULLER, R.C.: The evolution of photosynthetic carbon metabolism: the role of sequences ancillary to the Calvin cycle. - In: SCHOFFENIELS, E. (ed.): Biochemical Evolution and the Origin of Life. Molecular Evolution. Vol. 2. Pp. 259-273. North-Holland Publ. Comp., Amsterdam 1971.

9909 - FULLER, R.C., KIDDER, G.W., NUGENT, N.A., DEWEY, V.C., RIGOPOULOS, N.: The association and activities of pteridines in photosynthetic systems. - Photochem. Photobiol. *14*: 359-371, 1971.

9910 - GABIDZASHVILI, M.A.: [Characteristics of photosynthesis of some winter-vegetative herbaceous plants at temperatures near 0°C.] - Soobshch. Akad. Nauk Gruz. SSR *64*: 421-424, 1971. [In Georg., ab: E, R.]

9911 - GAENSSLEN, R.E.: Amine uptake in chloroplasts. - Diss. Abstr. int. B *32*: 1349-B, 1971. [Ps.]

9912 - GAFFRON, H.: Variable photosynthetic units, energy transfer and light-induced evolution of hydrogen in algae and bacteria. - In: BRODA, E., LOCKER, A., SPRINGER-LEDERER, H. (ed.): First European Biophysics Congress. Y XIX A/2. Pp. 19-22. Verlag Wiener med. Akad., Wien 1971.

9913 - GALE, J.: The effect of barometric pressure on photosynthesis and transpiration. - Israel J. Bot. *29*: 334-335, 9171.

9914 - GALOUX, A.: Flux et transferts d'énergie au niveau des écosystèmes forestiers. - In: DUVIGNEAUD, P. (ed.): Productivity of Forest Ecosystems. Pp. 21-40. Unesco, Paris 1971.

9915 - GALSTON, A.W.: Is the use of chlorophyll as an indicator of photosynthetic activity still valid? - Acta agron. Acad. Sci. hung. *20*: 439, 1971.

9916 - GAMAYUNOVA, M.S., KUCHERENKO, V.P., DUBROVSKAYA, A.A., LICHADEEV, G.I.: Vydelenie i ispol'zovanie v fotokhimicheskikh reaktsiyakh khloroplastov preparatov ferredoksina. [Isolation and utilization of ferredoxin preparations in photochemical reactions of chloroplasts.] - Fiziol. Biokhim. kul't. Rast. *3*: 619-621, 1971. [In R, ab: E.]

9917 - GANCHARYK, M.M., MARSHAKOVA, M.I., SHARSTSYANIKINA, A.V.: Da pytannya ab mekhanizme taksichnaya dzeyannya khlarydaǔ na fotasintez. [Mechanism responsible for the toxic effect of chlorides on photosynthesis.] - Vestsy Akad. Navuk belarus. SSR, Ser. biyal. Navuk *1971* (2): 36-38, 1971. [In Belorus.]

*9918 - GANCHARYK, M.M., SHARSTSYANIKINA, A.V.: Uplyǔ mineral'naga zhyǔlennya na nakaplenne pigmentaǔ khlarafilu i karatsinoidaǔ u listsi bul'by. [Effect of mineral nutrition on accumulation of chlorophylls and carotenoids in potato leaves.] - Vestsy Akad. Navuk belarus. SSR, Ser. biyal. Navuk *1967* (4): 5-9, 1967. [In Belorus.]

9919 - GANCHARYK, M.M., URBANOVICH, T.A.: Da pytannya ab uplyve azotu na fotasintetychny aparat i fotosintez bul'by. [Effect of nitrogen on the photosynthetic apparatus and photosynthesis of the potato.] - Vestsy Akad. Navuk belarus. SSR, Ser. biyal. Navuk *1971* (1): 5-7, 1971. [In Belorus.]

9920 - GANTT, E., EDWARDS, M.R., PROVASOLI, L.: Chloroplast structure of the *Cryptophyceae*. Evidence for phycobiliproteins within intrathylakoidal spaces. - J. Cell Biol. *48*: 280-290, 1971.

9921 - GAPONENKA, V.I., NIKALAEVA, G.M., SHAǓCHUK, S.M., LASITSKAYA, T.U.: Nakaplenie khlarafilaǔ *a* i *b* u postetyyaliravanykh prarostkakh kukuruzy, infil'travanykh detergentami. [Accumulation of chlorophylls *a* and *b* in post-etiolated maize seedlings infiltrated with detergents.] - Vestsy Akad. Navuk belarus. SSR, Ser. biyal. Navuk *1971* (6): 59-66, 138, 1971. [In Belorus., ab: R.]

9922 - GAPONENKO, V.I.: Obnovlenie khlorofilla v fotosinteziruyushchem apparate kak fiziologicheskiǐ protsess. [Chlorophyll turnover in the photosynthetic apparatus as a physiological process.] - In: Problemy Biosinteza Khlorofillov. Pp. 78-137. Nauka i Tekhnika, Minsk 1971. [In R.]

*9923 - GAPONENKO, V.I., NIKOLAEVA, G.N., SHEVCHUK, S.N., LOSITSKAYA, T.V., SHLYK, A.A.: Vliyanie tritona X-100 i dodetsilsul'fata na nakoplenie khlorofillov *a* i *b* v zeleneyushchikh prorostkakh kukuruzy. [Effect of Triton X-100 and dodecylsulphate on the accumulation of chlorophyll *a* and *b* in greening maize seedlings.] - In: Materialy Chetvertoǐ Biokhimicheskoǐ Konferentsii Pribaltiǐskikh Respublik i Belorusskoǐ SSR. Pp. 334-335. Vil'nyus 1970. [In R.]

9924 - GARAI, A.B., ROJIK, I., HORVÁTH, M.: Change in the pigment content of autumn wheat 1201 from Bánkút. - Acta biol. (Szeged) *17*: 119-122, 1971. [Chl.]

9925 - GARCIA, A., THORNBER, J.P., VERNON, L.P.: Subchromatophore fragments: *Chromatium, Rhodospirillum rubrum*, and *Rhodopseudomonas palustris*. - In: COLOWICK, S.P., KAPLAN, N.O. (ed.): Methods in Enzymology. Vol. 23. Pp. 305-320. Academic Press, New York - London 1971.

9926 - GAREWAL, H.S., SINGH, J., WASSERMAN, A.R.: Purification of chloroplast cytochrome b_{559}. - Biochem. biophys. Res. Commun. *44*: 1300-1305, 1971.

9927 - GARGAS, E.: "Sun-shade" adaptation in microbenthic algae from the Øresund. - Ophelia *9*: 107-112, 1971. [Ps.]

9928 - GÁSPAR, L.: Can the soluble protein content of non-assimilating tissues influence chloroplast function in assimilating tissues? - Acta agron. Acad. Sci. hung. *20*: 439-441, 1971.

9929 – GASSMAN, M.L., BOGORAD, L.: Inhibitors of bacterial δ-aminolevulinic acid synthetase in plant extracts. – Plant Physiol. *47* (Suppl.): 45, 1971.

9930 – GASSMANN, J., STRELL, I., BRANDL, F., STURM, H., HOPPE, W.: Structure determination of methylpheophorbide *A*. – Tetrahedron Lett. *48*: 4609-4612, 1971.

9931 – GATES, D.M.: The flow of energy in the biosphere. – Sci. Amer. *225* (3): 88, 92, 94, 96-100, 1971.

9932 – GAUDILLERE, J.P., COSTES, C.: Les spectres d'action de l'assimilation photosynthétique du gaz carbonique chez les plantes superieures. – Photosynthetica *5*: 272-316, 1971.

*9933 – GAUSMAN, H.W., ALLEN, W.A., CARDENAS, P., BOWEN, R.L.: Color photos, cotton leaves and soil salinity. – Photogrammetric Eng. *1970*: 454-459, 1970. [Chl.]

*9934 – GAUSMAN, H.W., ALLEN, W.A., CARDENAS, R., BOWEN, R.L.: Detection of foot root disease of grapefruit trees with infrared color film. – J. Rio Grande Valley hort. Soc. *24*: 36-42, 1970. [Chl.]

9935 – GAUSMAN, H.W., ALLEN, W.A., CARDENAS, R., RICHARDSON, A.J.: Effects of leaf nodal position on absorption and scattering coefficients and infinite reflectance of cotton leaves, *Gossypium hirsutum* L. – Agron. J. *63*: 87-91, 1971.

9936 – GAUSMAN, H.W., ALLEN, W.A., ESCOBAR, D.E., RODRIGUEZ, R.R., CARDENAS, R.: Age effects of cotton leaves on light reflectance, transmittance, and absorptance and on water content and thickness. – Agron. J. *63*: 465-469, 1971.

9937 – GAUSMAN, H.W., ALLEN, W.A., WIEGAND, C.L., ESCOBAR, D.E., RODRIGUEZ, R.R.: Leaf light reflectance, transmittance, absorptance, and optical and geometrical parameters for eleven plant genera with different leaf mesophyll arrangements. – In: Proceedings of the Seventh International Symposium on Remote Sensing of Environment. Pp. 1599-1625. University of Michigan, Ann Arbor, Mich. 1971.

*9938 – GAUSMAN, H.W., CARDENAS, R.: Effect of pubescence on reflectance of light. – In: Proceedings of the Fifth Symposium on Remote Sensing of Environment. Pp. 291-297. Univ. Michigan, Ann Arbor, Mich. 1968. [Chl.]

9939 – GAVALAS, N.A., CLARK, H.E.: On the role of manganese in photosynthesis: Kinetics of photoinhibition in manganese-deficient and 3-(4-chlorophenyl)-1, 1-dimethylurea-inhibited *Euglena gracilis*. – Plant Physiol. *47*: 139-143, 1971.

9940 – GEACINTOV, N.E., VAN NOSTRAND, F., POPE, M., TINKEL, J.B.: Magnetic field effect on the chlorophyll fluorescence in *Chlorella*. – Biochim. biophys. Acta *226*: 486-491, 1971.

9941 – GEHRING, U., ARNON, D.I.: Ferredoxin-dependent phenylpyruvate synthesis by cell-free preparations of photosynthetic bacteria. – J. biol. Chem. *246*: 4518-4522, 1971.

9942 – GEIS, J.W., TORTORELLI, R.L., BOGGESS, W.R.: Carbon dioxide assimilation of hardwood seedlings in relation to community dynamics in Central Illinois. I. Field measurements of photosynthesis and respiration. – Oecologia *7*: 276-289, 1971.

9943 – GEJ, B.: Changes in $^{14}CO_2$ absorption rates by the successive leaves in buckwheat and white mustard plants of various age. – Acta Soc. Bot. Pol. *40*: 599-614, 1971.

9944 – GELIN, C.: Primary production and chlorophyll *a* content of nanoplankton. in a eutrophic lake. – Oikos *22*: 230-234, 1971.

*9945 – van GEMERDEN, H.: Utilization of reducing power in growing cultures of *Chromatium*. – Arch. Mikrobiol. *64*: 111-117, 1968. [Chl.]

*9946 – van GEMERDEN, H.: On the ATP generation by *Chromatium* in darkness. – Arch. Mikrobiol. *64*: 118-124, 1968. [Chl.]

*9947 – GENCHEV, S.: Study into the content of plastid pigments in the anthers of male sterile and fertile forms of onions. – Dokl. Akad. sel'skokhoz. Nauk Bolg. *3*: 25-28, 1970.

*9948 - GENEVÈS, L.: Sur la structure et le comportement des plastes dans le tissu sporogène au cours de sa prolifération, chez *Hypnum rusciforme*. - Compt. rend. Acad. Sci. Paris, Sér. D *262*: 2215-2218, 1966.

*9949 - GEORGESCU, M., DOROBANŢU, N.: Reflectarea proceselor fiziologice asupra fertilităţii soiurilor Ceauş, Chasselas, Afuz-Ali altoite pe diferiţi portaltoi. [Fertility in Ceaus, Chasselas and Afuz-Ali on various root-stocks as affected by physiological processes.] - Lucrări ştiinţ. Inst. agron. "N. Balcescu" Ser. B, Hort. *10*: 413-428, 1967. [Ps; in Roum., ab: E, F, R.]

*9950 - GEORGIEV, Kh., GEORGIEV, D., MIKHAÏLOVA, S.: Pridvizhvane na radioaktivniya v"glerod C^{14} v s"tsvetiyata ot otdelni lista v zavisimost ot tyakhnoto raspolozhenie po st"bloto pri domata. [Transport of radioactive carbon ^{14}C from individual leaves to inflorescences in dependence on their position on the tomato stem.] - Gradinar. lozar. Nauka (Sofia) *3*: 609-615, 1966. [In Bulg., ab: F, R.]

9951 - GERLOFF, E.D., ORTMAN, E.E.: Physiological changes in barley induced by greenbug feeding stress. - Crop Sci. *11*: 174-175, 1971. [Ps, Chl.]

9952 - GERSTER, R.: Essai d'interprétation des cinétiques d'échange isotopique entre $C^{18}O_2$ et eau d'une feuille: Expérience à l'obscurité. - Planta *97*: 155-172, 1971.

9953 - GESNNER, F.: Die Photosynthese plasmolysierter Wasserpflanzen. - Ber. deut. bot. Ges. *84*: 267-274, 1971.

9954 - GETOV, G.K., ÏORDANOVA, S.Ts.: Spektral'noe raspredelenie fotovol'taicheskogo effekta v sloyakh khlorofilla. [Spectral distribution of the photovoltaic effect in chlorophyll layers.] - Dokl. bolg. Akad. Nauk *24*: 1027-1030, 1971. [In R.]

9955 - GEVORKYAN, A.G., BAZHANOVA, N.V.: O deïstvii sineï oblasti spektra na "violaksantinovyï tsikl" v list'yakh klevera. [Action of the blue spectral region on the "violaxanthin cycle" in clover leaves.] - Dokl. Akad. Nauk Arm. SSR *52*: 229-304, 1971. [In R, ab: Armen.]

9956 - GHORASHY, S.R.: The effect of leaf pubescence on leaf water potential, apparent photosynthetic rate and yield of three isogenic lines of soybeans (*Glycine max* (L.) MERRILL). - Diss. Abstr. Int. B *31*: 5117-B-5118-B, 1971.

9957 - GHORASHY, S.R., PENDLETON, J.W., BERNARD, R.L., BAUER, M.E.: Effect of leaf pubescence on transpiration, photosynthetic rate and seed yield of three near-isogenic lines of soybeans. - Crop Sci. *11*: 426-427, 1971.

9958 - GHORASHY, S.R., PENDLETON, J.W., PETERS, D.B., BOYER, J.S., BEUERLEIN, J.E.: Internal water stress and apparant photosynthesis with soybeans differing in pubescence. - Agron. J. *63*: 674-676, 1971.

*9959 - GHOSH, A.K., OLSON, J.M.: Effect of acid on the bacteriochlorophyll-protein complex from green bacteria. - Biophys. Soc. Abstr., 10 annu. Meet. *1966*: 24, 1966.

*9960 - GIBBON, D.P., LUPPI, G., MATTEI, F.: The agronomic potential of an environment and its measurement. First report on an experiment carried out at four research centres (two in U.K. and two in Italy) during 1967. - Ricerca scientifica *38*: 578-583, 1968. [Energy conversion efficiency.]

B9961 - GIBBS, M. (ed.): Structure and Function of Chloroplasts. - Springer-Verlag, Berlin - Heidelberg - New York 1971.

9962 - GIBBS, M.: Carbohydrate metabolism by chloroplasts. - In: GIBBS, M. (ed.): Structure and Function of Chloroplasts. Pp. 169-214. Springer-Verlag, Berlin - Heidelberg - New York 1971.

9963 - GIBBS, M.: Photophosphorylation and O_2 evolution, assessment. - In: HATCH, M.D., OSMOND, C.B., SLATYER, R.O. (ed.): Photosynthesis and Photorespiration. Pp. 428-429. Wiley-Interscience, New York - London - Sydney - Toronto 1971.

9964 - GIBBS, M.: Biosynthesis of glycolic acid. - In: HATCH, M.D., OSMOND, C.B., SLATYER, R.O. (ed.): Photosynthesis and Photorespiration. Pp. 433-441. Wiley-Interscience, New York - London - Sydney - Toronto 1971.

9965 - GIESE, A.C.: Photosensitization by natural pigments. - In: GIESE, A.C. (ed.):
Photophysiology. Vol. VI. Pp. 77-129. Academic Press, New York - London 1971.
[Chl.]

9966 - GIFFORD, R.M.: The light response of CO_2 exchange: on the source of differ-
ences between C_3 and C_4 species. - In: HATCH, M.D., OSMOND, C.B., SLATYER,
R.O. (ed.): Photosynthesis and Photorespiration. Pp. 51-56. Wiley-Interscience,
New York - London - Sydney - Toronto 1971.

9967 - GILES, K.L., SARAFIS, V.: On the survival and reproduction of chloroplasts
outside the cell. - Cytobios 4: 61-74, 1971.

*9968 - GILIS, M.B., RADCHENKO, N.P.: Vliyanie mikroelementov na rost, razvitie i
nekotorye biokhimicheskie osobennosti kukuruzy i sakharnoi svekly v usloviyakh
zapadnoi lesostepi Ukrainy. [Effect of microelements on the growth, development and some biochemical features of maize and sugar beet in the west forest-
steppe of Ukraine.] - In: Mikroelementy v Sel'skom Khozyaistve i Meditsine.
Vol. 3. Pp. 27-34. Naukova Dumka, Kiev 1967. [Growth analysis; in R.]

9969 - GILLER, Yu.E., ASOEVA, L.M., KAS'YANENKO, A.G.: Izuchenie vosstanovleniya
pigmentnoi sistemy khloroplastov mutantov *Arabidopsis thaliana* viridoal'bina
40/3 pod vliyaniem dobavki leitsina v pitatel'nuyu sredu. [Reconstruction
of the pigment system of chloroplasts of the mutant viridoalbina 40/3 of
Arabidopsis thaliana induced by the addition of leucine into the nutrient
medium.] - In: NASYROV, Yu.S. (ed.): Geneticheskie Aspekty Fotosinteza. Pp.
107-119. Donish, Dushanbe 1971. [In R, ab: E.]

9970 - GILLER, Yu.E., KAS'YANENKO, A.G., VAKHIDOVA, L.R., YUSUPOVA, G.A.: Sostav,
sostoyanie i fotokhimicheskaya aktivnost' pigmentov plastid zhiznesposob-
nykh khlorofil'nykh mutantov *Arabidopsis thaliana*. [Composition, state and
photochemical activity of plastid pigments of vital chlorophyll mutants of
Arabidopsis thaliana.] - In: ZALENSKII, O.V. (ed.): Fotosintez i Ispol'zovanie Solnechnoi Energii. Pp. 247-254. Nauka, Leningrad 1971. [In R, ab:
E.]

9971 - GILLER, Yu.E., STOLBOVA, A.V., VAKHIDOVA, L.R., KVITKO, K.V.: Kolichestven-
nyi sostav i sostoyanie osnovnykh pigmentov plastid mutantnykh form vodorosli *Chlorella*. [Quantitive and qualitative composition and state of basic
plastid pigments in mutant forms of *Chlorella*.] - Biofizika 16: 67-77, 1971.
[In R, ab: E.]

9972 - GILLESPIE, T.J.: Carbon dioxide profiles and apparent diffusivities in corn
fields at night. - Agr. Meteorol. 8: 51-57, 1971.

9973 - GILLESPIE, T.J., KING, K.M.: Night-time sink strengths and apparant diffusivities within a corn canopy. - Agr. Meteorol. 8: 59-67, 1971.

9974 - GIMMLER, H., AVRON, M.: On the mechanism of benzoquinone penetration and
photoreduction by whole cells. - Z. Naturforsch. 26 b: 585-588, 1971. [Ps.]

9975 - GIMMLER, H., NEIMANIS, S., EILMANN, I., URBACH, W.: Photophosphorylation
and photosynthetic $^{14}CO_2$-fixation *in vivo*. II. Comparison of cyclic and non-
cyclic photophosphorylation with photosynthetic $^{14}CO_2$-fixation during the
synchronous life cycle of *Ankistrodesmus braunii*. - Z. Pflanzenphysiol.
64: 358-366, 1971.

9976 - GIORDANO, P.M., MORTVEDT, J.J.: Effect of substrate Zn level on distribution of photo-assimilated C^{14} in maize and bean plants. - Plant Soil 35:
193-196, 1971.

9977 - GIRS, G.I.: Izmenenie soderzhaniya pigmentov v khvoe listvennitsy sibirskoi
v svyazi s vysotnoi zonal'nost'yu lesov. [Change in the level of pigments in
Siberian larch needles in connection with the high zonality of forests.] -
Ekologiya 2: 37-43,1971. [In R.]

9978 - GIVAN, C.V., LEECH, R.M.: Biochemical autonomy of higher plant chloroplasts
and their synthesis of small molecules. - Biol. Rev. Cambridge phil. Soc.
46: 409-428, 1971.

9979 - GLADYSHEV, A.I.: Produktivnost' fitomassy solodkovo-kiyakovoi formatsii v
poime Amudar'i. [Productivity of plant matter of *Glycyrrhiza glabra* L.-
Imperata cylindrica (L.) P.B. formation in flood plain of the Amadur'ya

river.] - Izv. Akad. Nauk turkmen. SSҡ, Ser. biol. Nauk *1971* (4): 33-39, 1971. [In R, ab: E.]

*9980 - GLAGOLEVA, T.A., CHULANOVSKAYA, M.V., ZALENSKIĬ, O.V.: O bioenergetike assi-miliruyushchikh kletok *Chlorella pyrenoidosa* CHICK. [Bioenergetics of assi-milatory cells of *Chlorella pyrenoidosa* CHICK.] - In: Fotosintez i Ispol'-zovanie SolnechnoĬ Radiatsii. Tezisy Dokladov. Pp. 44-47. Dushanbe 1967. [In R.]

*9981 - GLAGOLEVA, T.A., ZALENSKIĬ, O.V.: Vliyanie ingibitorov fotofosforilirova-niya na metabolizm ugleroda C^{14} u *Chlorella pyrenoidosa* CHICK. [The effect of photophosphorylation inhibitors on the ^{14}C metabolism in *Chlorella pyre-noidosa* CHICK.] - In: Tezisy Sektsionnykh SoobshcheniĬ Vtorogo Biokhimiches-kogo S"ezda. Pp. 13-14. Tashkent.1969. [In R.]

9982 - GLAGOLEVA, T.A., ZALENSKIĬ, O.V.: Ob otnositel'noĬ otsenke intensivnosti fotofosforilirovaniya i okislitel'nogo fosforilirovaniya u *Chlorella pyre-noidosa* CHICK. [Relative rates of photophosphorylation and oxidative phos-phorylation in *Chlorella pyrenoidosa* CHICK.] - In: ZALENSKII, O.V. (ed.): Fotosintez i Ispol'zovanie SolnechnoĬ Energii. Pp. 190-193. Nauka, Lenin-grad 1971. [In R, ab: E.]

9983 - GLASZIOU, K.T., BULL, T.A.: Feedback control of photosynthesis in sugarcane. - In: HATCH, M.D., OSMOND, C.B., SLATYER, R.O. (ed.): Photosynthesis and Photorespiration. Pp. 82-88. Wiley-Interscience, New York - London - Sydney - Toronto 1971.

9984 - GLAZER, A.N., COHEN-BAZIRE, G.: Subunit structure of the phycobiliproteins of blue-green algae. - Proc. nat. Acad. Sci. USA *68*: 1389-1401, 1971.

9985 - GLAZER, A.N., COHEN-BAZIRE, G., STANIER, R.Y.: Characterization of phycoe-rythrin from a *Cryptomonas* sp. - Arch. Mikrobiol. *80*: 1-18, 1971.

9986 - GLICKSON, J.D., PHILLIPS, W.D., McDONALD, C.C., POE, M.: PMR characteriza-tion of alfalfa and soybean ferredoxins: The existence of two ferredoxins in soybean. - Biochem. biophys. Res. Commun. *42*: 271-279, 1971.

*9987 - GLOGOV, L.V.: Intensivnost' fotosinteza kukuruzy. [Photosynthetic rate of maize.] - Izv. TSKhA *1967* (3): 13-27, 1967. [In R, ab: E.]

9988 - GMELIG MEYLING, H.D.: Grain yield of winter rye and winter wheat in relation to leaf number and leaf age. - Neth. J. agr. Sci. *19*: 250-256, 1971.

*9989 - GNILITSKAYA, A.B.: Effektivnost' predposevnoĬ obrabotki semyan kukuruzy sernokisloĬ med'yu pri razlichnykh sposobakh vneseniya mineral'nykh udo-breniĬ. [Effectivity of the pre-sowing treatment of maize seeds with $CuSO_4$ under various application of mineral fertilizers.]- In: Mikroelementy v Sel'skom KhozyaĬstve i Meditsine. Vol. 3. Pp. 81-86. Naukova Dumka, Kiev 1967. [Chl, Car; in R.]

9990 - GOCHOLASHVILI, M.M., ADEĬSHVILI, N.I., DZHASHI, R.G.: Vliyanie razlichnykh kompleksov vneshnikh usloviĬ proizrastaniya na rost, razvitie, morozoustoĬ-chivost' i drugie fiziologicheskie i biokhimicheskie svoĬstva limonnogo de-reva. [Effect of various complexes of environmental factors of growing on the growth, development, frost resistance and other physiological and bio-chemical properties of lemon.] - Subtrop. Kul't. *1967* (4): 76-91, 1967. [Ps; in R.]

9991 - GODNEV, T.N.: Vliyanie svetovogo i temperaturnogo faktora na biosintez khlorofilla v rastenii. [Effects of light and temperature on chlorophyll biosynthesis in plants.] - In: ZALENSKIĬ, O.V. (ed.): Fotosintez i Ispol'-zovanie SolnechnoĬ Energii. Pp. 123-126. Nauka, Leningrad 1971. [In R, ab: E.]

*9992 - GODNEV, T.Ṅ., DOMASH, V.I., AKULOVICH, N.K.: K voprosu o vliyanii fitokhro-mnoĬ sistemy na biosintez fotosinteticheskikh pigmentov nekotorykh rasteniĬ. [Effect of phytochrome system on the biosynthesis of photosynthetic pigments of some plants.] - In: Fotosintez i Pitanie RasteniĬ. Pp. 3-8. Nauka i Tekhnika, Minsk 1969. [In R.]

9993 - GODNEV, T.N., DOMASH, V.I., AKULOVICH, N.K., KHODASEVICH, E.V.: DeĬstvie vysokoenergeticheskikh impul'sov sveta na pigmentnuyu sistemu rasteniĬ.

[Action of high-energy light pulses on the pigment system of plants.] -
In: Svetoimpul'snaya Stimulyatsiya Rastenit. Pp. 222-227, 363. Nauka,
Moskva 1971. [In R.]

9994 - GODNEV, T.N., KHODASEVICH, E.V., ARNAUTOVA, A.I.: Izuchenie biosinteza
pigmentov u khvotnykh s primeneniem ugleroda-14. [Biosynthesis of pigments
in conifers studied using ^{14}C.] - In: Primenenie Izotopov i Yadernykh Iz-
luchenit v Sel'skom Khozyatstve. Pp. 130-134. Atomizdat, Moskva 1971. [In
R.]

9995 - GOEDHEER, J.C., GULYAEV, B.A.: Fluorescence polarisation of photosynthetic
pigments. - In: BRODA, E., LOCKER, A., SPRINGER-LEDERER, H. (ed.): Proceedings of the First European Biophysics Congress. Vol. 4. Pp. 43-47. Verlag Wiener med. Akad., Wien 1971.

9996 - GOÏSA, N.I., MITROFANOV, B.A., OKANENKO, A.S., KUTENKO, G.I., MAKARENKO,
K.I.: Issledovanie fotosinteza ozimot pshenitsy v usloviyakh razlichnot
vlagoobespechennosti. [Winter wheat photosynthesis under different water
supply.] - Fiziol. Biokhim. kul't. Rast. 3: 392-397, 1971. [In R, ab: E.]

*9997 - GOL'D, V.M.: Primenenie infrakrasnogo gazoanalizatora dlya opredeleniya
intensivnosti fotosinteza. [Use of infra-red gas analyser for determining
photosynthetic rate.] - In: OKUNTSOV, M.M. (ed.): Spetsial'nyt Praktikum
po Biokhimii i Fiziologii Rastenit. Pp. 126-132. Izdat. tomskogo Gosuniv.,
Tomsk 1966. [In R.]

*9998 - GOL'D, V.M.: Ob izmenenii pogloshcheniya khlorelly i khloroplastov fasoli
v ul'trafioletovot oblasti pod detstviem vidimot chasti spektra. [Changes
in absorption of Chlorella and bean chloroplasts in UV under the influence
of visible part of spectrum.] - Izv. sibir. Otd. Akad. Nauk SSSR, Ser.
biol.-med. Nauk 15 (3): 123-124, 1967. [In R.]

*9999 - GOL'D, V.M.: Spektr detstviya fotosinteza pri svetovom nasyshchenii. [Ac-
tion spectrum of light saturated photosynthesis.] - Inform. Byull. (Irkutsk) 3: 109-110, 1968. [In R.]

10000 - GOL'D, V.M.: Nekotorye problemy spetsifiki detstviya sinet chasti spektra
v protsesse fotosinteza. [Some pecularities of action of the blue part of
spectrum in the process of photosynthesis.] - In: Biokhimiya i Biofizika
Fotosinteza. Pp. 158-184. Irkutsk 1971. [In R.]

*10001 - GOL'D, V.M., BOTKINA, I.I.: K voprosu o regulyatornom detstvii sinego sve-
ta v protsesse fotozinteza. [Regulatory action of blue light in the photosynthetic process.] - Inform. Byull. sibir. Inst. Fiziol. Biokhim. Rast.
7: 31-32, 1970. [In R.]

*10002 - GOL'D, V.M., GRIGOR'EV, Yu.S., GAEVSKIÏ, N.A.: Svetoindutsiruemye perek-
hody fluorestsentsii khlorofilla A in vivo i v model'nykh sistemakh. [Light-
induced transients of fluorescence of chlorophyll a in vivo and in model
systems.] - Inform. Byull. sibir. Inst. Fiziol. Biokhim. Rast. 7: 28-30,
1970. [In R.]

10003 - GOL'D, V.M., GRIGOR'EV, Yu.S., GAEVSKIÏ, N.A.: Fotosintez·i dykhanie.
[Photosynthesis and respiration.] - In: GOL'D, V.M. (ed.): Sbornik Labo-
ratornykh Rabot po Fiziologii Rastenit. Pp. 7-95. Izdat. krasnoyarsk. Gosuniv., Krasnoyarsk 1971. [In R.]

*10004 - GOL'D, V.M., GRIGOR'EV, Yu.S., OSIPOV, V.I., BOTKINA, T.I.: Issledovaniya
v oblasti induktsionnykh fluorestsentsii i svyaz' ikh s effektivnost'yu
fotofosforilirovaniya v kletkakh khlorelly. [Induction phenomena in fluorescence and their relation to the efficiency of photophosphorylation in
Chlorella cells.] - Inform. Byull. (Irkutsk) 5: 60-62, 1969. [In R.]

10005 - GOL'D, V.M., GRIGOR'EV, Yu.S., OSIPOV, V.I., BOTKINA, T.I.: Detstvie kofaktorov tsiklicheskogo potoka elektronov na induktsionnye perekhody
fluorestsentsii i effektivnost' fotofosforilirovaniya v kletkakh khlorelly.
[Effect of cofactors of cyclic electron flow on inductive transfer of fluorescence and photophosphorylation efficiency in Chlorella cells.] - Biofizika 16: 643-649, 1971. [In R, ab: E.]

*10006 - GOL'D, V.M., KOL'TSOVA, V.G.: Detstvie kofaktorov tsiklicheskogo fotofos-

foriIirovaniya na biosintez zelenykh pigmentov. [Action of the cofactors
of cyclic photophosphorylation on the biosynthesis of green pigments.] -
Inform. Byull. (Irkutsk) 4: 76-77, 1969. [In R.]

*10007 - GOL'D, V.M., POLOKH, V.P.: O svyazi reaktsii tsiklicheskogo fotofosforili-
rovaniya s biosintezom sakharozy v list'yakh pshenitsy. [Relation to cyclic
photophosphorylation to saccharose synthesis in wheat leaves.] - Inform.
Byull. (Irkutsk) 5: 59-60, 1969. [In R.]

10008 - GOL'D, V.M., POLOKH, V.P., BOTKINA, I.I.: Deĭstvie vikasola na uglevodnyĭ
obmen v list'yakh vysshikh rasteniĭ. [Effect of vicasol on the metabolism
of carbohydrates in leaves of higher plants.] - Izv. sibir. Otd. Akad.
Nauk SSSR, Ser. biol. Nauki 1: 141-143, 1971. [In R, ab: E.]

10009 - GOLDSWORTHY, A.: A method for the rapid measurement of photosynthesis. -
J. exp. Bot. 22: 753-755, 1971.

*10010 - GOLDSWORTHY, P.R.: The sources of assimilate for grain development in tall
and short sorghum. - J. agr. Sci. 74: 523-531, 1970.

10011 - GOLDTHWAITE, J.J., BOGORAD, L.: A one-step method for the isolation and
determination of leaf ribulose-1,5-diphosphate carboxylase. - Anal. Biochem. 41: 57-66, 1971.

10012 - GOLOD, M.G., SEMYCHAEVS'KYĬ, V.D.: Stan protokhlorofilu vnutrishnikh obo-
lonok nasinnya garbuza in vivo. [State of protochlorophyll in the inner
envelopes of pumpkin seeds in vivo.] - Ukr. bot. Zh. 28: 12-17, 126, 1971.
[In Ukr., ab: E, R.]

*10013 - GONCHARIK, M.N., IVANCHENKO, V.M., LEGENCHENKO, B.I.: K voprosu o spetsi-
fichnosti vliyaniya ionov Cl na fotosinteticheskiĭ apparat rasteniĭ i ego
funktsii. [Specificity of the effect of Cl ions on the photosynthetic apparatus of plants and its functions.] - In: Tezisy Vsesoyuznogo Soveshchaniya po Soleustoĭchivosti Rasteniĭ. Pp. 780-785. Fan, Tashkent 1969.
[In R.]

*10014 - GONCHARIK, M.N., VLASENKO, N.E.: Vliyanie form mineral'nykh udobreniĭ na
fotosintez i kachestvo sakharnoĭ svekly. [Effect of forms of mineral nutrients on photosynthesis and quality of sugar beet.] - In: Fiziologo-
biokhimicheskie Issledovaniya Rasteniĭ. Pp. 93-102. Nauka i Tekhnika,
Minsk 1968. [In R.]

10015 - GONCHAROVA, N.V., EVSTIGNEEV, V.B.: Ob uchastii fenazinmetasul'fata v
fotofosforilirovanii na beskletochnykh preparatakh Chromatium minutissimum i vozmozhnom mehkanizme fotofosforilirovaniya. [Participation of
phenazine methosulfate in photophosphorylation in cell-free preparations
of Chromatium minutissimum and possible mechanism of photophosphorylation.]
- Biokhimiya 36: 311-321, 1971. [In R, ab: E.]

10016 - GOODCHILD, D.J.: The relationship between grana and stroma lamellae and
photosystems I and II in spinach chloroplasts. - In: HATCH, M.D., OSMOND,
C.B., SLATYER, R.O. (ed.): Photosynthesis and Photorespiration. Pp. 400-
405. Wiley-Interscience, New York - London - Sydney - Toronto 1971.

10017 - GOODCHILD, D.J.: Chloroplast structure: assessment. - In: HATCH, M.D.,
OSMOND, C.B., SLATYER, R.O. (ed.): Photosynthesis and Photorespiration.
Pp. 426-427. Wiley-Interscience, New York - London - Sydney - Toronto
1971.

10018 - GOODCHILD, D.J., PARK, R.B.: Further evidence for stroma lamellae as a
source of Photosystem I fractions from spinach chloroplasts. - Biochim.
biophys. Acta 226: 393-399, 1971.

*10019 - GOODENOUGH, U.W.: Chloroplast division and pyrenoid formation in Chlamy-
domonas reinhardi. - J. Phycol. 6: 1-6, 1970.

10020 - GOODENOUGH, U.W.: The effects of inhibitors of RNA and protein synthesis
on chloroplast structure and function in wild-type Chlamydomonas reinhardi. - J. Cell Biol. 50: 35-49, 1971.

10021 - GOODENOUGH, U.W., LEVINE, R.P.: The effects of inhibitors of RNA and protein synthesis on the recovery of chloroplast ribosomes, membrane orga-

nization, and photosynthetic electron transport in the ac-20 strain of
Chlamydomonas reinhardi. - J. Cell Biol. *50*: 50-62, 1971.

10022 - GOODENOUGH, U.W., STAEHELIN, L.A.: Structural differentiation of stacked
and unstacked chloroplast membranes: Freeze-etch electron microscopy of
wild-type and mutant strain of *Chlamydomonas*. - J. Cell Biol. *48*: 594-
619, 1971.

10023 - GOODENOUGH, U.W., TOGASAKI,,R.K., PASZEWSKI, A., LEVINE, R.P.: Inhibition
of chloroplast ribosome formation by gene mutation in *Chlamydomonas reinhardi*. - In: BOARDMAN, N.K., LINNANE, A.W., SMILLIE, R.M. (ed.): Autonomy
and Biogenesis of Mitochondria and Chloroplasts. Pp. 224-234. North-
Holland Publ. Comp., Amsterdam - London 1971.

10024 - GOODWIN, T.W.: Algal carotenoids. - In: GOODWIN, T.W. (ed.): Aspects of
Terpenoid Chemistry and Biochemistry. Pp. 315-356. Academic Press, London - New York 1971.

10025 - GOODWIN, T.W.: Biosynthesis. - In: ISLER, O. (ed.): Carotenoids. Pp. 577-
636. Birkhäuser Verlag, Basel - Stuttgart 1971. [Car.]

10026 - GOODWIN, T.W.: Biosynthesis by chloroplasts. - In: GIBBS, M. (ed.): Structure and Function of Chloroplasts. Pp. 215-276. Springer-Verlag, Berlin -
Heidelberg - New York 1971.

10027 - GOODWIN, T.W.: Biosynthesis of carotenoids and plant triterpenes. - Biochem. J. *123*: 293-329, 1971.

*10028 - GOPALAKRISHNAN, S., GOSWAMI, N.N.: Note on varietal differences in leaf-area
development and net assimilation rate in tossa jute (*Corchorus olitorius* L.).
- Indian J. agr. Sci. *40*: 552-555, 1970.

10029 - GORE, M.G., EVANS, R.B., HILL, H.M., ROGERS, L.J.: The effect of isonico-
tinyl hydrazide on the greening of maize and barley. - Biochem. J. *121*: 7
P-8P, 1971.

10030 - GORSHKOVA, A.A., SPIVAK, A.I.: Dnevnye i sezonnye izmeneniya intensivnosti
fotosinteza nekotorykh vidov stepnykh rasteniĭ yugo-vostochnogo Zabaĭkal'ya
[Diurnal and seasonal changes in photosynthetic rate of some species of
steppe plants of the south-eastern Zabaĭkal.] - In: Ekologo-biologicheskie
Osobennosti i Produktivnost' Lugopastbishchnykh Rasteniĭ Zabaĭkal'ya. Pp.
19-20. Ulan-Ude 1971. [In R.]

*10031 - GORYSHINA, T.K.: Dykhanie letnevegetiruyushchikh travyanistykh vidov le-
sostepnoĭ dubravy i ego sezonnaya dinamika. [Respiration of summer-vegeta-
ting herbaceous species of a forest-steppe oak forest and its seasonal dynamics.] - Ekologiya *1* (6): 20-31, 1970. [Ps; in R.]

10032 - GORYSHINA, T.K.: Sezonnaya dinamika fotosinteza i produktivnosti u nekotorykh letnevegetiruyushchikh travyanistykh rasteniĭ lesostepnoĭ dubravy.
[Seasonal dynamics of photosynthesis and productivity in certain summer-
green herbage plants of an oakwood in the forest-steppe zone.] - Bot. Zh.
56: 62-75, 1971. [In R, ab: E.]

*10033 - GORYSHINA, T.K., MITINA, M.B.: Ekologo-fiziologicheskie osobennosti ve-
sennikh i letnikh list'ev snyti *Aegopodium podagraria* L. i ikh rol' v produktivnosti travostoya lesostepnoĭ dubravy. [Ecophysiological properties of
spring and summer leaves of *Aegopodium podagraria* L. and their role in productivity of herbaceous layer in oak forest of the forest-steppe zone.] -
Vestn. Leningrad. Univ., Ser. biol. *25* (4): 71-77, 1970. [Ps; in R, ab: E.]

*10034 - GOSTIMSKIĬ, S.A.: Vozmozhnosti ispol'zovaniya mutantov vysshikh rasteniĭ
dlya izucheniya fotosinteza. [Possible use of mutants of higher plants in
photosynthetic studies.] - Nauch. Dokl. vyssh. Shkoly, biol. Nauki *13* (3):
134-135, 1970. [In R.]

10035 - GOVINDJEE, MOHANTY, P.: Chlorophyll *a* fluorescence in the study of photosynthesis. - Fluorescence News (Silver Spring, Md.) *6* (2): 1-7, 1971.

10036 - GOVINDJEE, PAPAGEORGIOU, G.: Chlorophyll fluorescence and photosynthesis:
Fluorescence transients. - In: GIESE, A.C. (ed.): Photophysiology. Vol. VI.
Pp. 1-46. Academic Press, New York - London 1971.

10037 - GRACE, J.: The directional distribution of light in natural and controlled environment conditions. - J. appl. Ecol. 8: 155-164, 1971.

10038 - GRAHAM, D., ATKINS, C.A., EVERSON, R.G.: Light-induced pH changes, photosynthesis and carbonic anhydrase. - Proc. austral. biochem. Soc. 4: 10, 1971.

10039 - GRAHAM, D., ATKINS, C.A., REED, M.L., PATTERSON, B.D., SMILLIE, R.M.: Carbonic anhydrase, photosynthesis, and light-induced pH changes. - In: HATCH, M.D., OSMOND, C.B., SLATYER, R.O. (ed.): Photosynthesis and Photorespiration. Pp. 267-274. Wiley-Interscience, New York - London - Sydney - Toronto 1971.

10040 - GRAHAM, D., GRIEVE, A.M., SMILLIE, R.M.: Phytochrome-mediated plastid development in etiolated pea stem apices. - Phytochemistry 10: 2905-2914, 1971.

10041 - GRAHAM, D., REED, M.L.: Carbonic anhydrase and the regulation of photosynthesis. - Nature - new Biol. 231: 81-83, 1971.

10042 - GRAHAM, D., SMILLIE, R.M.: Chloroplasts (and lamellae): algal preparations. - In: COLOWICK, S.P., KAPLAN, N.O. (ed.): Methods in Enzymology. Vol. 23. Pp. 228-242. Academic Press, New York - London 1971.

*10043 - GRALL, J.-R.: Détermination de la production de matière organique en Manche occidentale à l'aide du carbone 14. - Compt. rend. Acad. Sci. Paris, Sér. D 262: 2514-2517, 1966. [Plankton production.]

10044 - GRANICK, S.: Preparation and properties of Chlorella mutants in chlorophyll biosynthesis. - In: COLOWICK, S.P., KAPLAN, N.O. (ed.): Methods in Enzymology. Vol. 23. Pp. 162-168. Academic Press, New York - London 1971.

10045 - GRAZIANI, Y., LIVNE, A.: Dehydration, water fluxes, and permeability of tobacco leaf tissue. - Plant Physiol. 48: 575-579, 1971. [Ps.]

10046 - GRAZIANI, Y., LIVNE, A.: Water fluxes and permeability in tobacco leaf tissue. - Israel J. Bot. 20: 332-333, 1971. [Ps.]

10047 - GREBENSKIĬ, S.O., DUDOK, E.P.: Vliyanie gibberellina na rost, soderzhanie khlorofilla i dykhanie u prorostkov, podvergnutykh rentgenovskomu oblucheniyu. [Effect of gibberellin on growth, chlorophyll content and respiration in seedlings subjected to X-ray irradiation.] - Fiziol. Biokhim. kul't. Rast. 1: 44-47, 1969. [In R, ab: E.]

10048 - De GREEF, J., BUTLER, W.L., ROTH, T.F.: Greening of etiolated bean leaves in far red light. - Plant Physiol. 47: 457-464, 1971.

10049 - De GREEF, J., BUTLER, W.L., ROTH, T.F., FREDERICQ, H.: Control of senescence in Marchantia by phytochrome. - Plant Physiol. 48: 407-412, 1971. [Chl.]

10050 - De GREEF, J.A., CAUBERGS, R.: Chlorophyll formation and protochlorophyll regeneration during the early greening stages of etiolated seedlings of "Phaseolus vulgaris cultivar" Limburg. - Arch. int. Physiol. Biochim. 79: 412-414, 1971.

10051 - De GREEF, J.A., CAUBERGS, R., VERBELEN, J.P.: Elimination of the lag phase of chlorophyll synthesis in etiolated bean seedlings by a pre-illumination of the apex. - Plant Physiol. 47 (Suppl.): 44, 1971.

B10052 - GREGORY, R.P.F.: Biochemistry of Photosynthesis. - Wiley-Interscience, London - New York - Sydney - Toronto 1971.

10053 - GREGORY, R.P.F., RAPS, S., BERTSCH, W.: Are specific chlorophyll-protein complexes required for photosynthesis? - Biochim. biophys. Acta 234: 330-334, 1971.

10054 - GREUB, L.J., WEDIN, W.F.: Leaf area, dry-matter accumulation, and carbohydrate reserves of alfalfa and birdsfoot trefoil under a three-cut management. - Crop Sci. 11: 341-344, 1971. [Growth analysis.]

10055 - GREUB, L.J., WEDIN, W.F.: Leaf area, dry-matter production, and carbohydrate reserve levels of birdsfoot trefoil as influenced by cutting height. - Crop Sci. 11: 734-738, 1971. [Growth analysis.]

*10056 - GREVILLE, G.D.: A scrutiny of Mitchell's chemiosmotic hypothesis of res-

piratory chain and photosynthetic phosphorylation. - In: SANADI, D.R.
(ed.): Current Topics in Bioenergetics. Vol. 3. Pp. 1-78. Academic Press,
New York - London 1969.

*10057 - GRIGOR'EV, Yu.S., GOL'D, V.M., GAEVSKIĬ, M.A.: Izuchenie kharaktera in-
duktsionnykh perekhodov fluorestsentsii. [Character of induction transients
of fluorescence.] - In: I Kraevaya Konferentsiya po Fiziologii i Biokhimii
Rastenii. Pp. 7-9. Krasnoyarsk 1970. [In R.]

10058 - GRIGOROV, L.N., KONONENKO, A.A.: O temperaturnykh zavisimostyakh fotoin-
dutsirovaniya okislitel'no-vosstanovitel'nykh reaktsii vnutrikletochnykh
tsitokhromov *Rhodopseudomonas* sp. [Temperature dependence of photoinduct-
ion of redox reactions of intracellular cytochromes of *Rhodopseudomonas*
sp.] - Nauch. Dokl. vyssh. Shkoly, biol. Nauki *13* (3): 138, 1970. [In R.]

10059 - GRIGOROVICH, V.I., ZAKHAROVA, N.I., KUTYURIN, V.M., ROZONOVA, L.N., EL'-
PINER, I.E.: Raspredelenie khlorofilla i margantsa po chastitsam khloroplastov i ikh fotokhimicheskaya aktivnost'. [Distribution of chlorophyll
and manganese among chloroplast particles and their photochemical activity.] - Biofizika *16*: 260-264, 1971. [In R, ab: E.]

10060 - GRISHINA, G.S.: Primenenie amperometricheskogo metoda dlya opredeleniya
obmena kisloroda u vodnykh rastenii i suspenzii khloroplastov. [The use
of amperometric method for determining oxygen exchange in water plants
and chloroplast suspensions.] - In: Biofizicheskie Metody v Fiziologii
Rastenii. Pp. 34-43. Nauka, Moskva 1971. [In R.]

10061 - GROB, E.C., PFANDER, H., LEUENBERGER, U., SIGNER, R.: Die Trennung von
Carotinoidgemischen durch Gegenstromextraktion. - Chimia *25*: 332-333, 1971.

10062 - GROGAN, C.O., MUSGRAVE, R.B.: New efficiency for corn plants. - N.Y.
Food Life Sci. Quart. *4* (2-3): 27-28, 1971. [Ps.]

10063 - GROMET-ELHANAN, Z.: Inhibition of photophosphorylation and NAD^+ photoreduction in *Rhodospirillum rubrum* chromatophores by valinomycin and nonac-
tin. - In: QUAGLIARIELLO, E., PAPA, S., ROSSI, C.S. (ed.): Energy Transduction in Respiration and Photosynthesis. Pp. 579-592. Adriatica Editrice, Bari 1971.

10064 - GROMET-ELHANAN, Z.: Relationship between light-induced quenching of atebrin fluorescence and ATP formation in *Rhodospirillum rubrum* chromatophores. - FEBS Lett. *13*: 124-126, 1971.

10065 - GROMET-ELHANAN, Z., FELDMAN, H.: Non-cyclic electron transport and photophosphorylation in *Rhodospirillum rubrum* chromatophores. - Israel J. Chem.
9: 35BC, 1971.

10066 - GRONEBAUM-TURCK, K., WILLENBRINK, J.: Isolierung und Eigenschaften von
Mitochondrien aus Blättern von *Spinacia oleracea* und *Beta vulgaris*. -
Planta *100*: 337-346, 1971. [Also chloroplast isolation.]

10067 - GRÖNEGRESS, P.: The greening of chromoplasts in *Daucus carota* L. - Planta
98: 274-278, 1971.

10068 - GROSS, E.: Uncoupling of photophosphorylation inhibition of proton binding
by quaternary ammonium salts and zwitterionic buffers. - Arch. Biochem. Biophys. *147*: 77-84, 1971.

*10069 - GROSS, J.A., BECKER, M.J., KYLE, J.L.: Studies of photoactive "38 S" units
from spinach chloroplasts. - Biophys. Soc. Abstr. *1966*: 21, 1966.

10070 - GROUZIS, J.-P., PARIS-PIREYRE, N.: Action du calcium sur la photophosphorylation cyclique des chloroplastes isolés de *Spinacia oleracea* L. -
Compt. rend. Acad. Sci. Paris, Sér. D *272*: 234-237, 1971.

*10071 - GROZOV, D.N.: Fotosintez vinograda pod vliyaniem razlichnykh uslovii
mineral'nogo pitaniya. [Photosynthesis of vine under the effect of various
conditions of mineral nutrition.] - In: Fotosintez Sel'skokhozyaĭstvennykh
Rastenii Moldavii v Svyazi s Usloviyami Proizrastaniya. Pp. 26-41. Red.-
izdat. Otd. Akad. Nauk mold. SSR, Kishinev 1970. [In R.]

10072 - GRUODENE, Ya.P.: Vliyanie NRV na rost i urozhaĭ fasoli. [Effect of petroleum growth substance on the growth and yield of beans.] - In: NRV v Sel'-

skom Khozyaĭstve. Pp. 165-167. Elm, Baku 1971. [Chl; in R.]

10073 - GUBAR', G.D.: Fotosintez i svetovoĭ rezhim kak faktory effektivnosti mineral'nogo pitaniya. [Photosynthesis and light regime as factors of effectivity of mineral nutrition.] - Izv. Akad. Nauk latv. SSR *1971* (9): 6-18, 1971. [In R, ab: E.]

10074 - GUBAR', G.D., VOĬTSEKHOVICH, Z.V.: Aktivnost' fotosinteticheskogo apparata v zavisimosti ot obespechennosti rasteniĭ mineral'nym pitaniem. [Activity of the photosynthetic apparatus in relation to mineral nutrients supply to plants.] - Izv. Akad. Nauk latv. SSR *1971* (9): 37-45, 1971. [In R, ab: E.]

10075 - GUGUNAVA, N.A.: Produktivnost' fotosinteza blagorodnogo lavra v svyazi s periodichnost'yu ekspluatatsii. [Productivity of photosynthesis of *Laurus nobilis* L. in relation to periodicity of exploitation.] - Subtrop. Kul't. *1971* (3): 84-90, 1971. [In R.]

10076 - GUILLORY, R.J., FISHER, R.R.: Measurement of simultaneous synthesis of inorganic pyrophosphate and adenosine triphosphate. - Anal. Biochem. *39*: 170-180, 1971.

*10077 - GULLVÅG, B.M.: On the fine structure of the spores of *Equisetum fluviatile* var. *verticillatum* studied in the quiescent, germinated and non-viable state. - Grana palynologica *8*: 23-69, 1968. [Chloroplast.]

*10078 - GULLVÅG, B.M.: Primary storage products of some pteridophyte spores - a fine structural study. - Phytomorphology *19*: 82-92, 1969. [Chloroplasts.]

*10079 - GULYAEV, B.A., LITVIN, F.F.: Proizvodnye spektry pogloshcheniya khlorofilla i bakterial'nykh pigmentov v kletkakh fotosinteziruyushchikh organizmov pri temperature 20 ᵁ i -196 °C. [Derivative absorption spectra of chlorophyll and bacterial pigment in cells of photosynthesizing organisms at the temperatures 20 ° and -196 °C.] - Nauch. Dokl. vyssh. Shkoly, biol. Nauki *13* (3): 132-133, 1970. [In R.]

*10080 - GULYAEV, B.I.: Mnogokanal'nye ustanovki dlya izmereniya gazoobmena CO_2 u rasteniĭ. [Multichannel devices for determining CO_2 exchange in plants.]- In: Tezisy Dokladov Vsesoyuznogo Soveshchaniya po Unifikatsii Metodov i Priborov dlya Massovykh Izmereniĭ Intensivnosti Fotosinteza. Pp. 29-33. Nauch.-issled. Inst. Rastenievod. N.I. Vavilova, Leningrad - Pushkin 1970. [In R.]

*10081 - GULYAEV, B.I., MANUIL'SKIĬ, V.D.: O deĭstvii sveta na dykhanie rasteniĭ. [Action of light on plant respiration.] - Dokl. Akad. Nauk SSSR *193*: 724-727, 1970. [In R.]

10082 - GULYAEV, B.I., OKANENKO, A.S.: Raspredelenie fotosinteticheski aktivnoĭ radiatsii v posevakh. [Distribution of photosynthetically active radiation in stands.] - In: ZALENSKIĬ, O.V. (ed.): Fotosintez i Ispol'zovanie Solnechnoĭ Energii. Pp. 18-25. Nauka, Leningrad 1971. [In R, ab: E.]

10083 - GULYĬ, M.F.: Fiksatsiya CO_2 u mikroorganizmov i drugikh geterotrofnykh organizmov i ee fiziologicheskoe znachenie. [CO_2 fixation in microorganisms and other heterotrophic organisms and its physiological significance.] - Izv. Akad. Nauk SSSR, Ser. biol. *1971*: 724-742, 1971. [In R, ab: E.]

10084 - GUPTA, K.K., CHATTERJEE, S.K.: Studies on certain aspects of senescence in *Nicotiana plumbaginifolia* SPS. - Ann. Bot. *35*: 857-864, 1971. [Chl.]

10085 - GUPTA, R.K., WOOLLEY, J.T.: Spectral properties of soybean leaves. - Agron. J. *63*: 123-126, 1971.

10086 - GUSEĬNOV, B.Z., MASIEV, A.M., AKHUNDOVA, E.M.: Vliyanie NRV na soderzhanie nukleinovykh kislot i khlorofilla v list'yakh rasteniĭ. [Effect of petroleum growth substance on the nucleic acid and chlorophyll levels in plant leaves.] - In: NRV v Sel'skom Khozyaĭstve. Pp. 406-408. Elm, Baku 1971. [In R.]

*10087 - GUSEV, M.V., KORZHENEVSKAYA, T.G.: Sinezelenaya vodorosl' *Anacystis nidulans* - obligatnyĭ fototrof. [A blue-green alga *Anacystis nidulans* - an obligate phototrophic organism.] - Nauch. Dokl. vyssh. Shkoly, biol. Nauki *13* (5): 87-101, 1970. [Ps; in R.]

*10088 - GUSTAFSON, J.P., CURTIS, B.C., YOUNGMAN, V.E.: A chlorotic mutation in wheat. - Crop Sci. *10*: 665-667, 1970.

10089 - GYLDENHOLM, A.O., PALMER, J.M., WHATLEY, F.R.: Development of electron transport functions in greening chloroplasts. - In: QUAGLIARIELLO, E., PAPA, S., ROSSI, C.S. (ed.): Energy Transduction in Respiration and Photosynthesis. Pp. 368-374. Adriatica Editrice, Bari 1971.

10090 - HÄDER, D.P., NULTSCH, W.: Untersuchungen zur Abgrenzung der Arten *Phormidium uncinatum* und *Phormidium autumnale*. - Schweiz. Z. Hydrol. *33*: 566-577, 1971. [Chl, biliproteins.]

10091 - HADSELL, R.M.: Photosynthesis in *Rhodospirillum rubrum*: photophosphorylation quantum requirements and isolation of subchromatophore particles. - Diss. Abstr. int. B *31*: 3895-B-3896-B, 1971.

10092 - HAEHNEL, W., DÖRING, G., WITT, H.T.: On the reaction between chlorophyll-a_1 and its primary electron donors in photosynthesis. - Z. Naturforsch. *26 b*: 1171-1174, 1971.

10093 - HAGEMANN, R.: Struktur und Funktion der genetischen Information in den Plastiden. I. Die Bedeutung von Plastommutanten und die genetische Nomenklatur extranukleärer Mutationen. - Biol. Zbl. *90*: 409-418, 1971.

*10094 - HAGGAR, R.J.: Seasonal production of *Andropogon gayanus*. I. Seasonal changes in yield components and chemical composition. - J. agr. Sci. (Cambridge) *74*: 487-494, 1970. [Growth analysis.]

10095 - HALÁS, L.: Effect of habitat irradiance on the year's course of photosynthetic rate of *Prunus laurocerasus* L. leaf discs. - Photosynthetica *5*: 352-357, 1971.

10096 - HALL, A.E.: A hypothesis concerning the relationship between photosynthesis and photorespiration. - Plant Physiol. *47* (Suppl.): 10, 1971.

10097 - HALL, A.E.: A model of leaf photosynthesis and respiration. - Carnegie Inst. Year Book *70*: 530-540, 1971.

10098 - HALL, D.O., CAMMACK, R., RAO, K.K.: Role for ferredoxins in the origin of life and biological evolution. - Nature *233*: 136-138, 1971.

10099 - HALL, D.O., EDGE, H., KALINA, M.: The site of ferricyanide photoreduction in the lamellae of isolated spinach chloroplasts: a cytochemical study. - J. Cell Sci. *9*: 289-303, 1971.

*10100 - HALL, D.O., EVANS, M.C.W.: Iron-sulphur proteins. - Nature *223*: 1342-1384, 1969. [Ferredoxin in Ps.]

10101 - HALL, D.O., REEVES, S.G., BALTSCHEFFSKY, H.: Photosynthetic control in isolated spinach chloroplasts with endogenous and artificial electron acceptors. - Biochem. biophys. Res. Commun. *43*: 359-366, 1971.

*10102 - HALL, J.D., STILES, J.W., AWASTHI, Y., CRANE, F.L.: Membranifibrils on cristae and grana membranes. - Proc. Indiana Acad. Sci. *78*: 189-197, 1969. [Chloroplast.]

10103 - HALL, S.M., BAKER, D.A., MILBURN, J.A.: Phloem transport of ^{14}C-labelled assimilates in *Ricinus*. - Planta *100*: 200-207, 1971.

10104 - HALLDAL, P., FRENCH, C.S., LAWRENCE, M.: Automatic recordings of action spectra for partial reactions of photosynthesis. - Carnegie Inst. Year Book *69*: 670-678, 1971.

10105 - HAMBRIGHT, P.: The coordination chemistry of metalloporphyrins. - Coord. Chem. Rev. *6*: 247-268, 1971.

*10106 - HAMLIN, P.A., LAMBERT, J.L.: Determination of dissolved oxygen using photoreduced leuco phenothiazine dyes. - Anal. Chem. *43*: 618-620, 1971.

10107 - HANNAN, H.H., ANDERSON, B.T.: Predicting the diel oxygen minimum in ponds containing macrophytes. - Progress. Fish-Culturist *33*: 45-47, 1971. [Ps.]

*10108 - HANNAN, P.J., PATOUILLET, C.: Toxicity of metals to *Chlorella* - a surface phenomenon. - Develop. ind. Microbiol. *8*: 313-320, 1967. [Ps.]

*10109 - HANNAN, P.J., PATOUILLET, C.: Nutrient and pollutant concentrations as
         determinants in algal growth rates. - FAO Technical Conference on Marine
         Pollution and Its Effects on Living Resources and Fishing, FIR: MP/70/E-56.
         Pp. 1-7. Roma 1970. [Chl.]

 10110 - HANSEN, G.K.: Photosynthesis, transpiration and diffusion resistance in
         relation to water potential in leaves during water stress. - Acta Agr.
         scand. 21: 163-171, 1971.

 10111 - HANSEN, P.: ^{14}C-studies on apple trees. VII. The early seasonal growth
         in leaves, flowers and shoots as dependent upon current photosynthates
         and existing reserves. - Physiol. Plant. 25: 469-473, 1971.

 10112 - HANSON, A.D., EDELMAN, J.: Secretion of photosynthetic products by carrot
         tissue cultures. - Planta 98: 97-108, 1971.

 10113 - HANSON, W.D.: Selection for differential productivity among juvenile maize
         plants: Associated net photosynthetic rate and leaf area changes. - Crop
         Sci. 11: 334-339, 1971.

*10114 - HANWAY, J.J., RUSSELL, W.A.: Dry-matter accumulation in corn (Zea mays L.)
         plants: Comparisons among single-cross hybrids. - Agron. J. 61: 947-951,
         1969.

 10115 - HANWAY, J.J., WEBER, C.R.: Dry matter accumulation in eight soybean (Gly-
         cine max (L.) MERRILL) varieties. - Agron. J. 63: 227-230, 1971.

 10116 - HARADA, J., NAKAYAMA, H.: Chlorophyll degradation in leaf sections from tall
         and short varieties of rice. - J. agr. Sci. 76: 573-574, 1971.

 10117 - HARDER, D.E., MARTENS, J.W., McKENZIE, R.I.H.: Changes in chlorophyll and
         carotenoid content in oats associated with the expression of adult plant
         resistance to stem rust conferred by gene pg-11. - Can. J. Bot. 49: 1783-
         1785, 1971.

 10118 - HARDMAN, L.L., BRUN, W.A.: Effect of atmospheric carbon dioxide enrich-
         ment at different developmental stages on growth and yield components of
         soybeans. - Crop Sci. 11: 886-888, 1971.

 10119 - HARDT, H., MALKIN, S.: Kinetic studies on chemiluminescence of chloro-
         plasts induced by changes in ion concentration. - Photochem. Photobiol.
         14: 483-492,1971.

 10120 - HARDY, S.I., CASTELFRANCO, P.A., REBEIZ, C.A.: Effect of the hypocotyl
         hook on chlorophyll accumulation in excised cotyledons of Cucumis sativus
         L. - Plant Physiol. 47: 705-708, 1971.

 10121 - HARMS, W.R.: Estimating leaf-area growth in pine. - Ecology 52: 931-934,
         1971.

*10122 - HARMSWORTH, R.V., WHITESIDE, M.C.: Relation of cladoceran remains in lake
         sediments to primary productivity of lakes. - Ecology 49: 998-1000, 1968.

 10123 - HARNISCHFEGER, G.: Correlation between structure and function in isolated,
         disintegrating chloroplasts. - Diss. Abstr. int. B 31: 6412-B, 1971.

 10124 - HARRIS, G.P.: The ecology of corticolous lichens. II. The relationship
         between physiology and the environment. - J. Ecol. 59: 441-452, 1971.
         [Ps.]

 10125 - HARRIS, W.M.: Ultrastructural observations on the mesophyll cells of pine
         leaves. - Can. J. Bot. 49: 1107-1109, 1971. [Chloroplast.]

 10126 - HART, A.L., TREGUNNA, E.B.: Some aspects of environmental control of the
         photosynthetic apparatus in Gomphrena globosa. - In: HATCH, M.D., OSMOND,
         C.B., SLATYER, R.O. (ed.): Photosynthesis and Photorespiration. Pp. 413-
         418. Wiley-Interscience, New York - London - Sydney - Toronto 1971.

 10127 - HART, B.A., GIBSON, J.: Ribulose-5-phosphate kinase from Chromatium sp.
         strain D. - Arch. Biochem. Biophys. 144: 308-321, 1971.

 10128 - HART, R.H., LEE, D.R.: Age vs. net CO_2 exchange rate of leaves of coastal
         bermudagrass. - Crop Sci. 11: 598-599, 1971.

*10129 - HARTMAIR, V.: Enrichissement en sucres et accroissement du volume des baies: Mécanisme, facterus, rôle du feuillage pour le rendement de la qualité du raisin: Productivité du feuillage. - Bull. O.I.V. *440*: 1046-1056, 1967. [Ps, Chl.]

10130 - HARTSHORN, L.G., BISHOP, E.: A stable low-current source for electrode polarisation. - Analyst *96*: 885-886, 1971.

10131 - HASE, E.: Studies on the metabolism of nucleic acid and protein associated with the processes of de- and re-generation of chloroplasts in *Chlorella protothecoides*. - In: BOARDMAN, N.K., LINNANE, A.W., SMILLIE, R.M. (ed.): Autonomy and Biogenesis of Mitochondria and Chloroplasts. Pp. 434-446. North-Holland Publ. Comp., Amsterdam - London 1971.

10132 - HASKINS, R.H., CONSTABEL, F., NESBITT, L., GAMBORG, O.L.: Plastid development in albino plants obtained from brome grass cell cultures by embryogenesis. - Amer. J. Bot. *58*: 452, 1971.

10133 - HASPELOVÁ-HORVATOVIČOVÁ, A.: Pigmentveränderungen mehltauresistenter Gerste nach der Infektion. - Biológia (Bratislava) *26*: 19-26, 1971.

10134 - HASPELOVÁ-HORVATOVIČOVÁ, A.: Pozorovania z patologickej fyziológie listových pigmentov. [Observations from pathological physiology of leaf pigments.] - In: Zborn. Predn. Zjazdu Slov. Bot. Spoloč., Tisovec 1970. Pp. 137-149. Bratislava 1971. [In Slovak, ab: G.]

10135 - HATCH, M.D.: Mechanism and function of the C₄ pathway of photosynthesis. - In: HATCH, M.D., OSMOND, C.B., SLATYER, R.O. (ed.): Photosynthesis and Photorespiration. Pp. 139-152. Wiley-Interscience, New York - London - Sydney - Toronto 1971.

10136 - HATCH, M.D.: The C₄-pathway of photosynthesis. Evidence for an intermediate pool of carbon dioxide and the identity of the donor C₄-dicarboxylic acid. - Biochem. J. *125*: 425-432, 1971.

B10137 - HATCH, M.D., OSMOND, C.B., SLATYER, R.O. (ed.): Photosynthesis and Photorespiration. - Wiley-Interscience, New York - London - Sydney - Toronto 1971.

10138 - HAUPT, W.: Chloroplast movement as an example of oriented phytochrome response. - Advance. Sci. *27*: 341-346, 1971.

10139 - HAUPT, W.: Schwachlichtbewegung des *Mougeotia*-Chloroplasten im Blaulicht. - Z. Pflanzenphysiol. *65*: 248-265, 1971.

10140 - HAUSKA, G.A., McCARTY, R.E., BERZBORN, R.J., RACKER, E.: Partial resolution of the enzymes catalyzing photophosphorylation. VII. The function of plastocyanin and its interaction with a specific antibody. - J. biol. Chem. *246*: 3524-3531, 1971.

*10141 - HAYASHI, K.-I.: Efficiencies of solar energy conversion and relating characteristics in rice varieties. - Proc. Crop Sci. Soc. Jap. *38*: 495-500, 1969. [Growth analysis.]

10142 - HEALEY, F.P.: What can be the best basis to use in comparing rates of light saturated photosynthesis? - Acta agron. Acad. Sci. hung. *20*: 449-451, 1971.

10143 - HEALEY, F.P., MYERS, J.: The Kok effect in *Chlamydomonas reinhardi*. - Plant Physiol. *47*: 373-379, 1971.

10144 - HEATH, R.L.: Hydrazine as an electron donor to the water-oxidation site in photosynthesis. - Biochim. biophys. Acta *245*: 160-164, 1971.

10145 - HEATH, R.L.: Quantum yield of proton movements by isolated chloroplasts. - Plant Physiol. *47* (Suppl.): 32, 1971.

10146 - HEBER, U., KRAUSE, G.H.: Transfer of carbon, phosphate energy, and reducing equivalents across the chloroplast envelope. - In: HATCH, M.D., OSMOND, C.B., SLATYER, R.O. (ed.): Photosynthesis and Photorespiration. Pp. 218-225. Wiley-Interscience, New York - London - Sydney - Toronto 1971.

10147 - HEBER, U., TYANKOVA, L., SANTARIUS, K.A.: Stabilization and inactivation

of biological membranes during freezing in the presence of amino acids.
- Biochim. biophys. Acta *241*: 578-592, 1971. [Ps.]

10148 - HEDLEY, C.L., STODDART, J.L.: Light-stimulation of alanine aminotrans-
ferase activity in dark-grown leaves of *Lolium temulentum* L. as related
to chlorophyll formation. - Planta *100*: 309-324, 1971.

10149 - HEHL, M., KRANZ, A.R.: Endogene Stoffproduktionsrhythmen bei *Phaseolus
vulgaris*. - Ber. deut. bot. Ges. *84*: 551-558, 1971. [Ps.]

10150 - HEICHEL, G.H.: Confirming measurements of respiration and photosynthesis
with dry matter accumulation. - Photosynthetica *5*: 93-98, 1971.

10151 - HEICHEL, G.H.: Response of respiration of tobacco leaves in light and
darkness and the CO_2 compensation concentration to prior illumination
and oxygen. - Plant Physiol. *48*: 178-182, 1971.

10152 - HEICHEL, G.H.: Stomatal movements, frequencies, and resistances in two
maize varieties differing in photosynthetic capacity. - J. exp. Bot.
22: 644-649, 1971.

10153 - HEILPORN, V., LIMBOSCH, S.: Les effets du bromure d'éthidium sur *Acetabularia mediterranea*. - Biochim. biophys. Acta *240*: 94-108, 1971.
[Ps, Chl.]

10154 - HELDT, H.W., SAUER, F.: The inner membrane of the chloroplast envelope
as the site of specific metabolite transport. - Biochim. biophys. Acta
234: 83-91, 1971.

10155 - HELDT, H.W., SAUER, F., RAPLEY, L.: Metabolite transport in spinach chlo-
roplasts. - In: QUAGLIARIELLO, E., PAPA, S., ROSSI, C.S. (ed.): Energy
Transduction in Respiration and Photosynthesis. Pp. 1007-1008. Adriatica
Editrice, Bari 1971.

10156 - HELLEBUST, J.A.: Glucose uptake by *Cyclotella cryptica*: Dark induction
and light inactivation of transport system. - J. Phycol. *7*: 345-349,
1971. [Ps.]

10157 - HELLER, H.: Estimation of photosynthetically active leaf area in forests.
- In: ELLENBERG, H. (ed.): Integrated Experimental Ecology. Pp. 29-31.
Springer-Verlag, Berlin - Heidelberg - New York 1971.

10158 - HELLMUTH, E.O.: Eco-physiological studies on plants in arid and semi-
arid regions in western Australia. III. Comparative studies on photosynthesis, respiration and water relations of ten arid zone and two semi-
arid zone plants under winter and late summer climatic conditions. - J.
Ecol. *59*: 225-259, 1971.

10159 - HELLMUTH, E.O.: Eco-physiological studies on plants in arid and semi-
arid regions in western Australia. IV. Comparison of the field physiology of the host, *Acacia grasbyi* and its hemiparasite, *Amyema nestor* under
optimal and stress conditions. - J. Ecol. *59*: 351-363, 1971. [Ps.]

10160 - HELLMUTH, E.O.: The effect of varying air-CO_2 level, leaf temperature,
and illuminance on the CO_2 exchange of the dwarf pea, *Pisum sativum* L. var.
Meteor. - Photosynthetica *5*: 190-194, 1971.

10161 - HELMS, J.A., COBB, F.W. Jr., WHITNEY, H.S.: Effect of infection of *Ver-
ticicladiella wagenerii* on the physiology of *Pinus ponderosa*. - Phytopathology *61*: 920-925, 1971. [Ps.]

*10162 - HENNINGSEN, K.W., BOYNTON, J.E.: Formation of chlorophyll pigments and
plastid membranes in barley greening at low light intensity. - In: Abstr. XI International Botanical Congress. P. 89. Seattle, Wash. 1969.

10163 - HENNINGSEN, K.W., KAHN, A.: Photoactive subunits of protochlorophyll(ide)
holochrome. - Plant Physiol. *47*: 685-690, 1971.

*10164 - HENRY, M., LEICKNAM, J.-P.: Mise en évidence et comportement de la liai-
son entre des groupements polaires et l'atome de Mg de la chlorophylle
"*a*" et de quelques autres molécules possédant le cycle tetrapyrrolique.
- In: Colloques Internationaux du C.N.R.S. No. *191*: La Nature et les
Propriétés des Liaisons de Coordination. Pp. 317-333. Éditions CNRS,
Paris 1970.

10165 - HENRY-HISS, Y., SCHANTZ, R.: Les acides aminés libres ches *Gonium octonarium* (POCOCK) cultivé à l'obscurité et à la lumière. - Compt. rend. Acad. Sci. Paris, Sér. D *272*: 2181-2184, 1971. [Chl.]

10166 - HERNANDEZ, R.J., SCHAEDLE, M.: Photosynthesis by chloroplasts isolated from cottonwood *(Populus deltoides)* leaves. - Plant Physiol. *47* (Suppl.): 9, 1971.

10167 - HERRMANN, F.: Genetic control of pigment-protein complexes I and Ia of the plastid mutant en:alba 1 of *Antirrhinum majus*. - FEBS Lett. *19*: 267-269, 1971.

10168 - HERRMANN, F.: Struktur und Funktion der genetischen Information in den Plastiden II. Untersuchung der photosynthesedefekten Plastommutante alba-1 von *Antirrhinum majus*.L. - Photosynthetica *5*: 258-266, 1971.

10169 - HERRMANN, F., HAGEMANN, R.: Struktur und Funktion der genetischen Information in den Plastiden. III. Genetik, Chlorophylle und Photosyntheseverhalten der Plastommutante "Mrs. Pollock" und der Genmutante "Cloth of Gold" von *Pelargonium zonale*. - Biochem. Physiol. Pflanzen *162*: 390-409, 1971.

10170 - HERRMANN, R.G.: Anzahl und Anordnung der genetischen Einheiten (Chloroplastengenome) in Chloroplasten. - Ber. deut. bot. Ges. *83*: 359-361, 1971.

10171 - HERTZBERG, S., LIAAEN-JENSEN, S.: The constitution of aphanizophyll. - Phytochemistry *10*: 3251-3252, 1971.

10172 - HERTZBERG, S., LIAAEN-JENSEN, S., SIEGELMAN. H.W.: The carotenoids of blue-green algae. - Phytochemistry *10*: 3121-3127, 1971.

10173 - HESKETH, J.D., BAKER, D.N., DUNCAN, W.G.: Simulation of growth and yield in cotton: Respiration and the carbon balance. - Crop Sci. *11*: 394-398, 1971.

10174 - HESLEHURST, M.R.: The point quadrat method of vegetation analysis: a review. - Univ. Reading, Dept. Agr. Study *10*: 1-81, 1971.

10175 - HESLEHURST, M.R., WILSON, G.L.: Studies on the productivity of tropical pasture plants. III. Stand structure, light penetration, and photosynthesis in field swards of *Setaria* and green leaf *Desmodium*. - Aust. J. agr. Res. *22*: 865-878, 1971.

10176 - HEVESI, J.: Energy migration in dye-detergent systems. - Acta biochim. biophys. Acad. Sci. hung. *6*: 467-468, 1971.

10177 - HEYES, J.K., DALE, J.E.: A virescent mutant of *Phaseolus vulgaris*: photosynthesis and metabolic changes during leaf development. - New Phytol. *70*: 415-426, 1971.

10178 - HICKMAN, M.: Standing crops and primary productivity of the epipelon of two small ponds in North Somerset, U.K. - Oecologia *6*: 238-253, 1971.

10179 - HIEDEMANN-van WYK, D., GAMINI KANNANGARA, C.: Localization of ferredoxin in the thylakoid membrane with immunological methods. - Z. Naturforsch. *26 b*: 46-50, 1971.

10180 - HILL, A.C.: Vegetation: A sink for atmospheric pollutants. - APCA J. *21*: 341-346, 1971. [Ps.]

10181 - HILL, D.J.: Experimental study of the effect of sulphite on lichens with reference to atmospheric polution. - New Phytol. *70*: 831-836, 1971. [Ps.]

10182 - HILL, H.M., CALDERWOOD, S.K., ROGERS, L.J.: Conversion of lycopene to β-carotene by plastids isolated from higher plants. - Phytochemistry *10*: 2051-2058, 1971.

10183 - HILL, R.: Historical outline. - In: GIBBS, M. (ed.): Structure and Function of Chloroplasts. Pp.1-6. Springer-Verlag, Berlin - Heidelberg - New York 1971. [Ps.]

10184 - HILLER, R.G., ANDERSON, J.M., BOARDMAN, N.K.: Photooxidation of cytochrome *b*-559 in leaves and chloroplasts at room temperature. - Biochim. biophys. Acta *245*: 439-452, 1971.

10185 - HILLER, R.G., BOARDMAN, N.K.: Light driven redox changes of cytochrome *f* and the development of photosystems I and II during greening of bean leaves. - Biochim. biophys. Acta *253*: 449-458, 1971.

10186 - HILLIARD, J.H., GRACEN, V.E., WEST, S.H.: Leaf microbodies (peroxisomes) and catalase localization in plants differing in their photosynthetic carbon pathways. - Planta *97*: 93-105, 1971.

10187 - HILLIARD, J.H., WEST, S.H.: The association of chloroplast peripheral reticulum with low photorespiration rates in a photorespiring plant species. - Planta *99*: 352-356, 1971.

10188 - HILTON, J.L., JOHN, J.B. St., CHRISTIANSEN, M.N., NORRIS, K.H.: Interactions of lipoidal materials and a pyridazinone inhibitor of chloroplast development. - Plant Physiol. *48*: 171-177, 1971.

10189 - HINCHMAN, R.R.: The DNA of the oat plastid. - In: Argonne National Laboratory Annual Report 1971, ANL-7870. Pp. 79-81. Argonne, Ill. 1971.

10190 - HINCKLEY, T.M.: Effect of decreasing soil moisture on net assimilation, needle xylem sap pressure, and stomatal aperture in *Abies*. - Plant Physiol. *47* (Suppl.): 37, 1971.

10191 - HINSHIRI, H.M., PROCTOR, M.C.F.: The effect of desiccation on subsequent assimilation and respiration of the bryophytes *Anomodon viticulosus* and *Porella platyphylla*. - New Phytol. *70*: 527-538, 1971.

10192 - HIPKE, H.: Untersuchungen über den Einfluss äusserer Faktoren auf die Pigmentausstattung induzierter Mutanten von *Pisum sativum*. - Z. Pflanzenphysiol. *64*: 41-51, 1971.

10193 - HIYAMA, T., KE, B.: A further study of P430: A possible primary electron acceptor of Photosystem I. - Arch. Biochem. Biophys. *147*: 99-108, 1971.

10194 - HIYAMA, T., KE, B.: A new photosynthetic pigment, "P430": its possible role as the primary electron acceptor of photosystem I. - Proc. nat. Acad. Sci. USA *68*: 1010-1013, 1971.

10195 - HIYAMA, T., KE, B.: Laser-induced reactions of P700 and cytochrome *f* in a blue-green alga, *Plectonema boryanum*. - Biochim. biophys. Acta *226*: 320-327, 1971.

10196 - HO, L.C., MORTIMER, D.C.: The site of cyanide inhibition of sugar translocation in sugar beet leaf. - Can. J. Bot. *49*: 1769-1775, 1971. [Ps.]

10197 - HOCH, G.E., RANDLES, J.: On the interaction of Photosystems I and II in algal cells. - Photochem. Photobiol. *14*: 435-449, 1971.

10198 - HOCHAPFEL, A., HIVER, J.-A., VIOVY, R.: Orientation de la chlorophylle *a* dans un cristal liquide nématique. - Compt. rend. Acad. Sci. Paris, Sér. C *272*: 1265-1268, 1971.

10199 - HOCHAPFEL, A., HIVER, J.-A., VIOVY, R.: Orientation de la chlorophylle dans un cristal liquide nematique. - In: BRODA, E., LOCKER, A., SPRINGER-LEDERER, H. (ed.): Proceedings of the First European Biophysics Congress. Vol. 4. Pp. 155-160. Verlag Wiener med. Akad., Wien 1971.

*10200 - HOCHMAN, A., CARMELI, C.: Partial reactions of photophosphorylation in *Chromatium* strain D chromatophores. - Israel J. Chem. *8* (Suppl.): 163p, 1970.

10201 - HOCHMAN, A., CARMELI, C.: A coupling factor from *Chromatium* strain D chromatophores. - FEBS Lett. *13*: 36-40, 1971.

10202 - HOFFMAN, G.J.: Estimating leaf area from length measurements for Hybrid Granex onion. - Agron. J. *63*: 948-949, 1971.

10203 - HOFFMAN, G.J., PHENE, C.J.: Effect of constant salinity levels on water-use efficiency of bean and cotton. - Trans. ASAE *14*: 1103-1106, 1971. [Ps.]

10204 - HOFFMANN, P.: Der Anteil der einzelnen Organe an der photosynthetischen Produktivität der Gesamtpflanze bei *Arabidopsis thaliana* (L.) HEYNH. - *Arabidopsis* Inform. Serv. *1971* (8): 13-15, 1971.

10205 - HOFFMANN, P.: Symposium über die "Biochemie und Biophysik der Photosynthese", Irkutsk, 13.-20. Juli 1970. - Photosynthetica 5: 88, 1971.

10206 - HOFFMANN, P.: Wie lange bleiben Chloroplasten ausserhalb der Mutterzelle funktionstüchtig? - Biol. Rundschau 9: 106, 1971.

10207 - HOFFMANN, P., NISSEN, E.: Nomogramme zur Auswertung routinemässiger Chlorophyllbestimmungen. - Biol. Rundschau 9: 405-407, 1971.

10208 - HOLDSWORTH, M.: Carbon dioxide uptake by succulents. - Can. J. Bot. 49: 1520-1522, 1971.

10209 - HOLDSWORTH, R.H.: The isolation and partial characterization of the pyrenoid protein of Eremosphaera viridis. - J. Cell Biol. 51: 499-513, 1971.

10210 - HOLLIGAN, P.M.: Routine analysis by gas-liquid chromatography of soluble carbohydrates in extracts of plant tissues. I. A review of techniques used for the separation, identification and estimation of carbohydrates by gas-liquid chromatography. - New Phytol. 70: 239-269, 1971.

10211 - HOLLIGAN, P.M., DREW, E.A.: Routine analysis by gas-liquid chromatography of soluble carbohydrates in extracts of plant tissues. II. Quantitative analysis of standard carbohydrates, and the separation and estimation of soluble sugars and polyols from a variety of plant tissues. - New Phytol. 70: 271-297, 1971.

10212 - HOLT, D.A., YOUNGBERG, H.W.: Influence of sampling frequency of the precision of estimating accumulated solar radiation. - Agron. J. 63: 239-240, 1971.

10213 - HÖLZL, J.: Die Kristallformen von synthetischem und rekrystallisiertem β-Carotin. - Mikroskopie 27: 85-94, 1971.

10214 - HOMANN, P.H.: Actions of carbonylcyanide m-chlorophenylhydrazone on electron transport and fluorescence of isolated chloroplasts. - Biochim. biophys. Acta 245: 129-143, 1971.

10215 - HONDA, S.I., HONGLADAROM-HONDA, T., KWANYUEN, P., WILDMAN, S.G.: Interpretations on chloroplast reproduction derived from correlations between cells and chloroplasts. - Planta 97: 1-15, 1971.

10216 - HONEYCUTT, R.C.: Studies on photosynthetic electron transport. - Diss. Abstr. int. B 32: 3157-B, 1971.

*10217 - HOPKINS, D.L., HAMPTON, R.E.: Effects of tobacco etch virus infection upon the light reactions of photosynthesis in tobacco leaf tissue. - Phytopathology 59: 677-679, 1969.

*10218 - HOPKINS, D.L., HAMPTON, R.E.: Effects of tobacco etch virus infection upon the dark reactions of photosynthesis in tobacco leaf tissue. - Phytopathology 59: 1136-1140, 1969.

10219 - HORAK, A., HILL, R.D.: Coupling factor for photophosphorylation in bean etioplasts and chloroplasts. - Can. J. Biochem. 49: 207-209, 1971.

*10220 - HORI, Y., TATSUMI, M.: [Studies on the growth of vegetables in relation to light conditions. 1 The effects of light intensity and duration of supplementary illumination on the seedling growth.] - Bull. hort. Res. Sta., Ser. A (Hiratsuka) 7: 157-171, 1968. [Ps, Chl; in Jap., ab: E.]

*10221 - HORI, Y., TATSUMI, M., SHIRAISHI, K.: [Studies on the growth of vegetables in relation to light conditions. II The effects of prolonged illumination on the growth of vegetables.] - Bull. hort. Res. Sta., Ser. A (Hiratsuka) 7: 173-185, 1968. [Chl; in Jap., ab: E.]

10222 - HORIE, T., UDAGAWA, T.: Canopy photosynthesis of sunflower plants. Its measurements and modeling. - Bull. nat. Inst. agr. Sci. (Japan), Ser. A 1971 (18): 1-56, 1971.

10223 - HORIO, T., NISHIKAWA, K., HORIUTI, Y.: Adenosine triphosphatase: bacterial. - In: COLOWICK, S.P., KAPLAN, N.O. (ed.): Methods in Enzymology. Vol. 23. Pp. 650-654. Academic Press, New York - London 1971.

10224 - HORIO, T., YAMAMOTO, N., HORIUTI, Y., NISHIKAWA, K.: Enzymes catalyzing

exchange reactions: bacterial. - In: COLOWICK, S.P., KAPLAN, N.O. (ed.):
Methods in Enzymology. Vol. 23. Pp. 654-664. Academic Press, New York -
London 1971. [Ps.]

10225 - HORVÁTH, L., KISS, A.S., POZSÁR, B.: Az Agronit serkentö hatása a kuko-
rica levelének magnézium-tartalmára és a fotoszintetikus széndioxidfixá-
lásra. [Stimulating effect of Agronit on magnesium content and photosynthetic CO_2 fixation in maize leaves.] - Takarmánybázis, Takarmányter-
mesztési Kutató Int. (Iregszemcse) 11: 37-42, 1971. [In Hung., ab: E,
F, G, R.]

*10226 - HORVÁTH, M., LONTAI, L.: Change of pigment content, protein content and
that of the ribonuclease enzyme activity in intact plants and isolated
barley leaves. - Acta biol. (Szeged) 14: 47-55, 1968.

10227 - HORWITZ, B.A.: Bioelectric response of higher plants induced by photosynthetically active radiation. - Photosynthetica 5: 414-416, 1971.

10228 - HOSHINO, M., MATSUMOTO, F., OKUBO, T.: [Studies on the assimilation and
translocation of $^{14}CO_2$ in ladino clover. V. Relation between leaf age and
ability of assimilation and translocation.] - Proc. Crop Sci. Soc. Jap.
40: 468-473, 1971. [In Jap., ab: E.]

*10229 - HOSHINO, M., NISHIMURA, S., OKUBO, T.: [Studies on the assimilation and
translocation of $^{14}CO_2$ in ladino clover. II. Distribution of ^{14}C in various
fraction of assimilates in plants.] - Proc. Crop Sci. Soc. Jap. 35: 137-
141, 1966. [In Jap., ab: E.]

*10230 - HOSHINO, M., OIZUMI, H.: [Studies on the assimilation and translocation of
$^{14}CO_2$ in ladino clover. IV. Utilization of reserve ^{14}C-assimilates by the
plants in the early stages of regrowth.] - Proc. Crop Sci. Soc. Jap. 37:
82-86, 1968. [In Jap., ab: E.]

*10231 - HOWELL, S.H., MOUDRIANAKIS, E.N.: Ultrastructural analysis of chloroplast
lamellae active in Hill reaction. - Biophys. Soc. Abstr. 1966: 21, 1966.

*10232 - HOWLES, R.: Spur-type apples. - Exp. Rec. Dept. Agr. south Aust. 5: 9-11,
1970. [Chl.]

10233 - HOZYO, Y.: [The translocation of photosynthates.] - Proc. Crop Sci. Soc.
Jap. 40: 549-565, 1971. [In Jap.]

10234 - HSU, J.C., SMITH, B.N.: Preliminary studies of $^{13}C/^{12}C$ ratios in photorespiration and dark respiration in tobacco leaves and peanut and sunflower seedlings. - Plant Physiol. 47 (Suppl.): 11, 1971.

10235 - HSU, P., WALTON, P.D.: Relationships between yield and its components
and structures above the flag leaf node in spring wheat. - Crop Sci.
11: 190-193, 1971.

*10236 - HÜBEL, H.: Die ^{14}C-Methode zur Bestimmung der Primärproduktion des Phy-
toplanktons. - Limnologica (Berlin) 4: 267-280, 1966.

10237 - HUBÍK, E.: Příspěvek k poznání vlivu koncentrace CO_2 v ovzduší na tvorbu
sušiny obilnin. [Effect of CO_2 concentration in air on dry-matter incre-
ments in cereals.] - Rostlinná Výroba (Praha) 17: 501-509, 1971. [In
Czech, ab: E, R.]

10238 - HUDSON, J.P.: Horticulture in 2000 A.D. - In: WAREING, P.F., COOPER, J.P.
(ed.): Potential Crop Production. A Case Study. Pp. 187-201. Heinemann
educational Books, London 1971. [Productivity.]

10239 - HUGHES, A.P., COCKSHULL, K.E.: The effects of light intensity and carbon
dioxide concentration on the growth of *Chrysanthemum morifolium* cv. Bright
Golden Anne. - Ann. Bot. 35: 899-914, 1971. [Growth analysis.]

10240 - HUGHES, A.P., COCKSHULL, K.E.: A comparison of the effects of diurnal
variation in light intensity with constant light intensity on growth of
Chrysanthemum morifolium cv. Bright Golden Anne. - Ann. Bot. 35: 927-
932, 1971. [Growth analysis.]

10241 - HUGHES, A.P., COCKSHULL, K.E.: The variation in response to light intensity and carbon dioxide concentration shown by two cultivars of *Chry-*

santhemum morifolium grown in controlled environments at two times of
year. - Ann. Bot. *35*: 933-945, 1971. [Growth analysis.]

10242 - HUGHES, M.K.: Ground vegetation biocontent and net production in a
deciduous woodland. - Oecologia *7*: 127-135, 1971.

*10243 - HURD, R.G.: Leaf resistance in a glasshouse tomato crop in relation to
leaf position and solar radiation. - New Phytol. *68*: 265-273, 1969.

*10244 - HUTCHINSON, T.C.: Ecotype differentiation in *Teucrium scorodonia* with
respect to susceptibility to lime-induced chlorosis and to shade factors.
- New Phytol. *66*: 439-453, 1967.

*10245 - HUTCHINSON, T.C.: Lime-chlorosis as a factor in seedling establishment
on calcareous soils. I. A comparative study of species from acidic and
calcareous soils in their susceptibility to lime-chlorosis. - New Phytol.
66: 697-705, 1967.

10246 - HUXLEY, P.A.: Leaf volume: a simple method for measurement and some notes
on its use in studies of leaf growth. - J. appl. Ecol. *8*: 147-153, 1971.

10247 - ICHIMURA, M.: [*Tatsia japonica* leaf studies.] - Kagaku No Jikken *22* (8):
134-140, 1971. [Ps, Chl, Car; in Jap.]

10248 - IDSO, S.B.: A simple technique for the calibration of long-wave radiation
probes. - Agr. Meteorol. *8*: 235-243, 1971.

10249 - IDSO, S.B.: Transformation of a net radiometer into a hemispherical radiometer. - Agr. Meteorol. *9*: 109-121, 1971/1972.

*10250 - IHÁSZ, I.: Néhány zöld pillangós β-karotin és össztokoferol tartalmának
viszgálata. [Contents of β-carotene and total tocopherol in some green
leguminous plants.]- Agrokem. Talajtan *16*: 635-644, 1967. [In Hung., ab:
E, G, R.]

10251 - IKEDA, T.: Prolamellar body formation under different light and temperature conditions. - Bot. Mag. (Tokyo) *84*: 363-375, 1971.

10252 - IKEHARA, N., URIBE, E.G.: The relation of the alteration of Photosystem
II activity to the inhibition of energy transfer and the proton pump in
Tris-washed chloroplasts. - Arch. Biochem. Biophys. *147*: 717-727, 1971.

*10253 - IKRAMOVA, M.M., MITYAKINA, K.A., TROFIMOVA, E.P., SHUB, S.S.: Khimiches-
kaya kharakteristika solyanok Tadzhikistana. [Chemical characteristics
of Tadzhikistan Russian thistles.] - Dokl. Akad. Nauk tadzh. SSR *10* (2):
44-48, 1967. [Car; in R, ab: Tadzh.]

10254 - ILANI, A., BERNS, D.S.: The effect of ferric ion on phycocyanin fluorescence. - Biochem. biophys. Res. Commun. *45*: 1423-1430, 1971.

10255 - IMAI, H., YAMADA, Y., HARADA, T.: Comparative studies on the photosynthesis of higher plants (1) Differences of the metabolic patterns of photosynthesis between tomato leaves and maize leaves. - Soil Sci. Plant
Nutr. *17* (3): 110-114, 1971.

10256 - IMBAMBA, S.K., MOSS, D.N.: Effect of atrazine on physiological processes
in leaves. - Crop Sci. *11*: 844-848, 1971. [Ps, stomata.]

*10257 - IMPENS, I.: Energiebudget en diffusieweerstanden van een jong zonnebloemen-
gewas (*Helianthus annuus* L.). [Energy budget and diffusion resistance of a
young sunflower crop (*Helianthus annuus* L.).] - Meded. Fac. Landbouwwet.
(Gent) *35*: 211-230, 1970. [In Holl., ab: E.]

10258 - IMPENS, I., BEHAEGHE, T.: Dagverloop van de netto photosynthese van gras-
land. [Diurnal changes in the net photosynthesis of a grass stand.] -
Meded. Fac. Landbouwwet. (Gent) *36*: 569-575, 1971. [In Holl., ab: E.]

10259 - INGERSOLL, K.A.: Liquid filters for the visible and near infrared. - Appl.
Optics *10*: 2781-2783, 1971.

10260 - INOUÉ, H., NISHIMURA, M.: Reactions of photosystems I and II in spinach
chloroplast fragments treated with ethylene glycol. - Plant Cell Physiol.
12: 137-145, 1971.

10261 - INOUÉ, H., NISHIMURA, M.: Electron flow from hydrogen peroxide in photo-

system II-catalyzed oxidation-reduction reactions of spinach chloroplast
fragments. - Plant Cell Physiol. *12*: 739-747, 1971.

10262 - INOUÉ, H., WAKAMATSU, K., NISHIMURA, M.: Characterization of the electron-
transfer system between photoreaction center II and the site of O_2 evolution
in chloroplasts. - In: QUAGLIARIELLO, E., PAPA, S., ROSSI, C.C. (ed.):
Energy Transduction in Respiration and Photosynthesis. Pp. 565-578. Adriatica Editrice, Bari 1971.

10263 - INOUÉ, H., WAKAMATSU, K., NISHIMURA, M.: Light-induced absorbance changes
at 475-515 nm, and electron flow close to photosystem II in Tris-washed
chloroplast fragments. - Plant Cell Physiol. *12*: 457-460, 1971.

10264 - IONĂŞESCU, L.: Procese fiziologice la *Chlorella vulgaris* 157, în suspensii
agitate şi neagitate. [Some physiological processes of *Chlorella vulgaris*
157 in stirred and unstirred suspension.] - An. Univ. Bucur. Biol. veg.
20: 219-228, 1971. [Ps; in Roum., ab: E, R.]

*10265 - IONIŢĂ, M., GOIA, V.: Influenţa unor factori agrotehnici asupra conţinu-
tului în caroten la porumbul cultivat pentru nutreţ. [Influence of some
agrotechnical factors on the carotene content of maize grown for ensila-
ge.] - Lucr. Ştiint., Inst. Agron. Timişoara, Ser. Agron. *12*: 405-413,
1969. [In Roum., ab: E, R.]

10266 - IRELAND, H.M.M., BRADBEER, J.W.: Plastid development in primary leaves
of *Phaseolus vulgaris*. The effects of *D*-threo and *L*-threo chloramphenicol
on the light-induced formation of enzymes of the photosynthetic carbon
pathway. - Planta *96*: 254-261, 1971.

*10267 - IRIKI, Y., ANDŌ, M., INABA, A., KITAOKA, H., HORIUCHI, N.: [Photosynthe-
tic products of *Sorbus commixta*. I. The sugars of *S. commixta*.] - Bull.
Inst. natural Educ., Shiga Heights (Shinshu Univ.) *9*: 69-75, 1970. [In
Jap., ab: E.]

10268 - IRIKI, Y., HORIUCHI, N.: [Photosynthetic products of *Sorbus commixta* II.
Incorporation of radioactive carbon from $NaH^{14}CO_3$ into sugar by *S. com-
mixta* during photosynthesis.] - Bull. Inst. natural Educ., Shiga Heights
(Shinshu Univ.) *10*: 61-66, 1971. [In Jap., ab: E.]

10269 - IRVINE, J.E.: Photosynthetic rate in sugarcane: effect of virus diseases,
genetic disorders, and freezing. - Preprint: Proc. Int. Soc. Sugarcane
Technol. [14]th Congr. *1971*: 1-8, 1971.

10270 - IRVINE, J.E.: Photosynthesis in sugarcane varieties infected with strains
of sugarcane mosaic virus. - Physiol. Plant. *24*: 51-54, 1971.

10271 - IRVINE, J.E.: Photosynthesis and stomatal behavior in sugarcane leaves as
affected by light intensity and low air flow rates. - Physiol. Plant. *24*:
436-440, 1971.

10272 - ISHERWOOD, F.A., SELVENDRAN, R.R.: Changes in the phosphate compounds in
strawberry leaves during a dark-light-dark transition in relation to sucrose biosynthesis. - Phytochemistry *10*: 579-584, 1971.

10273 - ISHIMOTO, M., YAMASHITA, J.: [Development of chromatophores in photosynthetic bacteria. [- Protein, nucl. Acid Enzyme [Tampakushitsu, Kakusan,
Koso] *16*: 822-830, 1971. [In Jap.]

10274 - ISHITANI, H., SATO, K., SMIMIZU, S., FUKUI, S.: [Chemical and biochemical
studies on vitamin B_{12} and its related compounds. (XXX) Effects of cultural
conditions and carbon sources on the growth and corrinoid production of
Rhodospirillum rubrum.] - Vitamins (Kyoto) *44*: 159-167, 1971. [Chl: in
Jap., ab: E.]

10275 - ISHMUKHAMEDOVA, S.G., AKHMEDOV, B.: Potentsial'nyT fotosintez nekotorykh
sortov i mutantov khlopchatnika. [Potential photosynthesis in some cultivars and mutants of cotton.] - In: NASYROV, Yu.S. (ed.): Geneticheskie
Aspekty Fotosinteza. Pp. 243-254. Donish, Dushanbe 1971. [In R, ab: E.]

10276 - ISLER, O.: Introduction. - In: ISLER, O. (ed.): Carotenoids. Pp. 11-27.
Birkhäuser Verlag, Basel - Stutgart 1971. [Car.]

10277 - ISSINGER, O., MAASS, I., CLAUSS, H.: Photosynthese-Intensität der Stiegelregionen von *Acetabularia mediterranea*. - Planta *101*: 360-364, 1971.

*10278 - ITO, T.: [Carbon dioxide depletion within the plant canopy in growing vegetable crops.] - J. jap. Soc. hort. Sci. *39*: 185-192, 1970. [In Jap., ab: E.]

*10279 - ITO, T.: Absorption and distribution of radioactive phosphorus in tomato plant with respect to the carbon dioxide concentration in the atmosphere. - Tech. Bull. Fac. Hort., Chiba Univ. *1970* (18): 21-28, 1970. [Ps.]

10280 - ITO, T.: [Photosynthetic activity of vegetable plants and its horticultural significance. I. Examination of the methods in measuring the photosynthesis with assimilation chamber.] - J. jap. Soc. hort. Sci. *40*: 35-40, 1971. [In Jap., ab: E.]

10281 - ITO, T.: [Photosynthetic activity of vegetable plants and its horticultural significance. II. The time course of photosynthesis in the tomato plant as influenced by some external and internal factors, especially by water and starch contents in the leaf.] - J. jap. Soc. hort. Sci. *40*: 41-47, 1971. [In Jap., ab: E.]

10282 - ITO, T.: Photosynthetic activity of vegetable plants and its horticultural significance. III. Physiological responses of young tomato plant to very low carbon dioxide concentrations.] - J. jap. Soc. hort. Sci. *40*: 375-382, 1971. [In Jap., ab: E.]

10283 - ITO, T.: [Photosynthetic activity of vegetable plants and its horticultural significance. IV. Physiological aspects of low carbon dioxide experience in tomato and cucumber plants.] - J. jap. Soc. hort. Sci. *40*: 383-388, 1971. [In Jap., ab: E.]

10284 - ITOH, S., KATOH, S., TAKAMIYA, A.: Studies on the delayed light emission in spinach chloroplasts. II. Participation of primary electron donor and acceptor of Photoreaction II in producing the delayed light emission. - Biochim. biophys. Acta *245*: 121-128, 1971.

10285 - ITOH, S., MURATA, N., TAKAMIYA, A.: Studies on the delayed light emission in spinach chloroplasts. I. Nature of two phases in development of the millisecond delayed light emission during intermittent illumination. - Biochim. biophys. Acta *245*: 109-120, 1971.

10286 - IVANCHANKA, V.M., LYAGENCHANKA, B.I., GANCHARYK, M.M.: Ab zalezhnastsi intensiŭnastsi fotasintezy ad vodnaga rezhymu asimilyatsylnal tkanki. [Relationship of photosynthetic rate and water relations of the assimilatory tissue.] - Vestsi Akad. Navuk BSSR, Ser. biyal. Navuk *1971* (6): 10-14, 136, 1971. [In Belorus., ab: R.]

10287 - IVANCHENKO, V.M., KRUCHININA, S.S., DOROZHKINA, L.N.: Ob ob"eme khloroplastov, vydelennykh v razlichnykh sredakh. [Chloroplast volume as influenced by the isolation medium.] - In: Plastidnyl Apparat i Zhiznedeyatel'nost' Rastenil. Pp. 11-15. Nauka i Tekhnika, Minsk 1971. [In R.]

*10288 - IVANCHENKO, V.M., KRUCHININA, S.S., DOROZHKINA, L.N., GONCHARIK, M.N.: O kinetike ob"emnykh izmenenil khloroplastov v gradiente kontsentratsii. [Kinetics of volume changes of chloroplasts in a concentration gradient.] - Fiziol. Biokhim. kul't. Rast. *1*: 33-36, 1969. [In R, ab: E.]

10289 - IVANCHENKO, V.M., KRUCHININA, S.S., GONCHARIK, M.N.: Izmenenie ob"ema khloroplastov s pomoshch'yu membran-aktivnykh preparatov. [Changing chloroplast volume by means of membrane-active preparations.] - Dokl. Akad. Nauk BSSR *15*: 554-556, 1971. [In R.]

10290 - IVANCHENKO, V.M., LEGENCHENKO, B.I., GONCHARIK, M.N.: Vodnyl rezhim i fotosintez ovsa v usloviyakh razlichnol vlazhnosti torfyanol pochvy. [Water relations and photosynthesis in oat under various moisture of peat soil.] - In: Plastidnyl Apparat i Zhiznedeyatel'nost' Rastenil. Pp. 16-22. Nauka i Tekhnika, Minsk 1971. [In R.]

10291 - IVANCHENKO, V.M., MARSHAKOVA, M.I., URBANOVICH, T.A., MIKUL'SKAYA, S.A., DOROZHKINA, L.N., GONCHARIK, M.N.: Vliyanie aminofillina na reaktsiyu Khilla i fotofosforilirovanie. [Effect of aminophylline on the Hill reaction and photophosphorylation.] - Dokl. Akad. Nauk BSSR *15*: 742-744, 1971. [In R.]

10292 - IVANCHENKO, V.M., URBANOVICH, T.A., MARSHAKOVA, M.I., KRUCHININA, S.S., MIKUL'SKAYA, S.A., GONCHARIK, M.N.: Aminofillin - effektor strukturnogo i funktsional'nogo sostoyaniya khloroplastov. [Aminophylline - a factor of structural and functional state of chloroplasts.] - Dokl. Akad. Nauk BSSR *15*: 645-648, 666, 1971. [In R.]

*10293 - IVANOV, A.F., KRAVCHENKO, L.V.: Dlitel'nost' deĭstviya vnekornevoĭ pod-kormki mikroelementami na intensivnost' fotosinteza drevesnykh rasteniĭ. [Length of the action of spray dressing with trace elements on photosynthetic rate of trees.] - In: Floristicheskie i Geobotanicheskie Issledovaniya v Belorussii. Pp. 168-171. Nauka i Tekhnika, Minsk 1970. [In R.]

*10294 - IVANOV, I.D., DEMINA, N.S., SOTNIKOV, G.G., ZHELEVA, V.I.: Biologicheskoe vosstanovlenie molekulyarnogo azota. III. Obrazovanie ATF pri vosstanovlenii molekulyarnogo azota v intaktnykh kletkakh fotosinteziruyushcheĭ bakterii *Chromatium minutissimum*. [Biological reduction of molecular nitrogen. III. Formation of ATP during reduction of molecular nitrogen in intact cells of the photosynthesizing bacterium *Chromatium minutissimum*.] - Nauch. Dokl. vyssh. Shkoly, biol. Nauki *13* (9): 87-90, 1970. [In R.]

*10295 - IVANOV, O.V., BYKOV, O.D.: Konduktometricheskiĭ laboratorno-polevoĭ pribor dlya massovykh izmereniĭ fotosinteza rasteniĭ. [Conductimetric laboratory and field apparatus for routine determination of plant photosynthesis.] - In: Tezisy Dokladov Vsesoyuznogo Soveshchaniya po Unifikatsii Metodov i Priborov dlya Massovykh Izmereniĭ Intensivnosti Fotosinteza. Pp. 36-39. Nauch.-issled. Inst. Rastenievod. N.I. Vavilova, Leningrad - Pushkin 1970. [In R.]

10296 - IVANOVA, R.P., BOKAREV, K.S.: Izmenenie soderzhaniya sukhogo veshchestva i pigmentov v list'yakh kartofelya pri obrabotke nekotorymi proizvodnymi kholina i betaina [Changes in dry matter and pigments in potato leaves during treatment with some choline and betaine derivatives.] - Fiziol. Rast. *18*: 121-124, 1971. [In R, ab: E.]

*10297 - IVANOVSKAYA, N.P., NEKRASOV, L.I., CHASOVNIKOVA, L.V.: Adsorbtsionnye vzaimodeĭstviya syvorotochnogo al'bumina s monosloyami khlorofilla na poverkhnosti vody. [Adsorption interrelation between serum albumin and chlorophyll monolayers on water surface.] - Nauch. Dokl. vyssh. Shkoly, biol. Nauki *13* (3): 133-134, 1970. [In R.]

10298 - IVANOVSKAYA, N.P., NEKRASOV, L.I., CHASOVNIKOVA, L.V., KOBOZEV, N.I.: Vliyanie pH i temperatury na adsorbtsionnye vzaimodeĭstviya syvorotochnogo al'bumina s monosloyami khlorofillov *a* i *b* na granitse faz voda - vozdukh. [Effect of pH and temperature on adsorption interrelationships between serum albumin and chlorophyll *a* and *b* monolayers on the water - air interface.] - Zh. fiz. Khim. *45*: 155-156, 1971. [In R.]

10299 - IVNITSKAYA, V.E., ISMAÏLOV, S.I.: Vliyanie primeneniya NRV na pochvenno-mikrobiologicheskie usloviya pitaniya khlopchatnika i na khod fiziologicheskikh protsessov. [Effect of using petrol growth regulators on the soil-microbiological conditions of nutrition and on physiological processes in cotton.] - In: NRV v Sel'skom Khozyaĭstve. Pp. 427-430. ELM, Baku 1971. [Ps; in R.]

*10300 - IZVOSHCHIKOV, V.P., KONOVALENKO, V.V.: Pribor dlya ucheta fotosinteza v polevykh usloviyakh. [Apparatus for assessing photosynthesis in field conditions.] - In: Tezisy Dokladov Vsesoyuznogo Soveshchaniya po Unifikatsii Metodov i Priborov dlya Massovykh Izmereniĭ Intensivnosti Fotosinteza. Pp. 40-43. Nauch.-issled. Inst. Rastenievod. N.I. Vavilova, Leningrad - Pushkin 1970. [In R.]

10301 - JACKSON, J.B., CROFTS, A.R.: The kinetics of light induced carotenoid changes in *Rhodopseudomonas spheroides* and their relation to electrical field generation across the chromatophore membrane. - Europe. J. Biochem. *18*: 120-130, 1971.

10302 - JACOBI, G.: Subchloroplast fragments: sonication method. - In: COLOWICK, S.P., KAPLAN, N.O. (ed.): Methods in Enzymology. Vol. 23. Pp. 289-296. Academic Press, New York - London 1971.

10303 - JACOBSON, B.S., SWADER, J.A., DUGGER, W.M.: Inhibition of photosynthesis by acetazolamide. - Plant Physiol. 47 (Suppl.): 32, 1971.

10304 - JAIN, T.C.: Contribution of stem, laminae and ears to the dry-matter production of maize (Zea mays L.) after ear emergence. - Indian J. agr. Sci. 41: 579-583, 1971.

10305 - JAIN, T.C.: Physiological analysis of variation in growth and yield of zea mays due to differences in time of sowing. I - Pre-flowering period, growth characters, growth attributes and morphological components of growth. - Trop. Agriculturist 127: 117-132, 1971.

*10306 - JAIN, T.C., MISRA, D.K.: Effect of water stress on : I. Phsyiological activities of plants. - Indian J. Agron. 15: 36-40, 1970. [Ps.]

*10307 - JAKŠINA, A.M.: Atmung und organische Stoffbildung bei Quercus robur L. - Tagungsber. deut. Akad. Landwirtschaftswiss. Berlin 100: 211-219, 1968. [Ps.]

*10308 - JAMES, B.: Carbon dioxide measurement. - Agr. Meteorol. 7: 419, 1970.

10309 - JANÁČ, J., ČATSKÝ, J., JARVIS, P.G., BROWN, K.W., ECKARDT, F.E., FOCK, H., SCHAUB, H., BJÖRKMAN, O., GAUHL, E., PIETERS, G.A.: Infra-red gas analysers and other physical analysers. - In: ŠESTÁK, Z., ČATSKÝ, J., JARVIS, P.G. (ed.): Plant Photosynthetic Production: Manual of Methods. Pp. 111-197. Dr. W. Junk N.V. Publishers, The Hague 1971. [Ps.]

*10310 - JARVIS, P.G.: Comparative plant water relations. - Ann. arid Zone 6: 74-91, 1967. [Stomata.]

10311 - JARVIS, P.G.: The estimation of resistance to carbon dioxide transfer. - In: ŠESTÁK, Z., ČATSKÝ, J., JARVIS, P.G. (ed.): Plant Photosynthetic Production: Manual of Methods. Pp. 566-631. Dr. W. Junk N.V. Publishers, The Hague 1971.

10312 - JARVIS, P.G., ČATSKÝ, J., ECKARDT, F.E., KOCH, W., KOLLER, D.: General principles of gasometric methods and the main aspects of installation design. - In: ŠESTÁK, Z., ČATSKÝ, J., JARVIS, P.G. (ed.): Plant Photosynthetic Production: Manual of Methods. Pp. 49-110. Dr. W. Junk N.V. Publishers, The Hague 1971. [Ps.]

*10313 - JARVIS, P.G., ROSE, C.W., BEGG, J.E.: An experimental and theoretical comparison of viscous and diffusive resistances to gas flow through amphistomatous leaves. - Agr. Meteorol. 4: 103-117, 1967. [Model.]

10314 - JAŠA, B., ŘEZNÍČEK, V., MUZIKANTOVÁ, J.: Studium využití morforegulačních přípravků u letniček. [Utilization of morphoregulatory preparations in annual plants.] - Rostl. Výroba (Praha) 17: 1291-1297, 1971. [Ps; in Czech, ab: E, R.]

*10315 - JAYARAMI REDDY, A., RAO, I.M.: Influence of induced water stress on chlorophyll components of proximal and distal leaflets of groundnut plants. - Current Sci. 38: 118-119, 1969.

10316 - JEFFREE, C.E., JOHNSON, R.P.C., JARVIS, P.G.: Epicuticular wax in the stomatal antechamber of Sitka spruce and its effects on the diffusion of water vapour and carbon dioxide. - Planta 98: 1-10, 1971.

10317 - JENSEN, R.G.: Activation of CO_2 fixation in isolated spinach chloroplasts. - Biochim. biophys. Acta 234: 360-370, 1971.

10318 - JENSEN, R.G. FRANCKI, R.I.B., ZAITLIN, M.: Metabolism of separated leaf cells. I. Preparation of photosynthetically active cells from tobacco. - Plant Physiol. 48: 9-13, 1971.

10319 - JEŠKO, T., HEINRICHOVÁ, K., LUKAČOVIČ, A.: Increase in photosynthetic activity during the formation of the first node roots and first tiller in Sorghum saccharatum (L.) MOENCH. - Photosynthetica 5: 233-240, 1971.

10320 - JOHANNES, B., BRZEZINKA, H., BUDZIKIEWICZ, H.: Zur Photosynthese grüner

Pflanzen, VI. Isolierung von Diatoxanthin aus *Euglena gracilis*. - Z. Naturforsch. *26b*: 377-378, 1971.

*10321 - JOHNSON, C.E., BRAY, R.C., CAMMACK, R., HALL, D.O.: Mössbauer spectroscopy of the iron-sulfur proteins. - Proc. nat. Acad. Sci. USA *63*: 1234-1238, 1969. [Ferredoxins.]

10322 - JOHNSON, C.E., CAMMACK, R., RAO, K.K., HALL, D.O.: The interpretation of the EPR and Mössbauer spectra of two-iron, one-electron, iron-sulphur proteins. - Biochem. biophys. Res. Commun. *43*: 564-571, 1971.

10323 - JOHNSON, H.S.: NADP-malate dehydrogenase: photoactivation in leaves of plants with Calvin cycle photosynthesis. - Biochem. biophys. Res. Commun. *43*: 703-709, 1971. [Chl.]

10324 - JOHNSON, H.S., SLACK, C.R., HATCH, M.D., ANDREWS, T.J.: The CO_2 carrier between mesophyll and bundle sheath chloroplasts in C_4 pathway species. - In: HATCH, M.D., OSMOND, C.B., SLATYER, R.O. (ed.): Photosynthesis and Photorespiration. Pp. 189-195. Wiley-Interscience, New York - London - Sydney - Toronto 1971.

10325 - JOHNSTON, T.D., YORK, P.A.: Genetical investigations into photosynthetic rate in *Brassica*. 1. A simple technique for measuring photosynthesis of leaves of kale. - Euphytica *20*: 316-318, 1971.

10326 - JOLIOT, P., JOLIOT, A.: Hétérogénéité des accepteurs d'électrons de la réaction photochimique II en photosynthèse. - Compt. rend. Acad. Sci. Paris, Sér. D *272*: 2604-2607, 1971.

10327 - JOLIOT, P., JOLIOT, A., BOUGES, B., BARBIERI, G.: Studies of System II photocenters by comparative measurements of luminescence, fluorescence, and oxygen emission. - Photochem. Photobiol. *14*: 287-305, 1971.

10328 - JOLIVET, E.: Métabolisme des acides organiques et des acides aminés libres dans le tubercule de pomme de terre au cours de la rupture provoquée de son repos végétatif par la "rindite" IV. - Déroulement des réactions après l'apparition des germes. - Physiol. vég. *9*: 311-325, 1971. [Ps.]

*10329 - JOLIVET, E., NICOL, M., BAUDET, J., MOSSE, J.: Differences dans l'incorporation du $^{14}CO_2$ à la lumière chez de jeunes maïs normal et opaque-2. - In: Imp. Plant Protein Nucl. Tech., Proc. Symp. Pp. 391-401. IAEA, Vienna 1970.

10330 - JOLLIFFE, P.A.: Photosynthesis, photorespiration and related aspects of CO_2 exchange by wheat, corn and *Amaranthus edulis*. - Diss. Abstr. Int. B *31*: 7138-B-7139-B, 1971.

10331 - JONES, L.H.: Adaptive responses to temperature in Dwarf French beans, *Phaseolus vulgaris* L. - Ann. Bot. *35*: 581-596, 1971. [Ps, growth analysis.]

*10332 - JONES, R.: The leaf area of an Australian heathland with reference to seasonal changes and the contribution of individual species. - Aust. J. Bot. *16*: 579-588, 1968. [Growth analysis.1

*10333 - JONES, R.: Productivity studies on heath vegetation in southern Australia. The use of fertilizer in studies of production processes. - Folia geobot. phytotax. (Praha) *3*: 355-362, 1968. [Growth analysis.]

*10334 - JONES, R., GROVES, R.H., SPECHT, R.L.: Growth of heath vegetation. III. Growth curves for heaths in southern Australia: a reassessment. - Aust. J. Bot. *17*: 309-314, 1969. [Growth analysis.]

10335 - JORDAN, C.F.: Productivity of a tropical forest and its relation to a world pattern of energy storage. - J. Ecol. *59*: 127-142, 1971. [Ps.]

*10336 - JOSEPH, B., GAUR, B.K.: X-ray induced stimulation of growth in *Ocimum killimandscharicum*. - Stimulation News Lett. *1*: 34-38, 1970. [Chl.]

*10337 - JOSHI, G.V.: Photosynthesis under saline conditions. - In: Proceedings of Department of Atomic Energy Symposium on Radiation and Radioisotopes in Soil Studies and Plant Nutrition, Bangalore. Pp. 217-224. Bangalore 1970.

10338 - JOSHI, G.V., PATIL, S.: Effect of EDTA on photosynthesis in *Bryophyllum pinnatum*. - Indian J. exp. Biol. *9*: 476-477, 1971.

10339 - JOYARD, J., FOURCY, A.: Sur l'évolution de la teneur en manganèse des étioplastes isolés de maïs au cours de leur verdissement. - Compt. rend. Acad. Sci. Paris, Sér. D *273*: 572-575, 1971.

10340 - JOYARD, J., FOURCY, A.: Influence du traitement au Tris sur la teneur en manganèse des étioplastes isolés au cours de leur verdissement. - Compt. rend. Acad. Sci. Paris, Sér. D *273*: 805-808, 1971.

*10341 - JUNG, J., EL-FOULY, M.M.: Der Einfluss von Chlorcholinchlorid (CCC) auf den Gehalt des Weizens an Chlorophyll, Karotin sowie N, P, K und Mg im Verlauf des Wachstums. - Landw. Forsch. *19*: 29-34, 1966.

10342 - JUNGE, W., SCHMID, R.: The mechanism of action of valinocycin on the thylakoid membrane. Characterization of the electric current density. - J. Membrane Biol. *4*: 179-192, 1971.

10343 - JUPIN, H., GIRAUD, G.: Modification du spectre d'absorption dans le rouge lointain d'une diatomée cultivée en lumière rouge. - Biochim. biophys. Acta *226*: 98-102, 1971. [Chl.]

10344 - KABANOVA, Yu.G., OCHAKOVSKII, Yu.E.: Zavisimost' pervichnoĭ produktsii fitoplanktona ot biogennykh elementov i sveta. [Dependence of primary production of phytoplankton on biogenic elements and light.] - Dokl. Akad. Nauk SSSR *201*: 1227-1230, 1971. [Ps; in R.]

*10345 - KABYSH, V.A.: Formirovanie listovogo apparata u raznykh sortov yachmenya v zavisimosti ot hustoty poseva. [Formation of leaf apparatus in different barley varieties in dependence on crop density.] - Izv. Timiryaz. sel'.-khoz. Akad. *1969* (6): 27-33, 1969. [In R, ab: E.]

10346 - KACHARAVA, N.F.: [Primary effect and aftereffect of ultraviolet radiation.] - Soobshch. Akad. Nauk gruz. SSR *63*: 161-164, 1971. [Ps, Chl; in Georgian, ab: E,R.]

10347 - KADANÍKOVÁ, V., LAŠTŮVKA, Z.: The growth of maize and ion accumulation at varying stand density of water culture. - Scr. Fac. Sci. Nat. Univ. Purkynianae Brun. *1* (7): 197-213, 1971. [Chl.]

10348 - KAFALIEVA, D.N., ILIEV, V.R.: Separation and certain properties of chloroplast fragments obtained by digitonine treatment. - Dokl. bolg. Akad. Nauk *24*: 1691-1693, 1971.

10349 - KAGAWA, T., BEEVERS, H.: Relationship of leaf peroxisomes to photosynthetic metabolism. - Plant Physiol. *47* (Suppl.): 28, 1971.

10350 - KAHN, J.S.: ATP-ADP exchange enzyme from spinach chloroplasts. - In: COLOWICK, S.P., KAPLAN, N.O. (ed.): Methods in Enzymology. Vol. 23. Pp. 561-565. Academic Press, New York - London 1971.

10351 - KAHN, J.S.: Evidence for a two-directional hydrogen ion transport in chloroplasts of *Euglena gracilis*. - Biochim. biophys. Acta *245*: 144-150, 1971.

10352 - KAKHNOVICH, L.V., KLIMOVICH, A.S.: Fotosinteticheskiĭ apparat v zavisimosti ot intensivnosti sveta. [Effect of illuminance on photosynthetic apparatus.] - Fiziol. Rast. *18*: 893-897, 1971. [In R, ab: E.]

*10353 - KAKIE, T., SUGIZAKI, Y.: Diurnal changes in the starch and sugars of tobacco leaves. - Soil Sci. Plant Nutr. *16*: 201-203, 1970.

*10354 - KAKIE, T., SUGIZAKI, Y.: Starch and sugars of tobacco leaves during maturity stage. - Soil Sci. Plant Nutr. *17*: 27-36, 1970. [Ps.]

*10355 - KAKIE, T., SUGIZAKI, Y.: [Effect of phosphorus deficiency on the incorporation of $^{14}CO_2$ assimilated into carbohydrate in tobacco plants.] - Bull. Hatano Tobacco Sta. [Hatano Tabako Shikenjo Hokoku] *67*: 87-90, 1970. [In Jap., ab: E.]

10356 - KAKUNO, T., BARTSCH, R.G., NISHIKAWA, K., HORIO, T.: Redox components associated with chromatophores from *Rhodospirillum rubrum*. - J. Biochem. (Tokyo) *70*: 79-94, 1971.

10357 - **KALBE, L.**: Zur limnologischen Beurteilung von eutrophen Flachseen nach ihrer Biomasse. - Limnologica (Berlin) *8*: 311-320, 1971. [Ps, Chl.]

10358 - **KALBE, L., SCHULZE, H.-A.**: Zusammenhänge zwischen Primärproduktion, Trophie und Licht im Kummerower See (Mecklenburg). - Z. gesamte Hyg. Grenzgeb. *17*: 658-664, 1971. [Ps.]

10359 - **KAL"CHEVA, I.**: Deformatsiya i dvizhenie na khloroplastite pri osvetlyavane i zat"mnyavane na izolirani tseli lista ot nyakoi sortove tsarevitsa. [Deformation and movement of chloroplasts at illumination and darkening of isolated whole leaves of certain maize cultivars.] - Rasteniev"d. Nauki (Sofia) *8* (10): 11-18, 1971. [In Bulg., ab: E, R.]

10360 - **KALER, V.L., PODCHUFAROVA, G.M.**: Nekotorye sledstviya iz sopostavleniya skorostei sinteza protokhlorofillida i khlorofilla pri zelenenii etiolirovannykh rastenii. [Comparison of protochlorophyllide and chlorophyll synthesis rates at greening of etiolated plants.] - In: Plastidnyi Apparat i Zhiznedeyatel'nost' Rastenii. Pp. 54-61. Nauka i Tekhnika, Minsk 1971. [In R.]

10361 - **KALER, V.L., PODCHUFAROVA, G.M.**: Protokhlorofillid kak upravlyayushchii metabolit v biosinteze khlorofilla. [Protochlorophyllide as the controlling metabolite in chlorophyll biosynthesis.] - In: GONCHARIK, M.N. (ed.): Plastidnyi Apparat i Zhiznedeyatel'nost' Rastenii. Pp. 50-53. Nauka i Tekhnika, Minsk 1971. [In R.]

10362 - **KALFF, J.**: Nutrient limiting factors in an arctic tundra pond. - Ecology *52*: 655-659, 1971. [Ps.]

*10363 - **KALININA, L.M.**: Obrazovanie dlinnovolnovykh form khlorofilla na rannei stadii zeleneniya etiolirovannykh prorostkov. [Formation of long-wave forms of chlorophyll during the early phase of greening of etiolated seedlings.] - In: Tezisy IV Nauchnoi Konferentsii Molodykh Uchenykh po Sovremennym Problemam Biologii. Pp. 54-56. Minsk 1970. [In R.]

10364 - **KALLIO, P., HEINONEN, S.**: Influence of short-term low temperature on net photosynthesis in some subarctic lichens. - Rep. Kevo subarctic Res. Sta. *8*: 63-72, 1971.

*10365 - **KALMANSON, A.E., CHUMAKOV, V.M.**: K voprosu o prirode temnovogo signala EPR v khlorofillsoderzhashchikh fotosinteziruyushchikh ob"ektakh. [Dark electron spin resonance signal in chlorophyll-containing photosynthetizing objects.] - Biofizika *15*: 436-437, 1970. [In R.]

10366 - **KAMEYA, T., TAKAHASHI, N.**: Division of chloroplast *in vitro*. - Jap. J. Genet. *46*: 153-157, 1971.

10367 - **KAMP-NIELSEN, L.**: The effect of deleterious concentrations of mercury on the photosynthesis and growth of *Chlorella pyrenoidosa*. - Physiol. Plant. *24*: 556-561, 1971.

10368 - **KANCHAVELI, L.A., KALICHAVA, G.S.**: O vzaimosvyazi mezhdu fotosinteticheskoi aktivnost'yu i zabolevaniem rastitel'noi kletki. [Interrelation between photosynthetic activity end disease of plant cell.] - Soobshch. Akad. Nauk gruz. SSR *64*: 469-472, 1971. [In R, ab: E, Georg.]

10369 - **KANDAUROV, V.I., MOVCHAN, V.K.**: Fotosinteticheskii potentsial i produktivnost' sortov yarovoi pshenitsy v sukhostepnoi zone severa Kazakhstana. [Photosynthetic potential and productivity of spring wheat cultivars in arid steppe zone of North Kazakhstan.] - Sel'skokhoz. Biol. *6*: 16-21, 1971. [In R.]

10370 - **KANEMASU, E.T., FELTNER, K.C., VESECKY, J.F.**: Light interception and reflectance measurements with ozalid paper. - Crop Sci. *11*: 931-933, 1971.

10371 - **KANIUGA, Z., FRACKOWIAK, B.**: The relation between non-cyclic electron transport and photophosphorylation. - In: QUAGLIARIELLO, E., PAPA, S., ROSSI, C.S. (ed.): Energy Transduction in Respiration and Photosynthesis. Pp. 551-557. Adriatica Editrice, Bari 1971.

*10372 - **KANNANGARA, C.G.**: Development of carboxydismutase activity and photosynthesis during greening of dark grown barley. - In: Abstracts XI International Botanical Congress. P. 107. Seattle, Wash. 1969.

10373 - **KANNANGARA, C.G., HENNINGSEN, K.W., STUMPF, P.K., APPELQVIST, L.-Å., von WETTSTEIN, D.**: Lipid biosynthesis by isolated barley chloroplasts in relation to plastid development. - Plant Physiol. *48*: 526-531, 1971.

10374 - **KANNANGARA, C.G., STUMPF, P.K.**: The formation of fatty acyl thioesters during ^{14}C-1-acetate incorporation into long chain fatty acids by isolated spinach chloroplasts. - Biochem. biophys. Res. Commun. *44*: 1544-1551, 1971.

*10375 - **KANNANGARA, C.G., WOOLHOUSE, H.W.**: Changes in the enzyme activity of soluble protein fractions in the course of foliar senescence in *Perilla frutescens* (L.) BRITT. - New Phytol. *67*: 533-542, 1968. [Ps.]

10376 - **KAO, O., BERNS, D.S., MACCOLL, R.**: *C*-phycocyanin monomer molecular weight. - Europe. J. Biochem. *19*: 595-599, 1971.

10377 - **KAPLANOVÁ, M.**: Photochemical changes of chlorophyll *a* as reflected in absorption spectra. - Acta Univ. Carolinae, Math. Phys. *12*: 11-24, 1971.

10378 - **KAPLANOVÁ, M.**: Vliv rozpouštědel na optická spektra chlorofylu *a*. [Influence of solvents on chlorophyll *a* spectra.] - Sb. věd. Prací VŠCHT Pardubice *26*: 67-86, 1971. [In Czech, ab: E, G.]

10379 - **KAPLER, R., NEKRASOV, L.I.**: Spektry pogloshcheniya adsorbtsionnykh sloev khlorofilla na polimernykh plenkakh. [Absorption spectra of chlorophyll layers adsorbed in polymer films.] - Biofizika *16*: 206-213, 1971. [In R, ab: E.]

10380 - **KAPLER, R., NEKRASOV, L.I.**: Infrakrasnye spektry pogloshcheniya adsorbirovannogo na okisi alyuminiya khlorofilla *a* i *b*. [Infra-red absorption spectra of chlorophyll *a* and *b* adsorbed on aluminium oxide.] - Zh. fiz. Khim. *45*: 750-753, 1971. [In R.]

10381 - **KAPLER, R., NEKRASOV, L.I., IROSHNIKOVA, N.G., MAMLEEVA, N.A.**: Paramagnitnye svoĭstva adsorbtsionnykh sloev khlorofilla *a* i *b* na okisi alyuminiya. [Paramagnetic properties of adsorption layers of chlorophyll *a* and *b* on aluminium oxide.] - Biofizika *16*: 32-38, 1971. [In R, ab: E.]

10382 - **KARANOV, E.N.**: The influence of certain growth regulators and their interaction in delaying the destruction of chlorophyll in discs of *Raphanus sativa* at different ages of the leaves. - Dokl. bolg. Akad. Nauk *24*: 99-102, 1971.

10383 - **KARAPETYAN, N.V., KLIMOV, V.V.**: Ustanovka dlya izmereniya kinetiki indutsirovannykh svetom izmeneniĭ vykhoda fluorestsentsii u fotosinteziruyushchikh organizmov. [A device for measurement of the kinetics of the light-induced changes in fluorescence yield of photosynthetizing organisms.] - Fiziol. Rast. *18*: 223-228, 1971. [In R, ab: E.]

10384 - **KARAPETYAN, N.V., KLIMOV, V.V., KRAKHMALEVA, I.N., KRASNOVSKIĬ, A.A.**: Induktsiya fluorestsentsii khloroplastov i khromatoforov v vosstanovitel'nykh usloviyakh. [Induction of fluorescence in chloroplasts and chromatophores under reductive conditions.] - Dokl. Akad. Nauk SSSR *201*: 1244-1247, 1971. [In R.]

10385 - **KARAPETYAN, N.V., KLIMOV, V.V., LANG, F., KRASNOVSKIĬ, A.A.**: Issledovanie induktsii fluorestsentsii list'ev kukuruzy v anaerobnykh usloviyakh. [Fluorescence induction in maize leaves in anaerobiosis.] - Fiziol. Rast. *18*: 507-517, 1971. [In R, ab: E.]

10386 - **KARAPETYAN, N.V., KOL'TOVER, V.K., KRAKHMALEVA, I.N., KRASNOVSKIĬ, A.A.**: Deĭstvie inaktiviruyushchikh faktorov na signal EPR i vosstanovlenie iminoksil'nogo radikala v khromatoforakh purpurnykh bakteriĭ. [Effect of inactivating factors on ESR signal and reduction of iminoxyl radical in chromatophores of purple bacteria.] - Biofizika *16*: 1138-1141, 1971. [In R, ab: E.]

10387 - **KARLISH, S.J.D., AVRON, M.**: Energy transfer inhibition and ion movements in isolated chloroplasts. - Europe. J. Biochem. *20*: 51-57, 1971.

10388 - **KARPILOV, Yu.S., BIL', K.Ya., MALYSHEV, O.G., KARNAUKHOV, V.V.**: Osobennosti fotosistem khloroplastov kletok mezofilla i parenkhimnykh obkladok provodyashchikh puchkov list'ev odnodol'nykh i dvudol'nykh rasteniĭ. [Features of chloroplasts in cells of mesophyll and bundle sheaths in leaves of monocotyledonous and dicotyledonous plants.] - Dokl. Akad. Nauk SSSR *197*: 480-483, 1971. [In R.]

10389 - **KARPUSHKIN, L.T.**: Primenenie infrakrasnogo gazoanalizatora dlya izucheniya CO_2-gazoobmena rasteniĭ. [The use of infra-red gas analyser for studying carbon dioxide exchange in plants.] - In: Biofizicheskie Metody v Fiziologii Rasteniĭ. Pp. 44-71. Nauka, Moskva 1971. [In R.]

10390 - **KARUNAKARAN, K., KISS, I.S.**: Frequency of chlorophyll mutation in M_2 and M_3 as affected by drying back of ethyl methanesulphonate treated rice seeds. - Acta biol. Acad. Sci. hung. *22*: 471-473, 1971.

10391 - **KARUNAKARAN, K., KISS, I.S.**: M_1 chlorophyll chimeras induced by different and mutagens their M_2 chlorophyll mutation yields in rice. - Biol. Plant. *13*: 207-208, 1971.

10392 - **KARUNEN, P.**: Lipid and pigment patterns in germinating *Polytrichum commune* spores. - Phytochemistry *10*: 2811-2812, 1971.

10393 - **KASHIN, V.K.**: Vliyanie nikelya na intensivnost' fotosinteza i nekotorye pokazateli vodnogo rezhima kukuruzy. [Effect of nickel on photosynthetic rate and some characteristics of water relations in maize.] - In: Mikroelementy v Biosfere i Primenenie ikh v Sel'skom Khozyaĭstve i Meditsine Sibiri i Dal'-nego Vostoka. Pp. 235-239. Ulan-Ude 1971. [In R.]

10394 - **KASPERBAUER, M.J.**: Spectral distribution of light in a tobacco canopy and effects of end-of-day light quality on growth and development. - Plant Physiol. *47*: 775-778, 1971.

*10395 - **KASPRZYK, Z.**: Fotosynteza. [Photosynthesis.] - Podst. Probl. współcz. Tech. (Warszawa) *14*: 263-287, 1970. [In Pol.]

10396 - **KAS'YANENKO, A.G.**: Allel'nye khlorofil'nye mutatsii *Arabidopsis thaliana*. [Allelochlorophyll mutations of *Arabidopsis thaliana*.] - In: ZALENSKIĬ, O.V. (ed.): Fotosintez i Ispol'zovanie Solnechnoĭ Energii. Pp. 244-247. Nauka, Leningrad 1971. [Ps, Chl; in R, ab: E.]

10397 - **KAS'YANENKO, A.G., LOGINOV, M.A., NASYROV, Yu.S.**: Sravnitel'nyĭ analiz fotosinteticheskogo apparata khlorofil'nykh mutantov *Arabidopsis thaliana* v raznykh kompleksakh vneshnikh usloviĭ. [Comparative analysis of photosynthetic apparatus of chlorophyll mutants of *Arabidopsis thaliana* in various complexes of environmental conditions.] - In: NASYROV, Yu.S. (ed.): Geneticheskie Aspekty Fotosinteza. Pp. 120-132. Donish, Dushanbe 1971. [In R.]

10398 - **KAS'YANENKO, A.G., NASYROV, Yu.S., SMOLINA, E.A.**: Leĭtsinovye mutatsii *Arabidopsis thaliana*. [Leucin mutations in *Arabidopsis thaliana*.] - In: NASYROV, Yu.S. (ed.): Geneticheskie Aspekty Fotosinteza. Pp. 56-76. Donish, Dushanbe 1971. [Ps, Chl; in R.]

*10399 - **KAS'YANENKO, A.G., USMANOV, P.D.**: Indutsirovannye gamma-luchami Co^{60} allel'nye mutatsii *Arabidopsis thaliana*. [Allel mutations of *Arabidopsis thaliana* induced by gamma rays ^{60}Co] - Dokl. Akad. Nauk tadzh. SSR *11* (5): 69-73, 1968. [Chl, Car; in R, ab: Tadzh.]

10400 - **KATHJU, S., TEWARI, M.N., KAUSHIK, D.D.**: Effects of certain fluorine substituted phenoxy compounds on expansion, chlorophyll development and activities of phosphatases in excised cotyledons of *Tephrosia purpurea*. - Z. Pflanzenphysiol. *65*: 85-87, 1971.

10401 - **KATOH, S.**: Plastocyanin. - In: COLOWICK, S.P., KAPLAN, N.O. (ed.): Methods in Enzymology. Vol. 23. Pp. 408-413. Academic Press, New York - London 1971.

10402 - **KATOH, S., TAKAMIYA, A.**: Two modes of inhibition by 2-heptyl-4-hydroxyquinoline-N-oxide of electron transport associated with photosystem 2 in spinach chloroplasts. - Plant Cell Physiol. *12*: 479-492, 1971.

10403 - **KAUL, K., SABHARWAL, P.S.**: Effects of sucrose and kinetin on growth and chlorophyll synthesis in tobacco tissue cultures. - Plant Physiol. *47*: 691-695, 1971.

10404 - **KAUL, R., CROWLE, W.L.**: Relations between water status, leaf temperature, stomatal aperture, and productivity of some wheat varieties. - Z. Pflanzenzücht. *65*: 233-243, 1971.

10405 - KAWASHIMA, N., CHAN, P.H., SAKANO, K., WILDMAN, S.G.: Simple method for obtaining crystalline fraction I protein from different species of *Nicotiana*. - Plant Physiol. *47* (Suppl.): 10, 1971.

10406 - KAWASHIMA, N., WILDMAN, S.G.: Studies on fraction-1 protein. I. Effect of crystallization of fraction-I protein from tobacco leaves on ribulose diphosphate carboxylase activity. - Biochim. biophys. Acta *229*: 240-249, 1971.

10407 - KAZLOVA, A.P., GANCHARYK, M.M.: Uplyŭ khloru kaliĭnykh soleĭ na fotasintetychnuyu dzaĭnasts' grechki. [Effect of chlorine in potassium salts on photosynthetic activity of buckwheat.] - Vestsi Akad. Navuk belarus. SSR, Ser. biyal. Navuk *1971* (6): 30-33, 1971. [In Belorus.]

10408 - KE, B.: Carotenoproteins. - In: COLOWICK, S.P., KAPLAN, N.O. (ed.): Methods in Enzymology. Vol. 23. Pp. 624-636. Academic Press, New York - London 1971.

10409 - KE, B., CHANEY, T.H.: Spectral and photochemical properties of subchromatophore fractions derived from carotenoid-deficient *Chromatium* by Triton treatment. - Biochem. biophys. Acta *226*: 341-353, 1971.

10410 - KE, B., OGAWA, T., HIYAMA, T., VERNON, L.P.: Experimental determination of the molar differential extinction coefficient of P700. - Biochim. biophys. Acta *226*: 53-62, 1971.

10411 - KEERBERG, H., KEERBERG, O., PÄRNIK, T., VILL, J., VÄRK, E.: CO_2 assimilation by *Phaseolus* and *Aspidistra* leaves under varying density of blue and red radiant flux. - Photosynthetica *5*: 99-106, 1971.

10412 - KEERBERG, Kh., VYARK, E., KEERBERG, O., PYARNIK, T.: Deĭstvie spektral'nogo sostava sveta na vklyuchenie ^{14}C v amino- i organicheskie kisloty pri assimilyatsii $^{14}CO_2$ list'yami fasoli. [Effect of spectral composition of light on ^{14}C incorporation into amino and organic acids during assimilation of $^{14}CO_2$ by bean leaves.] - Izv. Akad. Nauk est. SSR, Biol. *20*: 350-353, 1971. [In R.]

*10413 - KEERBERG, O.F., PYARNIK, T.R.: Mnogokanal'naya ekspozitsionnaya kamera dlya issledovaniya fotosinteza v diskakh list'ev. [Multichannel exposure chamber for studying photosynthesis in leaf discs.] - In: Tezisy Dokladov Vsesoyuznogo Soveshchaniya po Unifikatsii Metodov i Priborov dlya Massovykh Izmereniĭ Intensivnosti Fotosinteza. Pp. 61-65. Nauch.-issled. Inst. Rastenievod. N.I. Vavilova, Leningrad - Pushkin 1970. [In R.]

*10414 - KEISTER, D.L., HEMMES, R.B.: Pyridine nucleotide transhydrogenase from *Chromatium*. - J. biol. Chem. *241*: 2820-2825, 1966.

10415 - KEISTER, D.L., MINTON, N.J.: ATP synthesis driven by inorganic pyrophosphate in *Rhodospirillum rubrum* chromatophores. - Biochem. biophys. Res. Commun. *42*: 932-939, 1971.

10416 - KEISTER, D.L., MINTON, N.J.: Effect of light on respiration in *Rhodospirillum rubrum* chromatophores. - In: QUAGLIARIELLO, E., PAPA, S., ROSSI, C.S. (ed.): Energy Transduction in Respiration and Photosytnehtsis. Pp. 375-384. Adriatica Editrice, Bari 1971.

10417 - KEISTER, D.L., MINTON, N.J.: Energy-linked reactions in photosynthetic bacteria. VI. Inorganic pyrophosphate driven ATP sythesis in *Rhodospirillum rubrum*. - Arch. Biochem. Biophys. *147*: 330-338, 1971.

10418 - KELLER, T.: Auswirkungen der Luftverunreinigungen auf die Vegetation. - Städtehygiene *22* (6): 130-136, 1971. [Ps.]

10419 - KELLER, T.: Der Einfluss der Stickstoffernährung auf den Gaswechsel der Fichte. - Allgem. Forst Jagdzeit. *142*: 89-93, 1971.

10420 - KELLER, T.: Gaseous exchange - a good indicator of nutritional status and fertilizer response of forest trees. - In: SAMISH, R.M. (ed.): Recent Advances in Plant Nutrition. Vol. 2. Pp. 669-678. Gordon and Breach Science Publishers, New York - London - Paris 1971.

10421 - KELLY, A.R., PATTERSON, L.K.: Model systems for photosynthesis. II. Concentration quenching of chlorophyll *b* fluorescence in solid solutions. - Proc. roy. Soc. London A *324*: 117-126, 1971.

10422 - KENDALL, D.R.: Improved Hersch cell system for rapid determination of oxygen in gases. - Anal. Chem. *43*: 944-947, 1971.

10423 - KENNEL, S.J., KAMEN, M.D.: Iron-containing proteins in *Chromatium*. I. Solubilization of membrane-bound cytochrome. - Biochem. biophys. Acta *234*: 458-467, 1971. [Chl.]

10424 - KENNEL, S.J., KAMEN, M.D.: Iron-containing proteins in *Chromatium*. II. Purification and properties of cholate-solubilized cytochrome complex. - Biochem. biophys. Acta *253*: 153-166, 1971.

10425 - KERESZTES, Á.: Light microscopic examination of chloroplast mutation in *Tradescantia* leaves. - Acta bot. Acad. Sci. hung. *17*: 379-389, 1971.

10426 - KERR, M.W., ROBERTSON, A.: Properties of phosphoenolpyruvate carboxylase isolated from maize leaves. - Biochem. J. *125*: 34P, 1971.

10427 - KERSHAW, K.A., HARRIS, G.P.: A technique for measuring the light profile in a lichen canopy. - Can. J. Bot. *49*: 609-611, 1971.

10428 - KESSLER, E., ZWEIER, I.: Physiologische und biochemische Beiträge zur Taxonomie der Gattung *Chlorella*. V. Die auxotrophen und mesotrophen Arten. - Arch. Mikrobiol. *79*: 44-48, 1971. [Car.]

10429 - KESTEMONT, P.: Productivité primaire des taillis simples et concept de nécromasse. - In: DUVIGNEAUD, P. (ed.): Productivity of Forest Ecosystems. Pp. 271-279. Unesco, Paris 1971.

10430 - KEYHANI, E., FLOYD, R.A., CHANCE, B.: Membrane structure and function. I. An electron microscope study of the alteration induced by laser irradiation on the chloroplast lamellar membranes. - Arch. Biochem. Biophys. *146*: 618-626, 1971.

10431 - KEYLOCK, M.J., KIRK, J.T.O., ROGERS, L.J.: Isolation, amino acid composition and physical properties of chloroplast thylakoid protein from *Euglena gracilis*. - Biochem. J. *121*: 14 P, 1971.

10432 - KHACHIDZE, O.T.: Vklyuchenie C^{14} radioaktivnoĭ uglekisloty v belki list'ev vinogradnoĭ lozy. [Incorporation of ^{14}C of radioactive carbon dioxide into the protein of grapevine leaves.] - Soobshch. Akad. Nauk gruz. SSR *64*: 449-452, 1971. [In R, ab: E, Georg.]

10433 - KHAILOV, K.M.: Vneshnemetabolichèskie svyazi makrofitov v pribrezhnykh morskikh fitotsenozakh. [Interrelations among seaweeds in coastal plant communities through external metabolites.] - Bot. Zh. *56*: 1557-1563, 1971. [Ps; in R, ab: E.]

10434 - KHALILOV, G.R., ISMAĬLOV, S.A.: Izuchenie doz, srokov i sposobov vneseniya NRV pod khlopchatnik. [Doses, terms and kinds of treatment of cotton by the petroleum-growth-regulators.] - In: NRV v Sel'skom Khozyaĭstve. Pp. 314-318, ELM, Baku 1971. [Ps; in R.]

10435 - KHAN, A.A., AKOSU, F.I.: The physiology of groundnut. I. An autoradiographic study of the pattern of distribution of ^{14}carbon products. - Physiol. Plant. *24*: 471-475, 1971.

10436 - KHARLOVA, M.G.: Osobennosti ekstinktsii pigmentnykh rastvorov khvoi kedra v godichnoĭ dinamike. [Peculiarities of absorbance of pigments solutions from cedar needles during a year.] - Inform. Byull. (Irkutsk) *9*: 53-54, 1971. [In R.]

*10437 - KHEIN, Kh.Ya.: Svetovye krivye istinnogo fotosinteza intaktnykh list'ev, poluchaemye radiometricheskim metodom. [Light curves of gross photosynthesis of intact leaves, obtained by the radiometric method.] - In: Tezĭsy Dokladov Vsesoyuznogo Soveshchaniya po Unifikatsii Metodov i Priborov dlya Massovykh Izmereniĭ Intensivnosti Fotosinteza. Pp. 112-114. Nauch.-issled. Inst. Rastenievod. N. I. Vavilova, Leningrad - Pushkin 1970. [In R.]

10438 - KHEIN, Kh.Ya., OSIPOVA, O.P., NICHIPOROVICH, A.A.: Izmeneniya svoĭstv fotosintetichèskogo apparata rasteniĭ *Vicia faba* L. pri smene svetovogo rezhima. [Changes in the properties of the photosynthetic apparatus of *Vicia faba* plants during a change in the light regime.] - Dokl. Akad. Nauk SSSR *200*: 244-247, 1971. [In R.]

10439 - **KHODASEVICH, E.V.**: Sostoyanie fonda pigmentov i plastid v ontogeneze lista khvoinykh. [Pigments and plastids during leaf ontogenesis in conifers.] - In: Problemy Biosinteza Khlorofillov. Pp. 173-198. Nauka i Tekhnika, Minsk 1971. [In R.]

10440 - **KHODOS, V.N., KOLOSHA, O.I.**: Nekotorye pokazateli fotosinteticheskogo metabolizma razlichnykh sortov ozimoi pshenitsy pri otritsatel'noi temperature. [Indices of photosynthetic metabolism in different varieties of winter wheat at temperatures below zero.] - Fiziol. Biokhim. kul't. Rast. *3*: 151-156, 1971. [In R.]

10441 - **KHOKHLOVA, V.A., SVESHNIKOVA, I.N., KULAEVA, O.N.**: Vliyanie fitogormonov na formirovanie struktury khloroplastov v izolirovannykh semyadolyakh tykvy. [Effect of phytohormones on the formation of chloroplast structure in isolated pumpkin cotyledons.] - Tsitologiya *13*: 1074-1079, 1971. [In R, ab: E.]

10442 - **KHOLDEBARIN, B., OERTLI, J.J.**: Effects of metabolic inhibitors on salt uptake and organic acid synthesis by leaf tissues in the light and in the dark. - Z. Pflanzenphysiol. *66*: 352-358, 1971. [Ps.]

10443 - **KHOLMOGOROV, V.E.**: Eksitonnaya priroda signala E.P.R. v kompleksakh kristallicheskogo khlorofilla s elektronnymi aktseptorami. [Exciton nature of EPR signal in complexes of crystalline chlorophyll with electron acceptors.] - Dokl. Akad. Nauk SSSR *200*: 402-405, 1971. [In R.]

*10444 - **KHROMOV, V.M., FEDOROV, V.D.**: Sezonne izmenenie pervichnoi produktivnosti v Belom more. [Seasonal changes of primary productivity in the White Sea.] - Vestn. mosk. Univ., Ser. VI-Biol. Pochvoved. *25* (5): 23-26, 1970. [In R.]

10445 - **KHUDAIRI, A.K., ARBOLEDA, O.P.**: Phytochrome-mediated carotenoid biosynthesis and its influence by plant hormones. - Physiol. Plant. *24*: 18-22, 1971.

*10446 - **KHRUSHUDYAN, P.A.**: Kornevaya nedostatochnost' kak osnovnaya prichina prezhdevremennogo usykhaniya ivovykh i topolevykh nasazhdenii sevanskikh pochvogruntov. [Root deficiency as a principal cause of precocious wilting of willow and poplar stands on Sevan soils and subsoils.] - Biol. Zh. Armenii *23* (7): 54-61, 1970. [Ps; in R, ab: Arm.]

*B10447 - **KICHIGIN, A.A.**: Karotin v Dikorastushchikh i Kul'turnykh Rasteniyakh Komi [Carotene in Wild and Cultivated Plants of the Komi Autonomous SSR.] - Komi knizhnoe Izdat., Syktyvkar 1970. [In R.]

10448 - **KIKNADZE, G.S.**: Izmenenie fluorestsentsii khlorofilla *in vivo*, svyazannye s razvitiem list'ev mnogoletnikh rastenii. [Changes in the fluorescence of chlorophyll *in vivo* connected with the development of leaves of perennial plants.] - Fiziol. Rast. *18*: 215-218, 1971. [In R.]

*10449 - **KILEN, T.C., ANDREW, R.H.**: Measurement of drought resistance in corn. - Agron. J. *61*: 669-672, 1969. [Chl.]

10450 - **KIMIMURA, M., KATOH, S., IKEGAMI, I., TAKAMIYA, A.**: Inhibitory site of carbonyl cyanide *m*-chlorophenylhydrazone in the electron transfer system of the chloroplasts. - Biochim. biophys. Acta *234*: 92-102, 1971.

*10451 - **KIMURA, M., MOTOTANI, I., HOGETSU, K.**: Ecological and physiological studies on the vegetation of Mt. Shimagare VI. Growth and dry matter production of young *Abies* stand. - Bot. Mag. (Tokyo) *81*: 287-296, 1968. [Growth analysis.]

10452 - **KINERSON, R. Jr., FRITSCHEN, L.J.**: Modeling a coniferous forest canopy. - Agr. Meteorol. *8*: 439-445, 1971.

10453 - **KINKLADZE, D.Kh., TARASASHVILI, K.M.**: [Fading of UV-irradiated solutions of plastid pigments in plants of various habitats.] - Soobshch. Akad. Nauk gruz. SSR *64*: 671-674, 1971. [In Georg., ab: E, R.]

10454 - **KIPNIS, E.A., LOBOTSKAYA, L.I., MATOSHKO, I.V.**: Pigmenty i fosforsoderzhashchie soedineniya v khloroplastakh list'ev sakharnoi svekly raznoi ploidnosti. [Pigments and phosphorus containing compounds in chloroplasts of sugar beet leaves of different ploidy.] - Vestsi Akad. Navuk belarus. SSR, Ser. biyal. Navuk *1971* (5): 98-99, 1971. [Chl, Car; in R.]

*10455 - **KIPRIN, V.I.**: Usovershenstvovanie gazoanalizatora GIP-9 dlya registratsii malykh izmenenii kontsentratsii CO_2 v vozdukhe. [Improvement of the infra-red gas analyser GIP-9 for recording small changes in CO_2 concentration in air.] - In: Tezisy Dokladov Vsesoyuznogo Soveshchaniya po Unifikatsii Metodov i Priborov dlya Massovykh Izmerenii Intensivnosti Fotosinteza. Pp. 44-47. Nauch.-issled. Inst. Rastenievod. N.I. Vavilova, Leningrad - Pushkin 1970. [In R.]

10456 - KIPRIN, V.I.: Usovershenstvovanie infrakrasnogo gazoananlizatora dlya registratsii gazoobmena rastenii. [An improvement of the infra-red gas analyser for recording gas exchange in plants.] - Sel'.-khoz. Biol. 6: 142-150, 1971. [In R, ab: E.]

10457 - KIRA, T., OGAWA, H.: Assessment of primary production in tropical and equatorial forests. - In: DUVIGNEAUD, P. (ed.): Productivity of Forest Ecosystems. Pp. 309-321. Unesco, Paris 1971.

10458 - KIRCHMANN, R.: Inhibition de l'accumulation des chlorophylles dans les feuilles primordiales de *Phaseolus vulgaris* après irradiation neutronique. - Physiol. Plant. 25: 249-252, 1971.

10459 - KIRCHMANN, R., BONOTTO, S., BRONCHART, R.: Accumulation des chlorophylles chez les feuilles primordiales de *Phaseolus vulgaris* L., irradiées en présence d'AET et de cystamine. II. Influence de l'AET et de la cystamine sur l'accumulation des chlorophylles. - Radiat. Bot. 11: 419-423, 1971.

10460 - KIRCHMANN, R., BONOTTO, S., van PUYMBROECK, S.: Accumulation des chlorophylles chez les feuilles primordiales de *Phaseolus vulgaris* L., irradiées en présence d'AET ou de cystamine. I. Pénétration, localisation et incorporation de la cystamine et de l'AET dans les feuilles primordiales. - Radiat. Bot. 11: 411-417, 1971.

*B10461 - KIRICHENKO, E.B.: Metody Issledovaniya Fotofosforilirovaniya. [Methods of Studying Photophosphorylation.] - Akad. Nauk SSSR, Inst. Fotosinteza, Pushchino-na-Oke 1970. [In R, ab: E.]

*10462 - KIRK, Dzh. T.O.: Avtonomiya plastid. [Plastid autonomy.] - In: Funktsional'-naya Biokhimiya Kletochnykh Struktur. Pp. 39-51. Nauka, Moskva 1970.

10463 - KIRK, J.T.O.: Chloroplast structure and biogenesis. - Annu. Rev. Biochem. 40: 161-169, 1971.

10464 - KISAKI, T., IMAI, A., TOLBERT, N.E.: Intracellular localization of enzymes related to photorespiration in green leaves. - Plant Cell Physiol. 12: 267-273, 1971.

10465 - KISAKI, T., YOSHIDA, N., IMAI, A.: Glycine decarboxylase and serine formation in spinach leaf mitochondrial preparation with reference to photorespiration. - Plant Cell Physiol. 12: 275-288, 1971.

10466 - KISS, B., KISS, A.S., BÉLA, P.: Az agronit és a pétisó eltérö hatása a klorofilltartalomra, a fotoszintetikus széndioxid-fixálásra és a klorofill fotokémiai aktivitására Bezostája-I búzafajta levelében. [Various effects of Agronit and Peti salts on chlorophyll content, photosynthetic CO_2 fixation and photochemical activity of chlorophyll in leaves of wheat cv. Bezostaya-1.] - Föiskolák Felsöfoku Tech. Evkönyve 1971: 93-98, 1971. [In Hung., ab: R.]

10467 - KJØSEN, H., LIAAEN-JENSEN, S.: Carotenoids in higher plants. 5. Total synthesis of lycoxanthin. - Acta chem. scand. 25: 1500-1502, 1971.

10468 - KLEIN, S., SHOCHAT, M.: The electron microscopic image of chloroplasts after osmium tetraoxide fixation in the presence of sorbitol. - J. Microscop. (Paris) 10: 117-120, 1971.

10469 - KLEINHOFS, A., SHUMWAY, L.K., SIDERIS, E.G.: Biochemical and ultrastructural characterization of chlorophyll-deficient barley mutants. - In: NILAN, R.A. (ed.): Barley Genetics. Vol. II. Pp. 194-200. Washington State Univ. Press, Pullman 1971.

10470 - KLEMME, B., KLEMME, J.-H., SAN PIETRO, A.: PPase, ATPase, and photophospholation in chromatophores of *Rhodospirillum rubrum*: Inactivation by phospholipase A; reconstitution by phospholipids. - Arch. Biochem. Biophys. 144: 339-342, 1971.

10471 - KLEMME, J.-H., GEST, H.: Regulation of the cytoplasmic inorganic pyrophos-
phatase of *Rhodospirillum rubrum*. - Europe. J. Biochem. *22*: 529-537, 1971.

10472 - KLEMME, J.-H., KLEMME, B., GEST, H.: Catalytic properties and regulatory diversity of inorganic pyrophosphates from photosynthetic bacteria. - J. Bacteriol. *108*: 1122-1128, 1971.

10473 - KLEPPER, L., FLESHER, D., HAGEMAN, R.H.: Generation of reduced nicotinamide
adenine dinucleotide for nitrate reduction in green leaves. - Plant Physiol.
48: 580-590, 1971. [Ps.]

10474 - KLIMOVICH, A.S., KAKHNOVICH, L.V.: Intensivnost' fotofosforilirovaniya v
zavisimosti ot yarusnosti list'ev i ikh vozrasta. [Photophosphorylation
rate dependent on leaf insertion level and age.] - Vestn. belorus. Univ.,
Ser. 2. *1971* (2): 38-40, 1971. [In R.]

10475 - KLOSE, H., SCHÄCKE, D.: Die jahres- und tageszeitlichen Änderungen des Sauer-
stoffgehaltes im Sacrow-Paretzer Kanal bei Potsdam und ihre Beziehungen zu
den Änderungen anderer Faktoren. - Limnologica (Berlin) *8*: 337-345, 1971.
[Ps, Chl.]

10476 - KLUGE, M.: Studies on CO_2 fixation by succulent plants in the light. - In:
HATCH, M.D., OSMOND, C.B., SLATYER, R.O. (ed.): Photosynthesis and Photo-
respiration. Pp. 283-287. Wiley-Interscience, New York - London - Sydney -
Toronto 1971.

10477 - KLUGE, M.: Veränderliche Markierungsmuster bei $^{14}CO_2$-Fütterung von *Bryophyllum tubiflorum* zu verschiedenen Zeitpunkten der Hell-Dunkelperiode. II.
Beziehungen zwischen dem Malatgehalt des Gewebes und dem Markierungsmuster
nach $^{14}CO_2$-Lichtfixierung. - Planta *98*: 20-30, 1971.

10478 - KLUGE, M.: Der CO_2-Austausch der Sukkulenten: Biochemische Grundlagen einer
ökologischen Anpassung. - Ber. deut. bot. Ges. *84*: 417-424, 1971 (1972).

10479 - KLUGE, M., OSMOND, C.B.: Pyruvate P_i dikinase in Crassulacean acid metabolism.
- Naturwissenschaften *58*: 414-415, 1971.

10480 - KNAFF, D.B., ARNON, D.I.: On two photoreactions in system II of plant photosynthesis. - Biochim. biophys. Acta *226*: 400-408, 1971.

10481 - KNAFF, D.B., McSWAIN, B.D.: Action spectra of the three light reactions in
photosynthesis. - Biochim. biophys. Acta *245*: 105-108, 1971.

*10482 - KNAUFT, R.L., ARDITTI, J.: Partial identification of dark $^{14}CO_2$ fixation
products in leaves of *Cattleya* (*Orchidaceae*). - New Phytol. *68*: 657-661, 1969.
[Ps.]

10483 - KNIPLING, E.B., SCHRODER, V.N., DUNCAN, W.G.: CO_2 evolution from Florida organic soils. - Soil Crop Sci. Soc. Fla. Proc. *30*: 320-326, 1971. [Ps.]

10484 - KNOBLOCH, K., ELEY, J.H., ALEEM, M.I.H.: Generation of reducing power in
bacterial photosynthesis. *Rhodopseudomonas palustris*. - Biochem. biophys.
Res. Commun. *43*: 834-839, 1971.

10485 - KNOBLOCH, K., ELEY, J.H., ALEEM, M.I.H.: Thiosulfate-linked ATP dependent
NAD^+ reduction in *Rhodopseudomonas palustris*. - Arch. Mikrobiol. *80*: 97-114,
1971.

10486 - KNOWLES, R.E., LIVINGSTON, A.L.: Carotenoid analysis by two-dimensional, two
adsorbent thin-layer chromatography. - J. Chromatogr. *61*: 133-141, 1971.

10487 - KNUCKLES, B.E., WITT, S.C., MILLER, R.E., BICKOFF, E.M.: Determination of
carotene and xanthophyll in alfalfa protein concentrates. - J. Ass. offic.
anal. Chemists *54*: 769-772, 1971.

10488 - KNYPL, J.S.: Control of protein and RNA synthesis by AMO-1618 and other growth
retardants in cucumber cotyledons. - Biochem. Physiol. Pflanzen *162*: 127-141,
1971.

10489 - KNYPL, J.S.: Effect of 2-chloroethylphosphonic acid and vanillin on chlorophyll, protein and RNA synthesis in detached cucumber cotyledons, and chlorophyll degradation in senescing leaf discs of kale. - Acta Soc. Bot. Pol.
40: 257-274, 1971.

10490 - KNYPL, J.S., MAZURCZYK, W.: Arrest of chlorophyll and protein breakdown in

senescing leaf discs of kale by cycloheximide and vanillin. - Current Sci.
40: 294-295, 1971.

10491 - **KOBAK, K.I., KUPCHENKO, G.S.**: Fotosinteticheskaya produktivnost', koeffitsient ispol'zovaniya fotosinteticheski aktivnoĭ radiatsii i uglekislotnyĭ
obmen v posevakh. [Photosynthetic productivity, PhAR utilization and carbon
dioxide exchange in canopies.] - In: ZALENSKIĬ, O.V. (ed.): Fotosintez i Ispol'zovanie SolnechnoĭEnergii. Pp. 44-50. Nauka, Leningrad 1971. [In R, ab: E.]

10492 - **KOCH, W., LANGE, O.L., SCHULZE, E.-D.**: Ecophysiological investigations on wild
and cultivated plants in the Negev desert. I. Methods: A mobile laboratory
for measuring carbon dioxide and water vapour exchange. - Oecologia *8*: 296-
309, 1971.

10493 - **KÖCHER, H., LEONARD, O.A.**: Translocation and metabolic conversion of ^{14}C-la-
beled assimilates in detached and attached leaves of *Phaseolus vulgaris* L.
in different phases of leaf expansion. - Plant Physiol. *47*: 212-216, 1971.

*10494 - **KOCHUBEĬ, S.M.**: O primenimosti razlichnykh fizicheskikh modeleĭ v raschetakh
perenosa energii v FSE. [Applicability of different physical models in calculations of energy transfer in photosynthesis.] - In: Problemy Biofotokhimii.
P.33. Izdat. moskov. gos. Univ., Moskva 1970. [In R.]

10495 - **KOCHUBEĬ, S.M., MANUIL'SKAYA, S.V., OSTROVSKAYA, L.K.**: Spektral'nye svoĭstva
fragmentov khloroplastov, obrazovannykh galaktolipazoĭ pri razlichnykh pH
sredy. [Spectral properties of chloroplast fragments produced by galactoli-
pase at various pH.] - In: Biokhimiya i Biofizika Fotosinteza. Pp. 71-74.
Irkutsk 1971. [In R.]

10496 - **KOCÚRIK, Š., VOJTUŇ, A.**: Vplyv Cl⁻ na obsah chlorofylu v listoch *Chenopodium
album* L. [Influence of Cl⁻ on chlorophyll content in the leaves of *Chenopodium album* L.] - Biológia (Bratislava) *26*: 77-80, 1971. [In Slovak, ab: E.]

10497 - **KOENIG, F.**: Konzentration einiger Lipide in den Chloroplasten von *Zea mays*
und *Antirrhinum majus*. - Z. Naturforsch. *26 b*: 1180-1187, 1971. [Chl.]

10498 - **KOHEL, R.J., BENEDICT, R.C.**: Description and CO_2 metabolism of aberrant and
normal chloroplasts in variegated cotton, *Gossypium hirsutum* L. - Crop Sci.
11: 486-488, 1971. [Chl.]

*10499 - **KOLOVU, M.**: Opredelyane intenzivnostta na fotosintezata na porednite lista
na pshenitsa sort YubileĭnaIII. [Photosynthetic rate in ordinal leaves of
the wheat cultivar YubileĭnaIII.] - Rasteniev"d. Nauki (Sofiya) *6* (1): 11-
19, 1969. [In Bulg., ab: E, R.]

*10500 - **KOLPINA, L.S., SIVTSEV, M.V.**: Vliyanie molibdenizirovanogo i marganizirovan-
nogo superfosfata na urozhaĭkukuruzy v usloviyakh Kryma. [Effect of super-
phosphate enriched in molybdenum and manganese on maize yield in Crimea.]
- In: Mikroelementy v Sel'skom Khozyaĭstvei Meditsine. Vol. 3. Pp. 71-74.
Naukova Dumka, Kiev 1967. [Chl; in R.]

10501 - **KOMISSAROV, D.A., ARTAMONOVA, T.A.**: Vliyanie stimulyatorov (NRV, MU) na rost
i nekotorye fiziologicheskie protsessy seyantsev sireni obyknovennoĭ. [Effect
of stimulators (petroleum growth substance, trace-nutrient fertilizer) on
the growth and physiological processes of common lilac seedlings.] - In: NRV
v Sel'skom Khozyaĭstve. Pp. 45-50. Elm, Baku 1971. [Ps; In R.]

10502 - **KOMISSAROV, G.**: Functional model of photosynthesis. - Sov. Sci. Rev. *2*: 285-
290, 1971.

10503 - **KOMOV, S.V., MOKRONOSOV, A.T.**: O dinamicheskikh kharakteristikakh pri otsenke
kachestva fotosinteticheskoĭsistemy. [Dynamic characteristics in evaluation
of the quality of a photosynthetic system.] - In: ZALENSKIĬ, O.V. (ed.): Fotosintez i Ispol'zovanie SolnechnoĭEnergii·. Pp. 149-156. Nauka, Leningrad
1971. [In R, ab: E.]

*10504 - **KONDRAT'EVA, E.N., KRASIL'NIKOVA, E.N., NOVIKOVA, L.M.**: Obrazovanie polisakharidov zelenymi fotosinteziruyushchimi bakteriyami. [Production of polysaccharides by green photosynthesizing bacteria.] - Mikrobiologiya *37*: 417-
424, 1968. [In R, ab: E.]

B10505 - **KONEV, S.V., VOLOTOVSKIĬ, I.D.**: Vvedenie v Molekularnuyu Fotobiologiyu. [An

Introduction to Molecular Photobiology.] - Nauka i Tekhnika, Minsk 1971.
[Also Ps; in R.]

*10506 - KONONENKO, A.A., GRIGOROV, L.N., VERKHOTUROV, V.N., ANDREITSEV, A.P., RUBIN,
A.B.: Nizkotemperaturnaya pristavka dlya spektrofotometricheskikh absorbtsi-
onnykh i lyuminestsentnykn issledovaniI. [Low temperature appliance for spectrophotometric absorption and luminescence studies.] - Nauch. Dokl. vyssh.
Shkoly, biol. Nauki 12 (4): 128-133, 1969. [Chl; in R.]

*10507 - KONONENKO, A.A., RUBIN, A.B., FOKHT, A.S., USPENSKAYA, N.Ya.: Issledovanie
kineticheskikh zakonomernosteI indutsirovannykh svetom reaktsiI vnutrikletochnykh tsitokhromov *Rhodopseudomonas* sp. [Kinetics of intracellular cytochrome
light-induced reactions in *Rhodopseudomonas* sp.] - Mol. Biol. (Moskva) 2:
807-817, 1968. [In R, ab: E.]

10508 - KONONENKO, A.A., VENEDIKTOV, P.S., LUKASHEV, E.P., MATORIN, D.H., RUBIN, A.B.:
Dokazatel'stvo sushchestvovaniya obshchego pervichnogo aktseptora dlya sistem tsiklicheskogo i netsiklicheskogo transporta elektronov u fotosinteziruyushchikh bakteriI *Ectothiorhodospira shaposhnikovii*. [Evidence for the existence of common primary electron acceptor of the cyclic and non-cyclic electron transport systems in photosynthesizing bacteria *Ectothiorhodospira shaposhnikovii*.] - Stud. biophys. 28: 15-22, 1971. [In R, ab: E.]

10509 - KONONENKO, A.A., VENEDIKTOV, P.S., LUKASHEV, E.P., RUBIN, A.B.: Fotoreaktsii
bakteriokhlorofilla v kletkakh *Ectothiorhodospira shaposhnikovii*. [Photore-
actions of bacteriochlorophyll in cells of *Ectothiorhodospira shaposhnikovii*.] - Stud. biophys. 28: 9-14, 1971. [In R, ab: E.]

*10510 - KONOVALOV, I.N., LERMAN, R.I., MIKHALEVA, E.N., MUKHINA, V.A.: Fiziologiches-
kie osobennosti sakharnoI svekly v LeningradskoI oblasti. [Physiological
characteristics of sugar beet in the Leningrad region.] - Tr. bot. Inst. Akad.
Nauk SSSR, Ser. IV - eksp. Bot. 18: 62-85, 1966. [Ps; in R, ab: E.]

10511 - KOPYLOV, N.I., TIMOSHENKO, S.E.: Intensivnost' fotosinteza i zakladka tsvet-
kovykh pochek u yabloni pri ogranichenii vysoty krony. [Change of photosynthetic rate and the formation of flower buds in apple-trees with limited
height of the crown.] - Sel'.-khoz. Biol. 6: 515-517, 1971. [In R, ab: E.]

10512 - KORESHKIN, A.I.: TeploprovodyashchiI mikrokalorimetr s medno-konstantanovymi
termobatareyami. [Thermoconducting microcalorimeter with copper-constantan
thermobatteries.] - Byul. eksp. Biol. Med. 72 (11): 123-124, 1971. [In R,
ab: E.]

10513 - KORNILOV, A.A., KOSTINA, V.S., TKACHENKO, L.G.: Ispol'zovanie metoda krakhmal'-
noI proby dlya kharakteristiki fotosinteticheskoI deyatel'nosti rasteniI v
posevakh. [Photosynthetic activity of plants in crops determined by the method of starch test.] - Fiziol. Rast. 18: 832-834, 1971. [In R.]

10514 - KORTSCHAK, H.P.: Sugar synthesis in mesophyll chloroplasts of sugarcane. -
In: HATCH, M.S., OSMOND, C.B., SLATYER, R.O.: (ed.): Photosynthesis and Pho-
torespiration. Pp. 255-258. Wiley-Interscience, New York - London - Sydney -
Toronto 1971.

*10515 - KORZH, B.V.: Ustanovka dlya izucheniya gazoobmena list'ev rasteniI v uslovi-
yakh nepreryvnogo i preryvistogo osveshcheniya. [Device for studying gas ex-
change of plant leaves under continuous and intermittent light.] - In: Tezisy
Dokladov Vsesoyuznogo Soveshchaniya po Unifikatsii Metodov i Priborov dlya
Massovykh IzmereniI Intensivnosti Fotosinteza. Pp. 48-51. Nauch.-issled. Inst.
Rastenievod. N.I. Vavilova, Leningrad - Pushkin 1970. [In R.]

10516 - KOTSUR, M.V., GULYAEV, B.I., MANUIL'S'KYÏ, V.D., OKANENKO, A.S.: Pro inten-
syvni svitlovi polya yak mutagennyĬ faktor. [Intensive light fields as a mutagenic factor.] - Dopovidi Akad. Nauk ukr. RSR, Ser. B 33: 1029-1030, 1971.
[Ps; in Ukr., ab: E.]

10517 - KOTSUR, N.V., GULYAEV, B.I., MANUIL'SKIÏ, V.D., OKANENKO, A.S.: O vliyanii
svetoimpul'snogo oblucheniya pyl'tsy na fotosintez i dykhanie list'ev kukuruzy pervogo pokoleniya. [Effect of light-impulse irradiation of pollen on
photosynthesis and respiration in leaves of maize of the first generation.]
- In: Svetoimpul'snaya Stimulyatsiya RasteniI. Pp. 335-340. Nauka, Moskva
1971. [In R.]

10518 - KOTSUR, N.V., KOTLYAR, V.Z., LYUBINSKIĬ, N.A., SHAKHOV, A.A.: O vliyanii ponizhennoĭ vlazhnosti pochvy na fiziologo-morfologicheskoe proyavlenie svetoimpul'snogo effekta u kukuruzy. [Effect of decreased soil moisture on physiological and morphological realization of the light-impulse effect in
maize.] - In: Svetoimpul'snaya Stimulyatsiya Rasteniĭ. Pp. 291-299. Nauka,
Moskva 1971. [Ps; in R.]

10519 - de KOUCHKOVSKY, Y.: Effect of ionic environment on chlorophyll fluorescence,
absorption spectrum and photochemical activity of photosynthesizing cells.
- In: QUAGLIARIELLO, E., PAPA, S., ROSSI, C.S. (ed.): Energy Transduction
in Respiration and Photosynthesis. Pp. 269-281. Adriatica Editrice, Bari 1971.

10520 - de KOUCHKOVSKY, Y.: Photosynthesis at the 8th International Congress of Biochemistry (Symposium No. 4: Biological Oxidation and Bioenergetics) in Lucerne, Switzerland, and at the International Colloquium on Bioenergetics
(Energy Transduction in Respiration and Photosynthesis) in Pugnochiuso, Italy,
September 1970. - Photosynthetica 5: 88-91, 1971.

10521 - KOUSALOVÁ, I.: Uplatnění růstové analýzy při studiu tvorby výnosu ozimé pšenice. [Use of growth analysis in the study of the yield formation in winter
wheat.] - Rostlinná Výroba (Praha) 17: 493-500, 1971. [In Czech, ab: E, R.]

*10522 - KOVALIK, A.I.: Vliyanie sovmestnogo primeneniya mineral'nykh udobreniĭ i sti-
mulyatorov rosta na fiziologicheskie protsessy rasteniĭ tomatov. [Effect of
the combined use of mineral fertilizers and growth stimulators on physiological processes of tomato plants.] - Tr. khar'kov. sel'skokhoz. Inst. 90 (Issledovaniya po Fiziologii i Biokhimii Rasteniĭ): 21-24, 1970. [Ps, Chl, Car;
in R.]

10523 - KOWALCZEWSKI, A., LACK, T.J.: Primary production and respiration of the phytoplankton of the rivers Thames and Kennet at Reading. - Freshwater Biol. 1:
197-212, 1971.

10524 - KOWALLIK, W.: Light stimulated respiratory gas exchange in algae and its relation to photorespiration. - In: HATCH, M.D., OSMOND, C.B., SLATYER, R.O.
(ed.): Photosynthesis and Photorespiration. Pp. 514-522. Wiley-Interscience,
New York - London - Sydney - Toronto 1971.

10525 - KRAAN, G.P.B., AMESZ, J.: Mechanisms of delayed and stimulated delayed fluorescence. - In: QUAGLIARIELLO, E., PAPA, S., ROSSI, C.S. (ed.): Energy
Transduction in Respiration and Photosynthesis. Pp. 611-620. Adriatica Editrice, Bari 1971. [Chl.]

10526 - KRAAYENHOF, R.: Fluorescence changes and binding of uncouplers in chloroplasts
and mitochondria. - In: QUAGLIARIELLO, E., PAPA, S., ROSSI, C.S. (ed.): Energy Transduction in Respiration and Photosynthesis. Pp. 649-657. Adriatica
Editrice, Bari 1971.

10527 - KRAAYENHOF, R., KATAN, M.B., GRUNWALD, T.: The effect of temperature on
energy-linked functions in chloroplasts. - FEBS Lett. 19: 5-10, 1971.

10528 - KRAĬNOVA, N.N., CHERNOV, I.A.: Vliyanie razlichnykh velichin pH, kontsentratsii NaHC^{14}O$_3$, temperatury i osveshchennosti na intensivnost' fotosinteza
izolirovannykh khloroplastov. [Effect of different pH values, concentrations
of NaH^{14}CO$_3$, temperature and illuminance on photosynthetic rate in isolated
chloroplasts.] - Fiziol. Rast. 18: 887-892, 1971. [In R, ab: E.]

*10529 - KRAKKAI, J.: Wechselbeziehungen zwischen dem Wirkungsmechanismus von Atrazin
und den Farbstoffen von Pflanzen. - Tagungsber. DAL Berlin 109: 119-126,
1970. [Chl.]

10530 - KRANZ, A.R.: Genphysiologie quantitativer Merkmale bei Arabidopsis thaliana
(L.) HEYNH. Teil 1: Spaltungsanalyse und Pigmentbiosynthese in quantitativen
Chlorophyll b-Mangelmutanten. - Theor. appl. Genet. 41: 45-51, 1971.

10531 - KRANZ, A.R.: Genphysiologie quantitativer Merkmale bei Arabidopsis thaliana
(L.) HEYNH. Teil 2: Modifikation primärer und sekundärer Genwirkungen durch
langwelligen Strahlung bei monogenen Chlorophyll b-Defektmutanten. - Theor.
appl. Genet. 41: 91-99, 1971.

10532 - KRASICHKOVA, G.V., GILLER, Yu. E.: Spektral'nye svoĭstva protokhlorofilla v
sostoyanii svyazi s belkom. [Spectral properties of protochlorophyll bound

with protein.]. - Dokl. Akad. Nauk tadzh. SSR *14* (4): 69-73, 1971. [In R, ab: Tadzh.]

10533 - KRASNOVSKIĬ, A.A., BRIN, G.P., ALIEV, Z.Sh.: Fotosensibilizirovannoe vydelenie kisloroda v vodnykh rastvorakh *n*-benzokhinona. [Oxygen production in water solutions of *n*-benzoquinone stimulated by light.] - Dokl. Akad. Nauk SSSR *199*: 952-955, 1971. [In R.]

10534 - KRASNOVSKIĬ, A.A., BYSTROVA, M.I., LANG, F.: Modelirovanie raznykh form protokhlorofillovykh pigmentov v tverdykh plenkakh i v list'yakh rasteniĭ, obrabotannykh δ-aminolevulinovoĭ kislotoĭ. [Modelling of various forms of protochlorophyll pigments in dry layers and plant leaves treated with δ-aminolevulinic acid.] - Dokl. Akad. Nauk SSSR *201*: 1485-1488, 1971. [In R.]

10535 - KRASNOVSKIĬ, A.A. Jr., SHUVALOV, V.A., LITVIN, F.F., KRASNOVSKII, A.A.: Issledovanie fosforestsentsii i zamedlennoĭ fluorestsentsii protokhlorofillovykh pigmentov. [Phosphorescence and delayed fluorescence of protochlorophyllic pigments.] - Dokl. Akad. Nauk SSSR *199*: 1181-1184, 1971. [In R.]

10536 - KRATKY, B.A., WARREN, G.F.: A rapid bioassay for photosynthetic and respiratory inhibitors. - Weed Sci. *19*: 658-661, 1971.

10537 - KRAUSE, G.H.: Indirekter ATP-Transport zwischen Chloroplasten und Zytoplasma während der Photosynthese. - Z. Pflanzenphysiol. *65*: 13-23, 1971.

10538 - KRAUSE, G.H., HEBER, U.: Reaction of oxygen with the electron transport chain of photosynthesis. - In: BRODA, E., LOCKER, A., SPRINGER-LEDERER, H. (ed.): Proceedings of the First European Biophysics Congress. Vol. 4. Pp. 79-84. Verlag wiener med. Akad., Wien 1971.

10539 - KRAVCHENKO, L.V.: Primenenie NRV pod drevesnye rasteniya. [Use of petroleum growth substance under trees.] - In: NRV v Sel'skom Khozyaĭstve. Pp. 459-463, Elm, Baku 1971. [Ps, Chl; in R.]

*10540 - KREICBERGS, O., KRŪMIŅŠ, V., VOLFMANIS, A.: Fotosintēzes intensitātes noteinkšana ar portatīvo iekārtu PPF-3. [Determination of photosynthetic rate by means of portable apparatus PPF-3.] - In: Fotosintēzes Pētīšana Sējumos. Pp. 83-99. Zinātne, Rīgā 1970. [In Latvian, ab: E.]

*10541 - KREĬTSBERG, O.E.: Primenenie kolorimetricheskogo metoda dlya sozdaniya portativnogo pribora dlya opredeleniya intensivnosti fotosinteza. [Use of colorimetric method for construction of a portable apparatus for determining photosynthetic rate.] - In: Tezisy Dokladov Vsesoyuznogo Soveshchaniya po Unifikatsii Metodov i Priborov dlya Massovykh Izmereniĭ Intensivnosti Fotosinteza. Pp. 52-55. Nauch.-Issled. Inst. Rastenievod. N.I. Vavilova, Leningrad - Pushkin 1970. [In R.]

10542 - KREĬTSBERG, O.E., KRISTKALNE, S.Kh., VITOLA, A.K., PAVULINA, D.A.: ATF-aznaya aktivnost' khloroplastov i regulyatsiya soderzhaniya khlorofilla. [ATPase activity in chloroplasts and control of chlorophyll content.] - Izv. Akad. Nauk latv. SSR *1971* (9): 69-79, 1971. [In R, ab: E.]

10543 - KREJCAREK, G.E., TURNER, L., DUS, K.: Investigation of photosynthetic cytochromes *C* by high resolution NMR spectroscopy. - Biochem. biophys. Res. Commun. *42*: 983-991, 1971.

10544 - KRENDELEVA, T.E., KAUROV, B.S., SAKHAROVA, T.A., RUBIN, A.B.: O svyzai perenosa elektronov ot iskusstvennogo donora k NADF, s protsessami fotofosforilirovaniya v khloroplastakh. [Relation between electron transport from artificial donor to NADP and processes of photophosphorylation in chloroplasts.] - Stud. biophys. *28*: 183-190, 1971. [In R, ab: E.]

10545 - KRIEDEMANN, P.E.: Crop energetics and horticulture. - HortScience *6*: 432-438, 1971. [Ps.]

10546 - KRIEDEMANN, P.E.: Photosynthesis and transpiration as a function of gaseous diffusive resistances in orange leaves. - Physiol. Plant. *24*: 218-225, 1971.

10547 - KRIEDEMANN, P.E., BUTTROSE, M.S.: Chlorophyll content and photosynthetic activity within woody shoots of *Vitis vinifera*(L.). - Photosynthetica *5*: 22-27, 1971.

10548 - KRIEDEMANN, P.E., CANTERFORD, R.L.: The photosynthetic activity of pear leaves

(*Pyrus communis* L.). - Aust. J. biol. Sci. *24*: 197-205, 1971.

10549 - **KRIEDEMANN, P.E., SMART, R.E.**: Effects of irradiance, temperature, and leaf water potential on photosynthesis of vine leaves. - Photosynthetica *5*: 6-15, 1971.

10550 - **KRINSKY, N.I.**: Function. - In: ISLER, O. (ed.): Carotenoids. Pp. 669-716. Birkhäuser Verlag, Basel - Stuttgart 1971. [Car.]

10551 - **KRISHNAMURTHY, K.**: Phytoplankton pigments in Porto Novo waters (India). - Int. Rev. gesamten Hydrobiol. *56*: 273-282, 1971. [Chl, Car.]

*10552 - **KRISTKALNE, S.**: Hlorofila koncentrācijas noteikšana lapās. [Determination of chlorophyll concentration in leaves.] - In: Fotosintēzes PētTšana Sējumos. Pp. 101-117. Zinātne, RTgā 1970. [In Latvian, ab: R.]

10553 - **KRISTKALNE, S., PURVĪTE, I.**: Slāpekļa vielu maiņa un fotosintētiskā aparāta aktivitāte. [Some aspects of nitrogen metabolism and the activity of photosynthetic apparatus.] - Izv. Akad. Nauk latv. SSR *1971*:(9): 63-68, 151-152, 1971. [In Latvian, ab: E, R.]

10554 - **KRIZEK, D.T., ZIMMERMAN, R.H., KLUETER, H.H., BAILEY, W.A.**: Growth of crabapple seedlings in controlled environments: Effect of CO_2 level, and time and duration of CO_2 treatment. - J. amer. Soc. hort. Sci. *96*: 285-288, 1971.

*10555 - **KR"STEVA, M.**: Vliyanie na fungitsidnite preparati v"rkhu nyakoi fiziologichni protsesi v tretiranite s tyakh yab"lkovi i praskoveni d"rveta. [Influence of fungicide treatment on some physiological processes in apple and peach trees.] - Gradinar. lozar. Nauka (Sofia) *4* (3): 31-37, 1967. [Chl; in Bulg., ab: E, R.]

*10556 - **KRZYSCH, G.**: Untersuchungen des CO_2-Umsatzes eines Beta-Rübenbestandes. - Angew. Bot. *43*: 275-290, 1969.

10557 - **KRZYWACKA, T.**: Wpływ powierzchni asymilacyjnej liścia flagowego i kłosa na plonowanie pszenicy. [Influence of the flag leaf and ear assimilation area on wheat yields.] - Hodowla Rośl. Aklim. Nasienn. *15*: 51-69, 1971. [Growth analysis; in Pol., ab: E, R.]

10558 - **KUBÍČEK, F.**: Leaf area estimation in *Quercus petraea* using linear regression equations. - Photosynthetica *5*: 426-428, 1971.

10559 - **KUBÍN, Š.**: Measurement of radiant energy. - In: ŠESTÁK, Z., ČATSKÝ, J., JARVIS, P.G. (ed.): Plant Photosynthetic Production: Manual of Methods. Pp. 702-765. Dr. W. Junk N.V. Publ., The Hague 1971.

10560 - **KUBOTA, F., AGATA, W., KAMATA, E.**: [Dry matter production of forage plants. 2. Calculation of the photosynthesis and dry matter production in forage plant communities.]- J. jap. Soc. Grassland Sci. *17*: 229-234, 1971. [In Jap., ab: E.]

10561 - **KUBOTA, F., AGATA, W., KAMATA, E.**: [Dry matter production of forage plants. 4. Influences of light extinction coefficient (k) on dry matter production in forage plant populations. - Theoretical analyses.] - J. jap. Soc. Grassland Sci. *17*: 234-249, 1971. [In Jap., ab: E.]

10562 - **KUDINOVA, L.M.**: Nekotorye osobennosti vodoobmena i fotosinteza u kleshcheviny pri vozdeTstvii zasukhi, mikroelementov i fosfora. [Some features of water relations and photosynthesis in castor bean under influence of drought, microelements and phosphorus.] - In: Nekotorye Voprosy Sovremennogo Estestvoznaniya. Pp. 312-317. Rostov-na-Donu 1971 (1972). [In R.]

10563 - **KUKUSHKIN, A.K., BLYUMENFEL'D, L.A.**: Ob odnom aspekte vzaimodeTstviya dvukh fotosistem. [Aspect of the interaction of two photosystems.] - Biofizika *16*: 932-933, 1971. [In R, ab: E.]

10564 - **KULL, U.**: Neuere Ergebnisse der Photosynthese-Forschung. - Jahresh. Ges. Naturkunde Württemberg *126*: 244-255, 1971.

*10565 - **KUMAKOV, V.A.**: Vliyanie ob"ema i sposoba vyborki na tochnost' opredeleniT chistoT produktivnosti fotosinteza v melkodelyanochnykh opytakh. [Effect of size and type of selection on the accuracy of determinations of net assimilation rate in experiments on small plots.] - In: Tezisy Dokladov Vsesoyuz-

nogo Soveshchaniya po Unifikatsii Metodov i Priborov dlya Massovykh Izmereniǐ
Intensivnosti Fotosinteza. Pp. 56-59. Nauch.-issled. Inst. Rastenievod. N.I.
Vavilova, Leningrad - Pushkin 1970. [In R.]

10566 - KUMAKOV, V.A., KAMENSKIĬ, V.V.: Izmeneniya v fotosinteticheskoǐ deyatel'nosti
prosa pri selektsii na yugo-vostoke SSSR. [Changes in the photosynthetic activity of millet during selection in the southeastern USSR.] - Dokl. Akad.
Nauk SSSR *199*: 724-726, 1971. [In R.]

10567 - KUNG, S.D., THORNBER, J.P.: Photosystem I and II chlorophyll-protein complexes of higher plant chloroplasts. - Biochim. biophys. Acta *253*: 285-289,
1971.

10568 - KÜNSTLE, E.: Der Jahresgang des CO_2-Gaswechsels von einjährigen Douglasien-
trieben in einem 20jährigen Bestand. - Allg. Forst- Jagd-Zeit. *142* (4): 105-
108, 1971.

10569 - KUO, T., BOERSMA, L.: Soil water suction and root temperature effects on
nitrogen fixation in soybeans. - Agron. J. *63*: 901-904, 1971. [Ps.]

*10570 - KUPERMAN, I.A., BOCHKOV, G.A.: Novaya laboratornaya ustanovka dlya issle-
dovaniya gazoobmena rasteniǐ. [A new laboratory device for studying plant gas
exchange.] - In: Tezisy Dokladov Vsesoyuznogo Soveshchaniya po Unifikatsii
Metodov i Priborov dlya Massovykh Izmereniǐ Intensivnosti Fotosinteza. P.
60. Nauch.-issled. Inst. Rastenievodstva N.I. Vavilova, Leningrad - Pushkin
1970. [In R.]

*10571 - KUPRIYANOVA, A.S., ALIEV, N.A., MEL'NIKOV, N.N., LUZHNOVA, M.I.: Zavisimost'
mezhdu stroeniem anilidov i ingibirovaniem imi reaktsii Khilla. [Relation
between the structure of anilides and their inhibitory effect on the Hill
reaction.] - Khim. sel'. Khoz. *5* (10): 29-31, 1967. [In R.]

10572 - KURSANOV, A.: Is there a correlation between the protein content and photosynthetic activity of leaves of both ill and healthy plants? - Acta agron.
Acad. Sci. hung. *20*: 448, 1971.

10573 - KURSANOV, A.L., BROVCHENKO, M.I.: Uroven' assimilyatov v SP listovoǐ plas-
tinki pri razlichnykh usloviyakh ottoka. [Level of assimilates in leaf-blade
free space at various efflux conditions.] - Fiziol. Rast. *18*: 1158-1164, 1971.
[In R, ab: E.]

10574 - KÜRSCHNER, K.: Vom atmosphärischen Kohlendioxid zum Holz. - Wiss. Fortschr.
21: 352-357, 1971. [Ps.]

*10575 - KURTYSH, G.P.: K voprosu o soderzhanii khlorofilla i azota v list'yakh yaro-
vykh i ozimykh pshenits. [Level of chlorophyll and nitrogen in the leaves
of spring and winter wheat.] - Tr. khar'kov. sel'skokhoz. Inst. *90* (Issledovaniya po Fiziologii i Biokhimii Rasteniǐ): 57-60, 1970. [In R.]

10576 - KUSHNIRENKO, M.D., MEDVEDEVA, T.N., KRYUKOVA, E.V.: Vodnyǐ rezhim i sostoyanie plastidnogo apparata rasteniǐ. [Water regime and state of plant plastid
apparatus.] - Fiziol. Biokhim, kul't. Rast. *3*: 563-568, 1971. [In R, ab: E.]

10577 - KUTYURIN, V.M.: Voda - istochnik kisloroda pri fotosinteze. [Water - oxygen
source in photosynthesis.] - Priroda *1971* (7): 24-31, 1971. [In R.]

10578 - KUTYURIN, V.M., ULUBEKOVA, M.V.: Ob osobom mekhanizme razlozheniya vody i
perekhode fotoreduktsii vodorosleǐ v fotosintez v dlinnovolnovoǐ oblasti
spektra (λ> 700 mμ). [Special mechanism of water splitting and inversion
of algae photoreduction into photosynthesis in the long-wave spectral region
(λ> 700 nm).] - Dokl. Akad. Nauk SSSR *201*: 480-482, 1971. [In R.]

10579 - KUZIN, I.S.: Fotosinteticheskaya reaktsiya nekotorykh plodovykh rasteniǐ na
fotoperiodicheskie vozdeǐstviya. [Photosynthetic reaction of some fruit trees
on photoperiod.] - Tr. tsentr. genet. Lab. Im. I. V. Michurina *12*: 261-273,
1971. [In R.]

*10580 - KUZ'MENKO, M.I.: Usvoenie 1-C^{14}-alanina i 5-C^{14}-glutaminovoǐ kisloty nekoto-
rymi sinezelenymi vodoroslyami. [Assimilation of 1-^{14}C-alanine and 5-^{14}C-
glutamic acid by some blue-green algae.] - Gidrobiol. Zh. *4* (3): 41-48, 1968.
[Chl, Car; in R.]

10581 - KUZ'MENKO, M.I.: Miksotrofizm u vodorosleǐ. [Mixotrophy in algae.] - Gidrobiol.
Zh. *7* (5): 111-123, 1971. [Ps; in R.]

10582 - KUZ'MIN, V.A., CHIBISOV, A.K., VINOGRADOV, A.P.: Fotosensibilizirovannoe
khlorofillom i ego proizvodnymi okislenie vody i nizshikh spirtov. [Oxidation
of water and lower alcohols photosensibilized by chlorophyll and its derivatives.] - Dokl. Akad. Nauk SSSR *197*: 129-132, 1971. [In R.]

10583 - KUZNETSOV, S.I., BEZLER, F.I.: Opyt sostavleniya balansa organicheskogo veshchestva v Rybinskom vodokhranilishche. [Organic matter balance in the Rybinsk
reservoir.] - Tr. Inst. Biol. vnutr. Vod Akad. Nauk SSSR *21* (24) - Biologiya
i Produktivnost' Presnovodnykh Organizmov: 66-74, 296, 1971. [Ps; in R.]

10584 - KUZNETSOV, S.I., ROMANENKO, V.I., KARPOVA, N.S., ROMANENKO, V.A.: Chislennost'
bakteriT i produktsiya organicheskogo veshchestva v Rybinskom vodokhranilishche v 1966 g. [Quantity of bacteria and production of organic matter in
the Rybinsk reservoir in 1966.] - Tr. Inst. Biol. vnutr. Vod Akad. Nauk SSSR
21 (24) - Biologiya i Produktivnost' Presnovodnykh Organizmov: 17-22, 295,
1971. [In R.]

10585 - KUZNETSOV, S.I., ROMANENKO, V.I., KARPOVA; N.S., ROMANENKO, V.A.: Chislennost
bakteriT i produktsiya organicheskogo veshchestva v Rybinskom vodokhranilishche v 1967 g. [Quantity of bacteria and production of organic matter in
the Rybinsk reservoir in 1967.] - Tr. Inst. Biol. vnutr. Vod Akad. Nauk SSSR
21 (24) - Biologiya i Produktivnost' Presnovodnykh Organizmov: 23-30, 295,
1971. [In R.]

*10586 - KUZNETSOV, V.S., CHERYATNIKOVA, T.L.: Gustota steblestoya i fotosintetiches-
kaya produktivnost' raznotipnykh sortov i gibridov kukuruza. [Stand density
and photosynthetic productivity of different cultivars and hybrids of maize.]
- Dokl. timiryaz. sel'skokhoz. Akad. (Moskva) *142*: 61-64, 1968. [In R.]

10587 - KUZNETSOVA, G.K.: Vliyanie kachestva sveta na fotosintez i soderzhanie alka-
loidov u paslena dol'chatogo i krestovnika rombolistnogo. [Effect of light
quality on photosynthesis and alkaloid content in nightshade and groundsel.]
- Fiziol. Rast. *18*: 18-22, 1971. [In R, ab: E.]

10588   KUZNETSOVA, L.G., POLEVAYA, V.S., DOMAN, N.G.: O fiksatsii $C^{14}O_2$ izoliro-
vannymi tkanyami rasteniT. [Fixation of $^{14}CO_2$ by isolated plant tissues.] -
Dokl. Akad. Nauk SSSR *200*: 1463-1465, 1971. [In R.]

10589 - KVÅLE, A.: The effect of different nitrogen levels of the trees on pigment
content, ground colour, and soluble solids of the apple cultivars Prins,
Gravenstein, James Grieve and Ingrid Marie. - Acta Agr. scand. *21*: 207-213,
1971.

10590 - KVĚT, J., MARSHALL, J.K.: Assessment of leaf area and other assimilating
plant surfaces. - In: ŠESTÁK, Z., ČATSKÝ, J., JARVIS, P.G. (ed.): Plant Photosynthetic Production. Manual of Methods. Pp. 517-555. Dr. W. Junk N.V.
Publishers, The Hague 1971.

10591 - KVĚT, J., NEČAS, J., ONDOK, J.P.: Metody růstové analýzy. [Methods of growth
analysis.] - Stud. inform. ÚVTI (Praha), zákl. Vědy Zeměd. *1971* (1): 1-110,
1971. [In Czech, ab: E, R.]

10592 - KVĚT, J., ONDOK, J.P.: The significance of biomass duration. - Photosynthetica *5*: 417-420, 1971.

10593 - KVĚT, J., ONDOK, J.P., NEČAS, J., JARVIS, P.G.: Methods of growth analysis.
- In: ŠESTÁK, Z., ČATSKÝ, J., JARVIS, P.G. (ed.): Plant Photosynthetic Pro-
duction. Manual of Methods. Pp. 343-391. Dr. W. Junk N.V. Publishers, The
Hague 1971.

10594 - KWOK, S.-Y., KAWASHIMA, N., WILDMAN, S.G.: Specific effect of ribulose-1,
5-diphosphate on the solubility of tobacco Fraction I protein. - Biochim.
biophys. Acta *234*: 293-296, 1971.

*10595 - LADYGIN, V.G., SEMENOVA, G.A., TAGEEVA, S.V.: Vzaimosvyaz' ul'trastruktury,
fotosinteticheskoT aktivnosti i soderzhaniya pigmentov u mutantov. I. Pig-
mentnyT sostav i fotosintez mutantov *Chlamydomonas reinhardi*. [Relationships
between ultrastructure, photosynthetic activity, and pigment content in mutants. I. Pigment complex and photosynthesis of mutants *Chlamydomonas reinhardi*.] -In: FRANK, G.M. (ed.): Biofizika ZhivoT Kletki. Vol. 1. Pp. 84-89.
Pushchino 1970. [In R.]

10596 - **LADYGINA, M.E., RUBIN, A.B.**: BiolyuminestsentnyĬ metod kolichestvennogo opredeleniya otdel'nykh komponentov adenilovoĬ sistemy. [Bioluminescent method
of quantitative determination of individual components of the adenylate system.] - In: Biofizicheskie Metody v Fiziologii RasteniĬ. Pp. 72-84. Nauka,
Moskva 1971. [ATP determination; in R.]

10597 - **LAETSCH, W.M.**: Chloroplast structural relationships in leaves of C_4 plants.
- In: HATCH, M.D., OSMOND, C.B., SLATYER, R.O. (ed.): Photosynthesis and Pho-
torespiration. Pp. 323-349. Wiley-Interscience, New York - London - Sydney -
Toronto 1971.

10598 - **LAGOUTTE, B., DURANTON, J.**: Physicochemical study of structural proteins of
chloroplast from *Zea mays* L. - Biochim. biophys. Acta *253*: 232-239, 1971.

*10599 - **LAGUN, L.P., KONANUCHENKO, N.V.**: Vliyanie gustoty posadki razlichnykh sortov
kartofelya na fotosinteticheskuyu deyatel'nost' rasteniĬ. [Effect of plant
density of various potato varieties on photosynthetic activity.] - In: Fiziologo-biokhimicheskie Issledovaniya RasteniĬ. Pp. 30-37. Nauka i Tekhnika,
Minsk 1970. [In R.]

10600 - **LAGUN, L.P., TALANOVA, K.S.**: Fotosinteticheskaya deyatel'nost' rasteniĬ na
razlichnykh agrofonakh. [Photosynthetic activity of plants grown in soils of
varying fertility.] - In: GONCHARIK, M.N. (ed.): PlastidnyĬ Apparat i Zhiznedeyatel'nost' RasteniĬ. Pp.34-41. Nauka i Tekhnika, Minsk 1971. [In R.]

10601 - **LAHDEOJA, M.J.**: Primary production of boreal coniferous forests. - In: DUVIG-
NEAUD, P. (ed.): Productivity of Forest Ecosystems. Pp. 187-188. UNESCO,
Paris 1971.

10602 - **LAING, W.A., FORDE, B.J.**: Comparative photorespiration in *Amaranthus*, soybean
and corn. - Planta *98*: 221-231, 1971.

10603 - **LAÏSK, A.**: Rol' diffuzii pri fotosinteze i transpiratsii lista. [The role of
diffusion in leaf photosynthesis and transpiration.] - In: ZALENSKIĬ, O.V.
(ed.): Fotosintez i Ispol'zovanie SolnechnoĬ Energii. Pp. 97-105. Nauka,
Leningrad 1971. [In R, ab: E.]

10604 - **LAÏSK, A., MOLDAU, Kh., NIL'SON, T., ROSS, Yu., TOOMING, Kh.**: O modelirovanii
produktsionnogo protsessa rastitel'nogo pokrova. [Modelling of the productive
process of the plant cover.] - Bot. Zh. *56*: 761-776, 1971. [In R, ab: E.]

10605 - **LAÏSK, A., OYA, V.**: Izmenenie soprotivleniya mezofilla osiny v otvet na bys-
troe obezvozhivanie lista. [Changes in resistance of aspen mesophyll as a
response to rapid leaf drying.] - Fiziol. Rast. *18*: 553-562, 1971. [Ps; in
R, ab: E.]

*10606 - **LAÏSK, A. Kh., OYA, V.M.**: Trekhkanal'naya gazometricheskaya apparatura dlya
detal'nogo issledovaniya CO_2-obmena lista. [Three-channel gasometric device
for fine study of CO_2 exchange of a leaf.] - In: Tezisy Dokladov Vsesoyuznogo
Soveshchaniya po Unifikatsii Metodov i Priborov dlya Massovykh IzmereniĬ
Intensivnosti Fotosinteza. Pp. 66-69. Nauch.-issled. Inst. Rastenievod. N.I.
Vavilova, Leningrad - Pushkin 1970. [In R.]

*10607 - **LAÏSK, A. Kh., OYA, V.M.**: Apparatura dlya izmereniya energii fotosinteza
lista. [Device for measuring energy of leaf photosynthesis.] - In: Tezisy
Dokladov Vsesoyuznogo Soveshchaniya po Unifikatsii Metodov i Priborov dlya
Massovykh IzmereniĬ Intensivnosti Fotosinteza. Pp. 70-73. Nauch.-issled. Inst.
Rastenievod. N.I. Vavilova, Leningrad - Pushkin 1970. [In R.]

10608 - **LAKE, J.V.**: The behaviour of plants in various gas mixtures. - Proc. roy.
Soc. London, Ser. B *179*: 177-188, 1971. [Ps.]

10609 - **LAKE, J.V.**: Comparison of natural and artificial sources of light. - Proc.
roy. Soc. London, Ser. B *179*: 189-192, 1971. [Ps.]

10610 - **LAMBERTSEN, G., BRAEKKAN, O.R.**: Method of analysis of astaxanthin and its
occurrence in some marine products. - J. Sci. Food Agr. *22*: 99-101, 1971.

10611 - **LAND, E.J., SYKES, A., TRUSCOTT, T.G.**: The *in vitro* photochemistry of biological molecules - II. The triplet states of β-carotene and lycopene excited
by pulse radiolysis. - Photochem. Photobiol. *13*: 311-320, 1971.

10612 - **LAND, J.B., WHITTINGTON, W.J., NORTON, G.**: Environment dependent chlorosis in a mutant plant of *Festuca pratensis*, HUDS. - Ann. Bot. *35*: 605-613, 1971.

10613 - LANDSBERG, J.J., JAMES, G.B.: Wind profiles in plant canopies: studies on
an analytical model. - J. appl. Ecol. *8*: 729-741, 1971. [Growth analysis.]

10614 - LANG, F., VOROB'EVA, L.M., KRASNOVSKIĬ, A.A.: Obrazovanie khlorofilla i formirovanie khloroplastov v zeleneyushchikh list'yakh normal'nykh i mutantnykh
rasteniĭ kukuruzy. [Chlorophyll synthesis and formation of chloroplast in
greening leaves of normal and mutant plants of maize.] - Mol. Biol. (Moskva)
5: 366-374, 1971. [In R, ab: E.]

10615 - LANG, W., POTRYKUS, I.: Lichtmikroskopische und pigmentanalytische Untersuchungen zum Ablauf der Petalenentwicklung von *Torenia baillonii* (*Scrophula-
riac*.). - Z. Pflanzenphysiol. *65*: 1-12, 1971. [Chl, Car.]

10616 - LANGE, O.L., SCHULZE, E.-D.: Measurement of CO_2 gas-exchange and transpira-
tion in the beach (*Fagus silvatica* L.). - In: ELLENBERG, H. (ed.): Integrated
Experimental Ecology. (Ecological Studies. Analysis and Synthesis, Vol. 2.).
Pp. 16-28. Springer-Verlag, Berlin - Heidelberg - New York 1971.

10617 - LAPINA, L.P., POPOV, B.A.: Posledeĭstvie sernokislogo i khloristogo natriya
na funktsional'nuyu aktivnost' i strukturu khloroplastov kukuruzy. [After-
effect of sodium sulfate and sodium chloride on functional activity and structure of maize chloroplasts.] - Fiziol. Rast. *18*: 409-414, 1971. [In R, ab: E.]

10618 - LAPINA, T.V., IVANISHCHEV, V.N.: Podavlenie protsessov fotosinteza patoranom.
[Inhibition of photosynthesis by patoran.] - Fiziol. Biokhim. kul't. Rast.
3: 426-429, 1971. [In R, ab: E.]

*10619 - LAPTEV, V.V.: Fitotrony dlya izuchenixa fotosinteza i dykhaniya posevov.
[Phytotrons for studying stand photosynthesis and respiration.] - In: Tezisy
Dokladov Vsesoyuznogo Soveshchaniya po Unifikatsii Metodov i Priborov dlya
Massovykh Izmereniĭ Intensivnosti Fotosinteza. Pp. 74-76. Nauch.-issled.
Inst. Rastenievod. N.I. Vavilova, Leningrad - Pushkin 1970. [In R.]

10620 - LARSSON, C., COLLIN, C., ALBERTSSON, P.-Å.: Characterization of three classes
of chloroplasts obtained by counter-current distribution. - Biochim. biophys.
Acta *245*: 425-438, 1971.

10621 - LASCELLES, J., WERTLIEB, D.: Mutant strains of *Rhodopseudomonas spheroides*
which form photosynthetic pigments aerobically in the dark. Growth characteristics and enzymic activities. - Biochim. biophys. Acta *226*: 328-340, 1971.

10622 - LATZKO, E., GIBBS, M., LABER, L.J.: Photosynthetic carbon metabolism studied
in leaves of spinach and maize. - In: BRODA, E., LOCKER, A., SPRINGER-LEDE-
RER, H. (ed.): Proceedings of the First European Biophysics Congress. Vol.
4. Pp. 97-107. Verlag Wiener med. Akad., Wien 1971.

10623 - LATZKO, E., LABER, L., GIBBS, M.: Transient changes in levels of some compounds in spinach and maize leaves. - In: HATCH, M.D., OSMOND, C.B., SLATYER,
R.O. (ed.): Photosynthesis and Photorespiration. Pp. 196-201. Wiley-Interscience, New York - London - Sydney - Toronto 1971.

B10624 - LÄUGER, P.: Die Photosynthese der grünen Pflanzen. - Universitätsverlag G.m.
b.H., Constance 1971.

*10625 - LAUSI, D.: Chlorophyll content observations of the near-surface scattering
layers in the Mediterranean. - G. bot. ital. *101*: 304, 1967.

*10626 - LAUSI, D.: Chlorophyll content observations of the near-surface scattering
in the Mediterranean sea. - G. bot. ital. *102*: 521-528, 1968.

10627 - LAVOREL, J.: Fluorescence and luminescence studies of *in vivo* chlorophyll
with a laser phosphoroscope. - Photochem. Photobiol. *14*: 261-275, 1971.

10628 - LAVRENTOVICH, D.I., OKANENKO, A.S., MITROFANOV, B.A., POCHINOK, Kh.N.: Osvo-
enie solnechnoĭ energii v chistykh, smeshannykh i poukosnykh posevakh. [Ab-
sorption of solar energy in pure, mixed and hay making crops.] - In: ZALENS-
KIĬ, O.V. (ed.): Fotosintez i Ispol'zovanie Solnechnoĭ Energii. Pp. 70-75.
Nauka, Leningrad 1971. [Ps; In R, ab: E.]

10629 - LAVRINENKO, D.D.: Izuchenie gazoobmena duba, buka i graba kak indikatora
vzaimodeĭstviya drevesnykh porod v chistykh i smeshannykh kul'turakh. [Gas exchange of the oak, beech, and hornbeam as an indicator of the interaction of trees

in pure and mixed stands.] - Lesoved. Agrolesomelior. *27*: 101-108, 1971. [In R.]

10630 - **LAWES, D.A., TREHARNE, K.J.**: Variation in photosynthetic activity in cereals and its implications in a plant breeding programme. I. Variation in seedling leaves and flag leaves. - Euphytica *20*: 86-92, 1971.

10631 - **LAYCOCK, M.V., CRAIGIE, J.S.**: Purification and characterization of cytochrome 553 from the chrysophycean alga *Monochrysis lutheri*. - Can. J. Biochem. *49*: 641-646, 1971.

10632 - **LEACH, C.K., CARR, N.G.**: Pyruvate:ferredoxin oxidoreductase and its activation by ATP in the blue-green alga *Anabaena variabilis*. - Biochim. biophys. Acta *245*: 165-174, 1971.

10633 - **LEBEDEV, S.I., KLYACHENKO, V.I.**: Vliyanie kal'tsiya i magniya na strukturu i aktivnost' khloroplastov yachmenya. [Effect of calcium and magnesium on structure and activity of barley chloroplasts.] - Fiziol. Biokhim. kul't. Rast. *3*: 487-493, 1971. [In R, ab: E.]

10634 - **LEBEDEV, S.I., KOMARNITSKIĬ, P.A.**: Fiziologo-biokhimicheskie izmeneniya v pochkakh chereshni v ontogeneze i ikh zimostoĭkost'. [Physiological and biochemical changes in mazzard cherry buds during development and their frost resistance.] - Fiziol. Rast. *18*: 184-191, 1971. [Chl; in R, ab: E.]

10635 - **LEBEDEVA, E.V., DMITRIEVA, L.V., AKANOV, E.N.**: Analiz kislorodnoĭ produktivnosti rastitel'nykh konveĭerov pri razlichnoĭ velichine shaga. [Oxygen production analysis by plant conveyers with different steps.] - Kosmich. Biol. Med. *5* (1): 25-29, 1971. [Ps; in R, ab: E.]

10636 - **LECLERC, J.-C.**: Influence de chocs osmotiques sur l'efficacité photosynthétique de lumières monochromatiques chez *Porphyridium* Sp (LEWIN). - Compt. rend. Acad. Sci. Paris. Şér. D *272*: 231-233, 1971.

10637 - **LEDEBOER, F.B., SKOGLEY, C.R., McKIEL, C.G.**: Soil heating studies with cool season turfgrasses. II. Effects of N fertilization and protective covers on performance and chlorophyll content. - Agron. J. *63*: 680-685, 1971.

10638 - **LEE, C.C., HARRIS, R.F., SYERS, J.K., ARMSTRONG, D.E.**: Adenosine triphosphate content of *Selenastrum capricornutum*. - Appl. Microbiol. *21*: 957-958, 1971.

10639 - **LEE, J.A., STEWART, G.R.**: Desiccation injury in mosses. I. Intra-specific differences in the effect of moisture stress on photosynthesis. - New Phytol. *70*: 1061-1068, 1971.

10640 - **LEE, K.-C.**: Some effects of water stress and ALAR tratment on water relations and photosynthesis in *Pisum sativum* L. seedlings. - Diss. Abstr. Int. B *32*: 2034-B, 1971.

10641 - **LEES, D.N., MAPP, B.J., REDMOND, A.M., ROGERS, L.J.**: Physiocochemical studies on chloroplast thylakoid protein from *Phaseolus vulgaris*. - Biochem. J. *121*: 6 P, 1971.

*10642 - **LEGLER, C.**: Beitrag zur Bestimmung des im Wasser gelösten Sauerstoffs. - Limnologica (Berlin) *4*: 291-301, 1966.

10643 - **LEHMUSLUOTO, P.O.**: Kasviplanktonin perustuotanto Itämeren alueella. [Phytoplankton primary production in the Baltic area.] - Luonnon Tutkija *1971*: 86-91, 1971. [In Fin., ab: E.]

10644 - **LEMASSON, C., BARBIERI, G.**: Effet de la longueur d'onde de préillumination sur la désactivation des formes oxydées du donneur d'électrons du Photosystème II. - Biochim. biophys. Acta *245*: 386-397, 1971.

10645 - **LEMOINE, B.**: Le pin maritime (*Pinus pinaster*) dans les Landes de Gascogne. - In: DUVIGNEAUD, P. (ed.): Productivity of Forest Ecosystems. Pp. 207-211. Unesco, Paris 1971. [Growth analysis.]

10646 - **LEMON, E., STEWART, D.W., SHAWCROFT, R.W.**: The sun's work in a cornfield. - Science *174*: 371-378, 1971. [Ps.]

*10647 - **LEPAJÕE, J., SAAR, A.**: Kaherealise odra sortide bioloogilisi ja tehnoloogilisi omadusi. [Biological and technological characteristics of cultivars of distichons barley.] - Izv. Akad. Nauk eston. SSR, Biol. *18*: 300-304, 1969. [Chl; in Eston., ab: G, R.]

10648 - LEPLEY, C.R.: Studies on the isolation and separation of bundle sheath and mesophyll cell chloroplasts and on carbon-dioxide uptake in maize leaves. - Diss. Abstr. int. B *32*: 37-B, 1971.

10649 - LEUSCHNER, D., BÜTTNER, R.: Druckänderungen am Infrarotabsorptionsanalyse-gerät (IRGA) bei pflanzenphysiologischen Messungen. - Arch. Gartenbau *19*: 379-385, 1971.

10650 - LEVINE, R.P.: Preparation and properties of mutant strains of *Chlamydomonas reinhardi*. - In: COLOWICK, S.P., KAPLAN, N.O. (ed.): Methods in Enzymology. Vol. 23. Pp. 119-129. Academic Press, New York - London 1971. [Ps.]

10651 - LEWANTY, Z., MALESZEWSKI, S., POSKUTA, J.: The effect of oxygen concentration on ^{14}C incorporation into products of photosynthesis of detached leaves of maize. - Z. Pflanzenphysiol. *65*: 469-472, 1971.

10652 - LEWIS, D.H., SMITH, D.C.: The autotrophic nutrition of symbiotic marine co-elenterates with special reference to hermatypic corals. I. Movement of photosynthetic products between the symbionts. - Proc. roy. Soc. Ser. B *178*: 111-129, 1971.

10653 - LI, Y.-S.: Calcium chloride and chloroplast fluorescence. - Diss. Abstr. int. B *32*: 3194-B, 1971.

10654 - LIAAEN-JENSEN, S.: Recent progress in carotenoid chemistry. - In: GOODWIN, T.W. (ed.): Aspects of Terpenoid Chemistry and Biochemistry. Pp. 233-254. Academic Press, London - New York 1971.

10655 - LIAAEN-JENSEN, S.: Isolation, reactions. - In: ISLER, O. (ed.): Carotenoids. Pp. 61-188. Birkhäuser Verlag, Basel - Stuttgart 1971. [Car.]

10656 - LIAAEN-JENSEN, S., JENSEN, A.: Quantitative determination of carotenoids in photosynthetic tissues. - In: COLOWICK, S.P., KAPLAN, N.O. (ed.): Methods in Enzymology. Vol. 23. Pp. 586-602. Academic Press, New York - London 1971.

10657 - LICHTENTHALER, H.K.: Die unterschiedliche Synthese der lipophilen Plastiden-chinone in Sonnen- und Schattenblättern von *Fagus silvatica* L. - Z. Natur-forsch. *26 b*: 832-842, 1971.

10658 - LICHTENTHALER, H.K., VERBEEK, L., BECKER, K.: Promotion of vitamin K_1-synthesis by naphthoquinones. - Phytochemistry *10*: 79-84, 1971. [Car.]

10659 - LIEB, J.A.: Biochemical function of *Euglena mutabilis* in acid mine drainage. - Diss. Abstr. int. B *32*: 1993-B, 1971. [Ps.]

10660 - LIEN, S., BANNISTER, T.T.: Multiple sites of DCIP reduction by sonicated oat chloroplasts: Role of plastocyanin. - Biochim. biophys. Acta *245*: 465-481, 1971.

10661 - LIEN, S., RACKER, E.: Partial resolution of the enzymes catalyzing photo-phosphorylation. VIII. Properties of silicotungstate-treated subchloroplast particles. - J. biol. Chem. *246*: 4298-4307, 1971.

10662 - LIEN, S., RACKER, E.: Preparation and assay of chloroplast coupling factor CF_1. - In: COLOWICK, S.P., KAPLAN, N.O. (ed.): Methods in Enzymology. Vol. 23. Pp. 547-555. Academic Press, New York - London 1971.

10663 - LIEN, S., SAN PIETRO, A., GEST, H.: Mutational and physiological enhancement of photosynthetic energy conversion in *Rhodopseudomonas capsulata*. - Proc. nat. Acad. Sci. USA *68*: 1912-1915, 1971.

10664 - LIETH, H.: The phenological viewpoint in productivity studies. - In: DUVIG-NEAUD, P. (ed.): Productivity of Forest Ecosystems. Pp. 71-84. Unesco, Paris 1971. [Growth analysis.]

10665 - LIETH, H.: Mathematical modelling for ecosystems analysis. - In: DUVIGNEAUD, P. (ed.): Productivity of Forest Ecosystems. Pp. 567-575. Unesco, Paris 1971.

19666 - LILJENBERG, C.: Phytol changes in dark grown leaves of barley after varying light treatments. - Physiol. Plant. *25*: 358-362, 1971.

10667 - LILLEY, R.McC., HOPE, A.B.: Chloride transport and photosynthesis in cells of *Griffithsia*. - Biochim. biophys. Acta *226*: 161-171, 1971.

10668 - LIN, D.C., NOBEL, P.S.: Control of photophosphorylation by Mg^{++}. - Plant Physiol. *47* (Suppl.): 9, 1971.

10669 - LIN, D.C., NOBEL, P.S.: Control of photosynthesis by Mg^{2+}. - Arch. Biochem. Biophys. *145*: 622-632, 1971.

10670 - LINDER, S.: Photosynthetic action spectra of Scots pine needles of different ages from seedlings grown under different nursery conditions. - Physiol. Plant. *25*: 58-63, 1971.

10671 - LIPSKAYA, G.A.: Ab spetsyfitsy dzeyannya adnol'kavaĭ kantsentratsyi kobal'tu na fotasintetychny aparat roznykh raslin. [Characteristics of the effect of the same concentration of cobalt on the photosynthetic apparatus of different plants.] - Vestsy Akad. Navuk belarus. SSR, Ser. biyal. Navuk *1971* (1): 14-20, 133-134, 1971. [In Belorus., ab: R.]

10672 - LITVIN, F.F., BELYAEVA, O.B.: Kharakteristika otdel'nykh reaktsiĭ i obshch-aya skhema biosinteza nativnykh form khlorofilla v etiolirovannykh list'yakh rasteniĭ. [Characteristics of individual reactions and a general scheme of the biosynthesis of native forms of chlorophyll in etiolated leaves.] - Bio-khimiya *36*: 615-622, 1971. [In R, ab: E.]

10673 - LITVIN, F.F., BELYAEVA, O.B.: Sequence of photochemical and dark reactions in the terminal stage of chlorophyll biosynthesis. - Photosynthetica *5*: 200-209, 1971.

10674 - LITVIN, F.F., GULYAEV, B.A., SINESHCHEKOV, V.A.: Spektral'nye kharakteristiki, otnositel'naya kontsentratsiya i koeffitsienty migratsii energii 10 nativ-nykh form khlorofilla *a*. [Spectral characteristics, relative concentration, and coefficients of energy migration of ten native forms of chlorophyll *a*.] - Dokl. Akad. Nauk SSSR *199*: 1428-1431, 1971. [In R.]

10675 - LITVIN, F.F., ZVALINSKIĬ, V.I.: Poluprovodimost' fotosinteticheskikh struktur i ee svyaz' s fotosintezom. [Semiconductance of photosynthetic structures and its connection with photosynthesis.] - Biofizika *16*: 420-430, 1971. [In R, ab: E.]

10676 - LIU, P.S.-K.: The influence of translocation on photosynthetic efficiency of *Phaseolus vulgaris* L. - Diss. Abstr. Int. B *32*: 118-B, 1971.

10677 - LIVINGSTON, A.L., KNOWLES, R.E., PAGE, J., KOHLER, G.O.: Turf grass dehydra-tion. - J. agr. Food Chem. *19*: 951-953, 1971. [Car.]

10678 - LIVNE, A.: Water stress and plant responses. - Israel J. Bot. *20*: 333-334, 1971. [Ps.]

*10679 - LOACH, K.: Shade tolerance in tree seedlings. II. Growth analysis of plants raised under artificial shade. - New Phytol. *69*: 273-286, 1970.

10680 - LOACH, P.A., BAMBARA, R.A., RYAN, F.J.: Identification of the major ultravio-let absorbance photochanges in photosynthetic systems. - Photochem. Photo-biol. *13*: 247-257, 1971.

10681 - LOCKSHIN, A., FALK, R.H., BOGORAD, L., WOODCOCK, C.L.F.: A coupling factor for photosynthetic phosphorylation from plastids of light- and dark-grown maize. - Biochim. biophys. Acta *226*: 366-382, 1971.

10682 - LOFTUS, M.E., CARPENTER, J.H.: A fluorometric method for determining chloro-phylls *a*, *b*, and *c*. - J. mar. Res. *29*: 319-338, 1971.

10683 - LOGAN, K.T.: Monthly variations in photosynthetic rate of jack pine provenan-ces in relation to their height. - Can. J. Forest Res. *1*: 256-261, 1971.

10684 - LOGAN, K.T., POLLARD, D.F.W.: Comparative investigations of dry matter pro-duction in young tree seedlings using classical growth analysis and gas ana-lysis techniques. - In: Petawawa Forest Experiment Station, Chalk River, Ontario. Information Report PS-X-25. 19pp. 1971.

10685 - LOGINOV, M.A.: O fotosinteze list'ev raznykh yarusov posevov sorgo. [Photo-synthesis of leaves in different strata of sorghum stands.] - In: ZALENSKIĬ, O.V. (ed.): Fotosintez i Ispol'zovanie Solnechnoĭ Energii. Pp. 80-82. Nauka, Leningrad 1971. [In R, ab: E.]

*10686 - LOGINOV, M.A., AKHMEDOV, A.Ya.: O temperaturnoĭ adaptatsii fotosinteza ras-teniĭ. [Temperature adaptation of plant photosynthesis.] - Dokl. Akad. Nauk tadzh. SSR *13* (6): 53-56, 1970. [In R, ab: Tadzh.]

*10687 - LOGINOV, M.A., NASYROV, Yu.S.: Ekologo-fiziologicheskiĭ analiz fotosinteza vysotnozameshchayushchikh vidov *Astragalus* zapadnogo Pamiro-Alaya. [Eco-physiological analysis of photosynthesis of *Astragalus* species in high altitudes of the Western Pamiro-Alay.] - Bot. Zh. *55*: 1171-1176, 1970. [In R.]

10688 - LOMMEN, P.W., SCHWINTZER, C.R., YOCUM, C.S., GATES, D.M.: A model describing photosynthesis in terms of gas diffusion and enzyme kinetics. - Planta *98*: 195-220, 1971.

10689 - LONG, D.W.: Metabolism of photosynthetically ^{14}C labeled sugars in developing soybean seeds. - Diss. Abstr. int. B *32*: 2035-B, 1971.

10690 - LING, E.B., COOKE, G.D.: A quantitative comparison of pigment extraction by membrane and glass-filters. - Limnol. Oceanogr. *16*: 990-992, 1971.

10691 - LOOMIS, R.S.: Is soluble protein a better physiological base than chlorophyll? - Acta agron. Acad. Sci. hung. *20*: 448-449, 1971.

10692 - LOOMIS, R.S., WILLIAMS, W.A., HALL, A.E.: Agricultural productivity. - Annu. Rev. Plant Physiol. *22*: 431-468, 1971. [Ps.]

*10693 - LORD, J.M., CODD, G.A., MERRETT, M.J.: The effect of light quality on glycolate formation and excretion in algae. - Plant Physiol. *46*: 855-856, 1970.

10694 - LORD, J.M., MERRETT, M.J.: The intracellular localization of glycollate oxidoreductase in *Euglena gracilis*. - Biochem. J. *124*: 275-281, 1971.

10695 - LORD, J.M., MERRETT, M.J.: The growth of *Chlorella pyrenoidosa* on glycollate. - J. exp. Bot. *22*: 60-69, 1971. [Ps.]

*10696 - LORENZEN, H., WEISBRICH, J.: Zur Wirkung eines MORPHAKTINS auf *Rhoeo spathacea* und *Cardamine chenopodifolia*. - Vortr. Gesamtgeb. Biol. deut. bot. Ges. *3*: 103-115, 1969. [Chl.]

10697 - LOSEV, A.P., SUBOCH, V.P., GURINOVICH, G.P.: Stroenie khlorofilla *a*, poluchaemogo fotokhimicheskim gidrirovaniem protokhlorofilla. [Structure of chlorophyll *a* produced by photochemical hydration of protochlorophyll.] - Dokl. Akad. Nauk belorus. SSR *15*: 550-553, 1971. [In R.]

10698 - LOSEVA, N.L.: Vliyanie intensivnosti sveta na skorost' fotosinteza vodorosli khlorelly. [The effect of light intensity on the rate of photosynthesis in the alga *Chlorella*.] - In: Sbornik Aspirantskikh Rabot, Kazan'. Univ. Estestv. Nauki - Biol. Pp. 125-128. Kazan! 1971. [In R.]

10699 - LOSSAINT, P., RAPP, M.: Le cycle du carbone dans les forêts de *Pinus halepensis*. - In: DUVIGNEAUD, P. (ed.): Productivity of Forest Ecosystems. Pp. 213-216. Unesco, Paris 1971. [Growth analysis.]

10700 - LOURENCE, F.J., PRUITT, W.O.: Energy balance and water use of rice grown in the Central Valley of California. - Agron. J. *63*: 827-832, 1971.

10701 - LOVELL, P., MOORE, K.: A comparative study of the role of the cotyledon in seedling development. - J. exp. Bot. *22*: 153-162, 1971. [Ps.]

*10702 - LOVELL, P.H.: Leaf expansion in the potato, *Solanum tuberosum*. - Physiol. Plant. *21*: 626-634, 1968. [Leaf area determination.]

10703 - LOVELL, P.H.: Translocation of photosynthates in tall and dwarf varieties of pea, *Pisum sativum*. - Physiol. Plant. *25*: 382-385, 1971.

10704 - LOWE, J., SLACK, C.R.: Inhibition of maize leaf phosphopyruvate carboxylase by oxaloacetate. - Biochim. biophys. Acta *235*: 207-209, 1971.

10705 - LOZIER, R., BAGINSKY, M., BUTLER, W.L.: Inhibition of electron transport in chloroplasts by chaotropic agents and the use of manganese as an electron donor to photosystem II.- Photochem. Photobiol. *14*: 323-328, 1971.

10706 - LOZOVA, G.I., UDALOVA, V.I.: Osoblyvosti stanu khlorofilu v pigmentvmishchuyuchykh kompleksakh z riznykh fotosystem. [Features of chlorophyll state in pigment complexes from different photosystems.] - Ukr. bot. Zh. *28*: 223-226, 1971. [In Ukr.]

*10707 - LUCATU, E., VASILACHE, M.: Quenching and polarization of the fluorescence in solutions of chlorophyll "*a*" containing quenchers. - J. Luminescence *3*: 132-136, 1970.

10708 - LUDLOW, M.M.: Analysis of the difference between maximum leaf net photosynthetic rates of C_4 grasses and C_3 legumes. - In: HATCH, M.D., OSMOND, C.B., SLATYER, R.O. (ed.): Photosynthesis and Photorespiration. Pp. 63-67. Wiley-Interscience, New York - London - Sydney - Toronto 1971.

10709 - LUDLOW, M.M., JARVIS, P.G.: Methods for measuring photorespiration in leaves. - In: ŠESTÁK, Z., ČATSKÝ, J., JARVIS, P.G. (ed.): Plant Photosynthetic Production. Manual of Methods. Pp. 294-315. Dr. W. Junk N.V. Publ., The Hague 1971.

10710 - LUDLOW, M.M., JARVIS, P.G.: Photosynthesis in Sitka spruce (*Picea sitchensis* (BONG.) CARR.). I. General characteristics. - J. appl. Ecol. *8*: 925-953, 1971.

10711 - LUDLOW, M.M., WILSON, G.L.: Photosynthesis of tropical pasture plants. I. Illuminance, carbon dioxide concentration, leaf temperature, and leaf-air vapour pressure difference. - Aust. J. biol. Sci. *24*: 449-470, 1971.

10712 - LUDLOW, M.M., WILSON, G.L.: Photosynthesis of tropical pasture plants. II. Temperature and illuminance history. - Aust. J. biol. Sci. *24*: 1065-1075, 1971.

10713 - LUDLOW, M.M., WILSON, G.L.: Photosynthesis of tropical pasture plants. III. Leaf age. - Aust. J. biol. Sci. *24*: 1077-1087, 1971.

10714 - LUDWIG, L.J., CANVIN, D.T.: An open gas-exchange system for the simultaneous measurement of the CO_2 and $^{14}CO_2$ fluxes from leaves. - Can. J. Bot. *49*: 1299-1313, 1971.

10715 - LUDWIG, L.J., CANVIN, D.T.: The rate of photorespiration during photosynthesis and the relationship of the substrate of light respiration to the products of photosynthesis in sunflower leaves. - Plant Physiol. *48*: 712-719, 1971.

10716 - LUE-KIM, H.: Hydrostatic pressure in relation to the synchronous culture of algae in open and closed systems. - Int. Rev. ges. Hydrobiol. *56*: 873-921, 1971. [Ps.]

10717 - LUK'YANENKO, I.A.: Snyatie perioda pokoya u svezheubrannykh klubneĭ kartofelya s pomoshch'yu fiziologicheski-aktivnykh organicheskikh veshchestv. [Dormancy interruption in field-fresh potato tubers by means of physiologically active organic compounds.] - In: NRV v Sel'skom KhozyaĭTstve. Pp. 473-478. ELM, Baku 1971. [Chl; in R.]

*10718 - LUK'YANYUK, V.I., GOLOVKOV, A.M.: Formirovanie urozhaya ozimoĭ pshenitsy v zavisimosti ot agropriemov. [Formation of winter wheat yields in dependence on agrotechnics.] - Dokl. TSKHA (Moskva) *147*: 5-11, 1969. [Net assimilation rate; in R.]

10719 - LÜRSSEN, K.: Zur Lokalisierung der Proteinsynthese bei der Entwicklung des Lamellarsystems ergrünender Etioplasten. - Z. Naturforsch. *26b*: 725-729, 1971.

10720 - LUTSENKO, G.N., SAAKOV, V.S.: Obnovlenie i kinetika vklyucheniya C^{14} v molekuly karotinoidov. [Restoration and kinetics of ^{14}C incorporation into carotenoid molecules.] - In: Biokhimiya i Biofizika Fotosinteza. Pp. 80-86. Irkutsk 1971. [in R.]

10721 - LUTSISHINA, E.G., GRODZINSKIĬ, D.M.: O primenimosti printsipov popadaniya i misheni k izucheniyu fotosinteticheski aktivnykh edinits khloroplastov. [On applicability of hit and target principles for studying photosynthetically active chloroplast units.] - Fiziol. Biokhim. kul't. Rast. *3*: 608-613, 1971. [in R, ab: E.]

10722 - LÜTTGE, U.: A comparative study of the coupling of ion uptake to light reactions in the leaves of C_3 and C_4 plants. - In: HATCH, M.D., OSMOND, C.B., SLATYER, R.O. (ed.): Photosynthesis and Photorespiration. Pp. 394-399. Wiley-Interscience, New York - London - Sydney - Toronto 1971.

10723 - LÜTTGE, U., BALL, E.: Light-independent uncoupler-sensitive ion uptake by green and by pale cells of variegated leaves of higher plants in relation to protein content and chloroplast integrity. - Z. Naturforsch. *26 b*: 158-161, 1971.

10724 - LÜTTGE, U., BALL, E., von WILLERT, K.: Gas exchange and ATP levels of green cells of leaves of higher plants as affected by FCCP and DCMU in *in vitro* experiments. - Z. Pflanzenphysiol. *65*: 326-335, 1971.

10725 - LÜTTGE, U., BALL, E., von WILLERT, K.: A comparative study of the coupling of ion uptake to light reactions in leaves of higher plant species having the C_3- and C_4-pathway of photosynthesis. - Z. Pflanzenphysiol. *65*: 336-350, 1971.

10726 - LUUKKANEN, O., RÄSÄNEN, P.K., YLI-VAKKURI, P.: Neulasten väri myöhemmän kasvun ja lannoitusvaikutuksen ilmaisijana. [The use of needle colour in predicting growth and response to fertilization.] - Silva fenn. *5*: 297-313, 1971. [Ps, Chl; in Finn., ab: E.]

10727 - LUUKANAN, O., RÄSÄNEN, P.K., YLI-VAKKURI, P.: The use of needle color in predicting growth and response to fertilization. - In: IUFRO Congress Working Group 6, Section 21: Tree Physiology as Related to Site Factors. Pp. 62-71. Gainesville, Fla. 1971. [Chl, growth analysis.]

10728 - LUXMOORE, R.J., MILLINGTON, R.J., ASTON, A.R.: Modified tube solarimeters for additive and net measurements of visible, infrared and solar radiation. - Agron. J. *63*: 329-331, 1971.

10729 - LUXMOORE, R.J., MILLINGTON, R.J., MARCELLOS, H.: Soybean canopy structure and some radiant energy relations. - Agron. J. *63*: 111-114, 1971. [Growth analysis.]

10730 - LYAKHNOVICH, Ya.P.: Vliyanie polnogo zatemneniya na pigmentnyi apparat nekotorykh odnokletochnykh zelenykh vodoroslei. [Influence of complete darkening on pigment apparatus of some unicellular green algae.] - In: Problemy Biosinteza Khlorofillov. Pp. 227-246. Nauka i Tekhnika, Minsk 1971. [In R.]

*10731 - LYAKHNOVICH, Ya.P., GODNEV, T.N.: O vliyanii vysokoi temperatury na sostoyanie pigmentnykh kompleksov u nekotorykh mutantov khlorelly. [Effect of high temperature on the state of chloroplast pigments in some maize mutants.] - In: Fiziologo-biokhimicheskie issledovaniya Rastenii. Pp. 6-11. Nauka i Tekhnika, Minsk 1970. [In R.]

10732 - LYAKHNOVICH, Ya.P., GODNEV, T.N.: O pigmentnom sostave nekotorykh mutantov khlorelly. [Pigment composition of some *Chlorella* mutants.] - In: Plastidnyi Apparat i Zhiznedeyatel'nost'Rastenii. Pp. 3-10. Nauka i Tekhnika, Minsk 1971. [In R.]

*10733 - LYASHCHENKO, I.F., ULITCHEVA, I.I.: Vliyanie γ-luchei i khimicheskikh mutagenov na chastotu reversii khlorofil'nykh mutantov podsolnechnika. [Effect of γ-rays and chemical mutagens on reversion frequency of chlorophyll mutants in sunflower.] - Tsitol. Genet. *4*: 553-559, 1970. [In R, ab: E.]

10734 - LYTTLETON, J.W.: Studies of chloroplasts from *Amaranthus* and maize after isolation in aqueous systems. - In: HATCH, M.D., OSMOND, C.B., SLATYER, R.O. (ed.): Photosynthesis and Photorespiration. Pp. 232-239. Wiley-Interscience, New York - London - Sydney - Toronto 1971.

10735 - LYTTLETON, J.W., BALLANTINE, J.E.M., FORDE, B.J.: Development and environment studies on chloroplasts of *Amaranthus lividus*. - In: BOARDMAN, N.K., LINNANE, A.W., SMILLIE, R.M. (ed.): Autonomy and Biogenesis of Mitochondria and Chloroplasts. Pp. 447-452. North-Holland Publ. Comp., Amsterdam - London 1971.

*10736 - LYUBIMOVA, E.E.: Fotosinteticheskaya deyatel'nost' pastbishchnykh trav v zavisimosti ot urovnya mineral'nogo pitaniya. [Photosynthetic activity of pasture grasses in dependence on level of mineral nutrients.] - Dokl. TSKHA (Moskva) *147*: 257-261, 1969. [In R.]

10737 - LYUTOVA, M.I., KISLYUK, I.M.: Posledeistvie nagrevaniya i okhlazhdeniya list'ev na fotosintez i fotokhimicheskie reaktsii. [Aftereffect of heating and cooling of leaves on photosynthesis and photochemical reactions.] In: ZALENSKII, O.V. (ed.): Fotosintez i Ispol'zovanie Solnechnoi Energii. Pp. 193-199. Nauka, Leningrad 1971. [In R, ab: E.]

10738 - MacCOLL, R., BERNS, D.S., KOVEN, N.L.: Effect of salts on *C*-phycocyanin. - Arch. Biochem. Biophys. *146*: 477-482, 1971.

10739 - MacCOLL, R., LEE, J.J., BERNS, D.S.: Protein aggregation in *C*-phycocyanin.

Studies at very low concentrations with the photoelectric scanner of the ul-
tracentrifuge. - Biochem. J. *122*: 421-426, 1971.

10740 - **MACHOLD, O.**: Die Wirkung von Chloramphenicol und Streptomycin auf Lamellar-
proteine der Chloroplasten von *Vicia faba*. - Exp. Cell Res. *65*: 466-467, 1971.

10741 - **MACHOLD, O.** Lamellar proteins of green and chlorotic chloroplasts as affected
by iron deficiency and antibiotics. - Biochim. biophys. Acta *238*: 324-331,
1971.

10742 - **MACHOLD, O., MEISTER, A., ADLER, K.**: Spektroskopische Eigenschaften von elek-
trophoretisch getrennten Chlorophyll-Protein-Komplexen der Photosysteme I
und II aus *Vicia faba* und *Chlorella pyrenoidosa*. - Photosynthetica *5*: 160-
165, 1971.

10743 - **MacRAE, J.C.**: Quantitative measurement of starch in very small amounts of
leaf tissue. - Planta *96*: 101-108, 1971.

10744 - **MADSEN, E.**: Cytological changes due to the effect of carbon dioxide concentration on the accumulation of starch in chloroplasts of tomato leaves. -
Roy. vet. agr. Univ. Copenhagen Yearbook (Kong. Vet.-Landbohøjsk. Årsskr.)
1971: 191-194, 1971.

10745 - **MADSEN, E.**: The effect of carbon dioxide concentration on the photosynthetic
rate in tomato leaves. - Roy. vet. agr. Univ. Copenhagen Yearbook (Kong. Vet.-
Landbohøjsk. Årsskr.) *1971*: 195-200, 1971.

*10746 - **MADZHAROVA, D., BENBASAM, E.**: Nyakoi biologicheski osobenosti na repichkite
pri parnikovo proizvodstvo. [Some biological peculiarities of radishes grown
in hotbeds.] - Gradinar. lozar. Nauka (Sofia) *4* (2): 75-88, 1967. [Chl; in
Bulg., ab: E, R.]

10747 - **MAGALHÃES, A.C., MONTOJOS, J.C.**: Effect of solar radiation on the growth
parameters and yield of two varieties of common beans (*Phaseolus vulgaris* L.).
- Turrialba *21*: 165-168, 1971. [Growth analysis.]

10748 - **MAGALHÃES, A.C., MONTOJOS, J.C., MIYASAKA, S.**: Effect of dry organic matter
on growth and yield of beans (*Phaseolus vulgaris* L.). - Exp. Agr. *7*: 137-
143, 1971. [Growth analysis.]

10749 - **MAGNE, F.**: Sur la présence vraisemblable de polysaccharides dans les globules osmiophiles des plastes. - Compt. rend. Acad. Sci. Paris, Sér. D *273*:
340-343, 1971. [Chloroplast.]

10750 - **MAHRINGER, W., MOTSCHKA, O.**: Verbesserungen an Sternpyranometern und Strah-
lungsbilanzmessern. - Arch. Meteorol. Geophys. Bioklimatol., Ser. B *19*: 149-
156, 1971.

10751 - **MAKARIEV, Z.**: Metod za opredelyane obema na krushovi plodove, bez da se otk"-
svat ot plodnite klonki. [Method for determining the volume of pear fruits
without picking them from the fruit branchlets.] - Gradinar. lozar. Nauka
(Sofia) *8* (2): 25-29, 1971. [In Bulg., ab: E, R.]

10752 - **MAKARLA, U.K., MAKHMADBEKOVA, L.M., NASYROV, Yu.S.**: Transportnye formy i metabolizm C^{14}-produktov fotosinteza v chereshkakh khlopchatnika. [Transport forms
and metabolism of ^{14}C-photosynthates in cotton petioles.] - Dokl. Akad. Nauk
tadzh. SSR *14* (7): 64-67, 1971. [In R, ab: Tadzh.]

10753 - **MAKARLA, U.K., PINKHASOV, Yu.I., MAKHMADBEKOVA, L.M.**: Transport assimilyatov
iz lista v plodovye organy khlopchatnika. [Translocation of photosynthates
from a leaf into the fruit producing organs of cotton.] - Dokl. Akad. Nauk
tadzh. SSR *14* (5): 58-61, 1971. [In R, ab: Tadzh.]

10754 - **MAKAROV, A.D., KARTASHOV, I.M.**: Nekotorye osobennosti okisleniya i vosstanovleniya NADF khloroplastami. [NADP oxidation and reduction by chloroplasts.]
- Biofizika *16*: 467-471, 1971. [In R, ab: E.]

10755 - **MAKAROV, A.D., MAL'YAN, A.N., OPANASENKO, V.K.**: Vzaimodeĭstvie adenozinfos-
fatov s khloroplastami v reaktsii fotofosforilirovaniya. [Interaction of adenosine phosphates with chloroplasts in the course of photophosphorylation.]
- Biofizika *16*: 1125-1128, 1971. [In R, ab: E.]

10756 - **MAKHARINETS, S.N.**: Fotoustoĭchivost' pigmentov i fotokhimicheskaya aktivnost' khloroplastov pshenitsy. [Photoresistance of pigments and photochemical activity of wheat chloroplasts.] - Tr. kishinev. sel'.-khoz. Inst. *85*: 36-40, 77, 1971. [In R.]

10757 - MAKHMADBEKOVA, L.M.: Deĭstvie kinetina na peredvizhenie assimilyatov-C^{14} u pigmentnykh mutantov gorokha. [Action of kinetin on the translocation of ^{14}C photosynthates in pigment mutants of pea.] - In: NASYROV, Yu.S. (ed.): Geneticheskie Aspekty Fotosinteza. Pp. 207-214. Donish, Dushanbe 1971. [In R, ab: E.]

10758 - MAKHMADBEKOVA, L.M., PINKHASOV, Yu.I., BATALOVA, A.G., NASYROV, Yu. S.: Raspredelenie pogloshchennogo pri fotosinteze ugleroda C^{14} u khlopchatnika v ontogeneze. [Ontogenetic changes in the distribution of carbon-^{14}C absorbed in photosynthesis by cotton leaves.] - In: ZALENSKIĬ, O.V. (ed.): Fotosintez i Ispol'zovanie Solnechnoĭ Energii. Pp. 114-116. Nauka, Leningrad 1971. [In R, ab: E.]

10759 - MAKSIMOVA, E.V., NEUPOKOEVA, N.K.: Vliyanie regulyatorov rosta na urozhaĭ semyan sal'vii nizkoĭ i astry kitaĭskoĭ. [Effect of growth regulators on the yield of *Salvia splendens* and *Callistephus chinensis* seeds.] - Khim. sel'. Khoz. *9*: 605-606, 1971. [Ps; in R.]

*10760 - MALCHEV, E.: Izmeneniya na khlorofilite pri sterilizatsiya na zelen fasul. I. s"obshchenie. [Changes in chlorophyll during sterilization of green beans. I. communication.] - Nauch. Tr. Vissh. Inst. Khranit. Vkusova Prom. (Plovdiv) *17* (Pt. 2): 15-18, 1970. [In Bulg., ab: E, R.]

10761 - MALKIN, R.: Photoreduction of NADP by photosystem II chloroplast fragments lacking P700 and cytochrome *f*. - Plant Physiol. *47* (Suppl.): 33, 1971.

10762 - MALKIN, R.: Photosystem II activity of chloroplast fragments lacking P_{700}. - Biochim. biophys. Acta *253*: 421-427, 1971.

10763 - MALKIN, R., BEARDEN, A.J.: Primary reactions of photosynthesis: Photoreduction of a bound chloroplast ferredoxin at low termperature as detected by EPR spectroscopy. - Proc. nat. Acad. Sci. USA *68*: 16-19, 1971.

10764 - MALKIN, S.: Fluorescence induction studies in isolated chloroplasts. III. On the electron-transport equilibria in the pool of electron acceptors of Photosystem II. - Biochim. biophys. Acta *234*: 415-427, 1971.

10765 - MALLERY, C.H., RICHARDSON, N.: Disc gel electrophoresis of biliproteins. - Plant Cell Physiol. *12*: 997-1001, 1971.

10766 - MALOFEEV, V.M.: Odnovremennaya registratsiya intensivnosti fluorestsentsii i fotosinteticheskoĭ aktivnosti lista v statsionarnom sostoyanii. [Simultaneous recording of fluorescence intensity and photosynthetic activity of a leaf under steady-state conditions.] - In: Biofizicheskie Metody v Fiziologii Rastenĭi. Pp. 85-105. Nauka, Moskva 1971. [In R.]

10767 - MALOFEEV, V.M., SHATILOV, I.S., ABISALOV, R.S., VAULIN, A.V.: Fotosinteticheskaya aktivnost' vskhodov ozimoĭ pshenitsy v osenniĭ period. [Photosynthetic activity of winter wheat shoots in autumn.] - Dokl. mosk. sel'.-khoz. Akad. K.A. Timiryazeva *162*: 226-231, 1971. [In R.]

10768 - MALOFEEV, V.M., SHATILOV, I.S., ABISALOV, R.S., VAULIN, A.V.: Dinamika fotosinteza i dykhaniya ovsa (polevye issledovaniya). [Dynamics of photosynthesis and respiration of oats (field studies).] - Dokl. mosk. sel'.-khoz. Akad. K.A. Timiryazeva *162*: 232-236, 1971. [In R.]

10769 - MALONE, T.C.: The relative importance of nannoplankton and netplankton as primary producers in tropical oceanic and neritic phytoplankton communities. - Limnol. Oceanogr. *16*: 633-639, 1971. [Chl.]

*10770 - MAMAEV, A.S., NIKOLAEVSKIĬ, V.S.: Nekotorye osobennosti povrezhdaemosti sernistym gazom prorostkov sosny obyknovennoĭ. [Some peculiarities of damage of pine seedlings by sulphur dioxide.] - Tr. Inst. Ekol. Rast. Zhivot. ural. fil. Akad. Nauk SSSR *62*: 203-207, 1968. [Chl, Car; in R.]

*10771 - MAMUSHINA, N.S.: Deĭstvie temperatury na biosintez polisakharidov u raznykh shtammov khlorelly. [The effect of temperature on the polysaccharides biosynthesis by different strains of *Chlorella*.] - In: Materialy V rabochego Soveshchaniya po Voprosu Krugovorota Veshchestv v Zamknutoĭ Sisteme na Osnove

Zhiznedyeatel'nosti Nizshikh Organizmov. Pp. 176-178. Naukova Dumka, Kiev
1968. [In R.]

*10772 - MAMUSHINA, N.S., ZALENSKIĬ, O.V.: Vliyanie temperatury na metabolizm ugleroda
C^{14} pogloshchennogo pri fotosinteze u raznykh shtammov khlorelly. [The influence of temperature on the metabolism of ^{14}C absorbed during photosynthesis by different strains of *Chlorella*.] - In: Fotosintez i Ispol'zovanie Solnechnoĭ Radiatsii. Tezisy Dokladov. Pp. 59-60. Dushanbe 1967. [In R.]

*10773 - MAMUSHINA, N.S., ZALENSKIĬ, O.V.: Deĭstvie temperatury na dinamiku rasprede-
leniya C^{14} sredi organicheskikh veshchestv posle fotosinteza u razlichnykh
shtammov khlorelly. [The effect of temperature on the distribution of ^{14}C
assimilated during photosynthesis by different strains of *Chlorella*.] - In:
Tezisy Dokladov na II vsesoyuznom Biokhimicheskom S"ezde. Sektsiya 19 "Problemy Fotosinteza". Pp. 79-80. Tashkent 1969. [In R.]

10774 - MAMUSHINA, N.S., ZALENSKIĬ, O.V.: Deĭstvie temperatury na fotosinteticheskiĭ
metabolizm ugleroda razlichnykh shtammov khlorelly. [Effect of temperature
on ^{14}C metabolizm absorbed during photosynthesis by different *Chlorella*
strains.] - In: ZALENSKIĬ, O.V. (ed.): Fotosintez i Ispol'zovanie Solnechnoĭ
Energii. Pp. 177-181. Nauka, Leningrad 1971. [In R, ab: E.]

*10775 - MANGENOT, G., MANGENOT, S.: Les corps pigmentés des chromoplastes de *Neottia*
nidus avis. - Compt. rend. Acad. Sci. Paris, Sér. D *262*: 2456-2457, 1966.
[Chl.]

10776 - MANKEVICH, V.I.: Uplyŭ ploidnastsi na nekatoryya biyakhimichnyya asablivastsi
Lychnis chalcedonica L., *Dianthus barbatus* L. i *Saponaria officinalis* L.
[Effect of ploidy on some biochemical features of *Lychnis chalcedonica, Dian-
thus barbatus*, and *Saponaria officinalis*.] - Vestsi Akad. Navuk belarus. SSR,
Ser. biyal. Navuk *1971* (6): 55-58, 137-138, 1971. [Chl, Car; in Belorus.,
ab: R.]

*10777 - MANN, J.D., JAWORSKI, E.G.: Comparison of stresses which may limit soybean
yields. - Crop Sci. *10*: 620-624, 1970.

10778 - MANSFIELD, T.A.: Stomatal behaviour following treatment with auxin-like substances and phenylmercuric acetate. - New Phytol. *66*: 325-330, 1967. [Ps.]

*10779 - MANSFIELD, T.A., WILLMER, C.M.: Stomatal responses to light and carbon dioxi-
de in the Hart's-tongue fern, *Phyllitis scolopendricum* NEWM. - New Phytol.
68: 63-66, 1969. [Chloroplasts in guard cells.]

10780 - MANSUROV, N.I., MECHISLAVSKIĬ, Yu. A., KRIVKINA, N.P., MEDVEDOVSKAYA, L.E.:
Nekotorye fiziologo-biokhimicheskie osobennosti geterozisnykh gibridov khlop-
chatnika. [Physiological and biochemical features of heterosis cotton hybrids.]
- Dokl. Akad. Nauk tadzh. SSR *14* (1): 72-75, 1971. [Chl; in R, ab: Tadzh.]

10781 - MANTAI, K.E., HIND, G.: On the mechanism and stoichiometry of the oxidation
of hydrazine by illuminated chloroplasts. - Plant Physiol. *48*: 5-8, 1971.

10782 - MAR, T., GOVINDJEE: Thermoluminescence in spinach chloroplasts and in *Chlorella*. - Biochim. biophys. Acta *226*: 200-203, 1971.

*10783 - MARČENKO, E.: The effect of illumination regimen on temperature-induced and
spontaneous bleaching in *Euglena gracilis*. - Acta bot. croat. *29*: 27-32, 1970.

*10784 - MARČENKO, E.: Veränderungen im Chromatophoren-Feinbau von *Netrium digitus*
(*Desmidiaceae*) bei Lichtabschluss. - Acta bot. croat. *29*: 33-38, 1970.

10785 - MARCHANT, R.H.: On the process of induction of adenosinetriphosphate hydrolysis by chloroplasts with cysteine as activating thiol. - Arch. Biochem.
Biophys. *147*: 502-508, 1971.

10786 - MARCHETTI, S.E., BARON, F.J.: Response by chloroplast suspensions to their
direct treatment with kinetin. - Plant Physiol. *47* (Suppl.): 49, 1971.

10787 - MARGALEF, R.: Una campaña oceanográfica del "Cornide de Saavedra" en la región
de afloramiento del noroeste africano. [An oceanographic field trip of the
"Cornide de Saavedra" in the region of the northwest African coastal upwel-
lings.] - Invest. pesq. *35* (Suppl.): 1-39, 1971. [Ps, Chl; in Span.]

10788 - MĂRGINEANU, L.: Probleme actuale în studiul fotosintezei şi respiraţiei celu-
lare.] Actual problems in studying cell photosynthesis and respiration.] -

Prog. şti. 7: 224-231, 1971. [In Roum., ab: E, F, G, R.]

10789 - MARGULIES, M.M.: Concerning the sites of synthesis of proteins of chloroplast ribosomes and of fraction I protein (ribulose-1,5-diphophate carboxylase). - Biochem. biophys. Res. Commun. 44: 539-545, 1971.

*B10790 - MARKS, G.S.: Heme and Chlorophyll. Chemical, Biochemical and Medical Aspects. - D. van Nostrand Comp., London - Princeton - New Jersey - Toronto - Melbourne 1969.

10791 - MÁRKUS, L., HORVÁTH, L., POZSÁR, B.: A fotoszintetikus széndioxid fixálás intenzitásának fagypont alatti meghatározása a fagytűrőképesség jellemzésére, intenzív búzafajtáknál. [Sub-freezing point determination of photosynthetic carbon dioxide fixation rate for characterizing frost-hardiness in intensive wheat varieties.] - Agrobotanika 11: 61-66, 1969 (1971). [In Hung., ab: E.]

10792 - MAROTI, I., GABNAI, É.: Separation of chlorophylls and carotenoids by thin-layer chromatography. - Acta biol. (Szeged) 17: 67-77, 1971.

10793 - MARÓTI, M.: Do soluble proteins give a true picture of the intensity of carbon dioxide fixation after a photosynthetic activity of longer duration? - Acta agron. Acad. Sci. hung. 20: 445-446, 1971.

10794 - MARSHALL, J.K.: Drag measurements in roughness arrays of varying density and distribution. - Agr. Meteorol. 8: 269-292, 1971. [Model experiments.]

10795 - MARSHALL, N., OVIATT, C.A., SKAUEN, D.M.: Productivity of the benthic micro-flora of shoal estuarine environments in Southern New England. - Int. Rev. ges. Hydrobiol. 56: 947-955, 1971.

10796 - MARSHO, T.V., KOK, B.: Detection and isolation of P700. - In: COLOWICK, S.P., KAPLAN, N.O. (ed.): Methods in Enzymology. Vol. 23. Pp. 515-522. Academic Press, New York - London 1971.

10797 - MARTIN, E.S., MEIDNER, H.: Endogenous stomatal movements in *Tradescantia virginiana.* - New Phytol. 70: 923-928, 1971. [Resistances.]

10798 - MARTIN, J.K.: ^{14}C-labelled material leached from the rhizosphere of plants supplied with $^{14}CO_2$. - Aust. J. biol. Sci. 24: 1131-1142, 1971. [Ps.]

10799 - MARTIN, T., CIOFU, A., GEORGESCU, M., CHIRILĂ, R., PAUNEŢ, M.: Cercetări as-upra absorbţiei microelementelor prin hrănire extraradiculară şi a influenţei acestora asupra viţei de vie. [Microelement absorption by extra-radicular feeding and their influence upon vine.] - Lucrări ştiinţ. Inst. agron. "N. Bălcescu", Ser. B 14: 315-326, 1971. (1973). [Ps; in Roum., ab: E, F, R.]

*10800 - MARTIN, T., GEORGESCU, M., CIOFU, A.: Influenţa hrănirii extraradiculare cu bor, cobalt, mangan, şi molibden asupra proceselor de metabolism şi de rodire la Riesling italian. [Influence of extra-radicular nutrition with B, Co, Mn, Mo on metabolism of fruit-bearing processes of the Italian Riesling variety.] - Lucrări ştiinţ. Inst. agron. "N. Bălcescu", Ser. B, Hort. 9: 149-162, 1966. [Ps; in Roum., ab: E, F, R.]

10801 - MARTINEZ, H.: Gradientes en la planta madura de *Wedelia glauca*: I. Hetero-geneidad de las hojas de acuerdo a su posicion en el tallo. [Gradients in the mature plant of *Wedelia glauca*: I. Heterogeneity of the leaves according to their position on the stem.] - Rev. agron. noroeste Argent. 8: 239-252, 1971. [Ps; in Span., ab: E.]

10802 - MARTINEZ NADAL, N.G., PEREZ, J.W.: Chromatographic separation of pigments in *Spirulina maxima* (*Algae: Cyanophyceae*). - Caribb. J.Sci. 11: 211-214, 1971. [CC, TLC.]

10803 - MARTY, D.: Peroxysomes (= microbodies) associés aux plastes des zones non chlorophylliennes des feuilles panachées de *Coleus blumei* BENTH. - Compt. rend. Acad. Sci. Paris, Sér. D 273 : 860-863, 1971.

10804 - MARTYNENKO, G.N., NAUMENKO, A.S., NARYADOVOÏ, V.L.: Formula dlya opredeleniya ploshchadi listovoï poverkhnosti vinograda sorta Shasla. [Equation for deter-mining leaf surface area in vine cv. Chasla.] - Dokl. VASKHNIL 1971 (8): 16-17, 1971. [In R.]

10805 - MASHTAKOV, S.M., VOLYNETS, A.P., KORNELYUK, V.N.: Sovmestnoe deïstvie neko-torykh flavonoidov i abstsizovoï kisloty na otdel'nye formy rosta. [Combined

effect of flavonoids and abscisic acid on various forms of growth.] - Fiziol.
Rast. *18*: 802-807, 1971. [Chl; in R, ab: E.]

*10806 - MASLOVA, T.G., POPOVA, I.A., SAPOZHNIKOV, D.I.: Prevrashchenie ksantofillov
pri osveshchenii rasteniĭ svetom raznogo spektral'nogo sostava. [Xanthophyll
transformation in plants illuminated with light of different spectra.] - In:
Tezisy 3 Biokhimicheskoĭ Konferentsii. Vol. 1. Pp. 19-20. Minsk 1968. [In R.]

*10807 - MASUCH, G.: Stärkebildende Plastiden im Elektronenmikroskop. Amyloplasten
der Gartenkresse. - Mikrokosmos *59*: 110-115, 1970.

*10808 - MASYUK, N.P.: Otsinka prydatnosti ropy Saks'kykh vodoĭm dlya vyroshchuvannya
karotynonosnykh vodorosteĭ. [Estimation of salt-water usefulness of Saki reservoirs for cultivating carotene-bearing algae.] - Ukr. bot. Zh. *24* (4):
37-43, 1967. [In Ukr., ab: E, R.]

10809 - MASYUK, N.P., RADCHENKO, M.I.: Kolichestvennoe soderzhanie pigmentov v kletkakh nekotorykh vidov *Dunaliella* TEOD. v usloviyakh, blagopriyatnykh dlya
ikh razmnozheniya. [Content of pigments in cells of the species *Dunaliella*
under conditions favourable for their reproduction.] - Gidrobiol. Zh. *7* (6):
31-41, 1971. [In R, ab: E.]

*10810 - MATCHETT, R.W., POLLOCK, B.M., ROBERTSON, D.W.: The "Grandpa" gene: a chlo-
rophyll mutation in *Hordeum* species. - J. Hered. *59*: 279-282, 1968.

10811 - MATHIESON, A.C., BURNS, R.L.: Ecological studies of economic red algae. I.
Photosynthesis and respiration of *Chondrus crispus* STACKHOUSE and *Gigartina
stellata* (STACKHOUSE) BATTERS. - J. exp. mar. Biol. Ecol. *7*: 197-206, 1971.

10812 - MATHIS, P.: Les caroténoïdes vous protègent contre l'oxygène. - Recherche
(Paris) *2* (8): 76-77, 1971.

*10813 - MATSUBARA, H., SASAKI, R.M., CHAIN, R.K.: Spinach ferredoxin. I. Amino acid
composition and terminal sequences. - J. biol. Chem. *243*: 1725-1731, 1968.

10814 - MATSUDA, Y., KIKUCHI, T., ISHIDA, M.R.: Studies on chloroplast development
in *Chlamydonomas reinhardtii* I. Effect of brief illumination on chlorophyll
synthesis. - Plant Cell Physiol. *12*: 127-135, 1971.

*10815 - MATSUIKE, K., MORINAGA, T., SASAKI, T.: The optical characteristics of the
water in the three oceans. Part IV. An attempt to the approximate figures
of seasonal solar energy reached to and penetrated in the water of the three
oceans. - J. oceanograph. Soc. Jap. *26*: 52-60, 1970.

*10816 - MATSUSHIMA, S., MATSUZAKI, A., TOMITA, T.: [Analysis of yield-determining
process and its application to yield-prediction and culture improvement of
lowland rice. CI. On a method for expressing the leaf colour of rice plants
under field conditions (1).] - Proc. Crop Sci. Soc. Jap. *39*: 231-236, 1970.
[Chl; in Jap., ab: E.]

*10817 - MATSUSHIMA, S., SUZUKI, H., WADA, G., MATSUZAKI, A., ABE, H.: [Analysis of
yield-determining process and its application to yield prediction and culture
improvement of lowland rice. LXXIX. An experiment to maximizing yield based
on ideal plant types by using a circulating irrigation method applied with
nitrate nitrogen.] - Proc. Crop Sci. Soc. Jap. *38*: 1-10, 1969. [Chl, growth
analysis; in Jap., ab: E.]

10818 - MATSUYAMA, M., SAIJO, Y.: Studies on biological metabolism in a meromictic
lake Suigetsu. - J. oceanograph. Soc. Jap. *27*: 197-206, 1971. [Ps.]

*10819 - MATTEI, F.: Ricerche sui rapporti tra radiazione solare, accrescimento e
traspirazione di alcune specie erbacee coltivate. [Relation of solar radiation, growth and transpiration in some species of cultivated herbs.] - Ric.
Sci. (Roma) *37*: 598-607, 1967. [Ps; in Ital.]

*10820 - MAURINA, H.: Hlorofila satura, fotosintēzes intensitātes un citu fiziologis-
ko procesu izmainas dažādos augu organogenēzes etapos. [Changes in chlorophyll content, photosynthetic rate and other physiological processes at
different stages of plant organogenesis.] - In: Fotosintēzes Pētīšana Sējumos. Pp. 119-128. Zinātne, Riga 1970. [In Latv., ab: E.]

10821 - MAUZERALL, D., MALLEY, M.: The light-induced increase in fluorescence yield
in *Chlorella* is complete in 60 nanoseconds. - Photochem. Photobiol. *14*: 225-
227, 1971.

*10822 - MAXWELL, M.A.B., BIDWELL, R.G.S.: Synthesis of asparagine from $^{14}CO_2$ in wheat seedlings. - Can. J. Bot. *48*: 923-927, 1970.

10823 - MAYER, F.: Light-induced chloroplast contraction and movement. - In: GIBBS, M. (ed.): Structure and Function of Chloroplasts. Pp. 35-49. Springer-Verlag, Berlin - Heidelberg - New York 1971.

*10824 - MAYER, H.: Synthesis in rhodoxanthin and related compounds. - Tidsskr. Kjemi, Bergv., Metallurgi *26*: 131, 1966.

10825 - MAYER, H., ISLER, O.: Total syntheses. - In: ISLER, O. (ed.): Carotenoids. Pp. 325-575. Birkhäuser Verlag, Basel - Stuttgart 1971. [Car.]

10826 - MAYNE, B.C., EDWARDS, G.E., BLACK, C.C.: Light reactions in C_4 photosynthesis. - In: HATCH, M.D., OSMOND, C.B., SLATYER, R.O. (ed.): Photosynthesis and Photorespiration. Pp. 361-371. Wiley-Interscience, New York - London - Sydney - Toronto 1971.

10827 - MAYNE, B.C., EDWARDS, G.E., BLACK, C.C. Jr.: Spectral, physical, and electron transport activities in the photosynthetic apparatus of mesophyll cells and bundle sheath cells of *Digitaria sanguinalis* (L.) SCOP. - Plant Physiol. *47*: 600-605, 1971.

10828 - MAYNE, B.C., HOBBS, L.J., KETTERING, C.F.: Flash activation of chemically induced luminescence of chloroplasts. - Plant Physiol. *47* (Suppl.): 31, 1971.

10829 - McCARTY, R.E.: Chloroplast preparation deficient in coupling factor 1. - In: COLOWICK, S.P., KAPLAN, N.O. (ed.): Methods in Enzymology. Vol. 23. Pp. 251-253. Academic Press, New York - London 1971.

10830 - McCARTY, R.E.: Preparation and properties of phosphorylating subchloroplast particles. - In: COLOWICK, S.P., KAPLAN, N.O. (ed.): Methods in Enzymology. Vol. 23. Pp. 302-305. Academic Press, New York - London 1971.

10831 - McCARTY, R.E., FUHRMAN, J.S., TSUCHIYA, Y.: Effects of adenine nucleotides on hydrogen-ion transport in chloroplasts. - Proc. nat. Acad. Sci. USA *68*: 2522-2526, 1971.

10832 - McCULLOUGH, E.C., PORTER, W.P.: Computing clear day solar radiation spectra for the terrestrial ecological environment. - Ecology *52*: 1008-1015, 1971.

10833 - McDONALD, D.J.: The effect of temperature and light on the rate of photosynthesis of 20 rice varieties and hybrids. - Diss. Abstr. int. B *32*: 1309-B, 1971.

10834 - McEVOY, F.A., LYNN, W.S.: Effect of anions on chloroplast particles. - Biochem. biophys. Res. commun. *44*: 1388-1393, 1971.

10835 - McFEETERS, R.F.: Purification and properties of chlorophyllase from *Ailanthus altissima* (tree-of-heaven). - Diss. Abstr. int. B *31*: 7098, 1971.

10836 - McFEETERS, R.F., CHICHESTER, C.O., WHITAKER, J.R.: Purification and properties of chlorophyllase from *Ailanthus altissima* (tree-of-heaven). - Plant Physiol. *47*: 609-618, 1971.

10837 - McGAHEE, C.F., DAVIS, G.J.: Photosynthesis and respiration in *Myriophyllum spicatum* L. as related to salinity. - Limnol. Oceanogr. *16*: 826-829, 1971.

*10838 - McINTIRE, C.D., WULFF, B.L.: A laboratory method for the study of marine benthic diatoms. - Limnol. Oceanogr. *14*: 667-678, 1969. [Ps, Chl. Car.]

10839 - McLAUGHLIN, S.B.,Jr.: The effects of fluoride on processes related to photosynthesis and respiration of several species of forest trees. - Diss. Abstr. int. B *31*: 5840-B - 5841-B, 1971.

10840 - MEDINA, E.: Relationships between nitrogen level, photosynthetic capacity, and carboxydismutase activity in *Atriplex Patula* leaves. - Carnegie Inst. Year Book *69*: 655-662, 1971.

10841 - MEDINA, E.: Effect of nitrogen supply and light intensity during growth on the photosynthetic capacity and carboxydismutase activity of leaves of *Atriplex patula* ssp. *hastata*. - Carnegie Inst. Year Book *70*: 551-559, 1971.

10842 - MEEKS, J.C., CASTENHOLZ, R.W.: Growth and photosynthesis in an extreme thermophyle, *Synechococcus lividus* (*Cyanophyta*). - Arch. Mikrobiol. *78*: 25-41, 1971.

10843 - MEÏLANOV, I.S., ZOLOTOVITSKIÏ, Ya.M., BENDERSKIÏ, V.A.: Fotoeffekt na granitse sloev khlorofilla s rastvorom elektrolita. [Photoeffect on the border of
chlorophyll layers with electrolyte solution.] - Biofizika 16: 415-419, 1971.
[In R, ab: E.]

10844 - MELANDRI, B.A., BACCARINI-MELANDRI, A., GEST, H., SAN PIETRO, A.: Studies
on resolution of the photophosphorylating system of the photosynthetic bacterium *Rhodopseudomonas capsulata*. - In: QUAGLIARIELLO, E., PAPA, S., ROSSI,
C.S. (ed.): Energy Transduction in Respiration and Photosynthesis. Pp. 593-
608. Adriatica Editrice, Bari 1971.

10845 - MELANDRI, B.A., BACCARINI-MELANDRI, A., SAN PIETRO, A., GEST, H.: Inter-
changeability of phosphorylation coupling factors in photosynthetic and respiratory energy conversion. - Science 174: 514-516, 1971.

10846 - MELLOR, G.E., TREGUNNA, E.B.: The localization of nitrate-assimilating enzymes in leaves of plants with the C_4-pathway of photosynthesis. - Can. J.
Bot. 49: 137-142, 1971.

10847 - MENGEL, K.: Photosynthese und CO_2-Assimilation der höheren Pflanze. - Stärke
23: 375-379, 1971.

10848 - MENKE, W., RUPPEL, H.-G.: Molekulargewicht, Grösse und Gestalt von Proteinen
der Thylakoidmembran. - Z. Naturforsch. 26b: 825-831, 1971.

10849 - MENKE, W., SCHÖLZEL, E.: Molecular weight of polypeptides of the thylakoid
membrane. - Z. Naturforsch. 26 b: 378-379, 1971.

10850 - MENSHUTKIN, V.V., PRIKHOD'KO, T.I.: Model'noe issledovanie vertikal'nogo
raspredeleniya i produktsiya fitoplanktona. [Analogue investigation of phytoplankton vertical distribution and production.] - Gidrobiol. Zh. 7 (2):
5-10, 1971. [In R, ab: E.]

10851 - MEREZHKO, A.I., KUZ'MENKO, M.I., VELICHKO, I.M.: Vzaimootnosheniya razlichnykh vidov vodorosleĭ u vysshikh vodnykh rasteniĭ, obuslovlennye ikh meta-
bolitami. [Relationships among different species of algae and green water
plants induced by their metabolites.] - In: Letuchie Biologicheski Aktivnye
Soedineniya Biogennogo Proiskhozhdeniya. Pp. 143-152. Izdat. mosk. Univ.
Moskva 1971. [Ps; in R, ab: E.]

10852 - MEREZHKO, A.I., SHIYAN, P.N., STRELKO, D.V.: O karotinoidakh rdesta pronzen-
nolistnogo (*Potamogeton perfoliatus*). [Carotenoids of the pond-weed *Potamogeton perfoliatus*.] - Gidrobiol. Zh. (Kiev) 7 (5): 93-95, 1971. [In R.]

10853 - MESQUITA, J.F.: Inhibition de la morphogénèse des chloroplastes dans les
racines du *Lupinus albus* L. exposées à la lumière. - Bol. Soc. broteriana
45: 363-371, 1971.

10854 - MESQUITA, J.F.: Altérations ultrastructurales des plastes au cours de verdissement expérimental des racines du *Lupinus albus* L. - Portug. Acta biol.,
Sér. A 12: 33-52, 1971.

10855 - METZNER, H., FISCHER, K.: Energy conservation in photosynthesis models. -
In: BRODA, E., LOCKER, A., SPRINGER-LEDERER, H. (ed.): Proceedings of the
First European Biophysics Congress. Vol. 4. Pp. 27-29. Verlag Wiener med.
Akad., Wien 1971.

10856 - MEYER, K.A., McCORMICK, J.F.: Seasonal fluctuation of phytoplankton composition, diversity, and production in a freshwater lake. - J. Elisha Mitchell
sci. Soc. 87: 127-138, 1971.

10857 - MEYER, M.M. Jr., SPLITTSTOESSER, W.E.: The utilization of carbohydrate and
nitrogen reserves by *Taxus* during its spring growth period. - Physiol. Plant.
24: 306-314, 1971.

10858 - MEYER, T.E., BARTSCH, R.G., KAMEN, M.D.: Cytochrome c_3. A class of electron
transfer heme proteins found in both photosynthetic and sulfate-reducing
bacteria. - Biochim. biophys. Acta 245: 453-464, 1971.

10859 - MEYER, T.E., SHARP, J.J., BARSCH, R.G.: Isolation and properties of rubredoxin
from the photosynthetic green sulfur bacteria. - Biochim. biophys. Acta 234:
266-269, 1971.

10860 - **MICHEL, J.-P.**: Acide aspartique et capacité en oxygène photosynthétique chez le maïs. Action du rouge lointain. - Compt. rend. Acad. Sci. Paris, Sér. D *272*: 2455-2458, 1971.

10861 - **MIGINIAC-MASLOW, M.**: Etude de la photophosphorylation endogène des chloroplastes isolés d'épinard. - Biochim. biophys. Acta *234*: 353-359, 1971.

10862 - **MIGINIAC-MASLOW, M., CHAMPIGNY, M.L.**: Relations entre l'assimilation photosynthétique du CO_2 et la photophosphorylation des chloroplastes isolés. II. L'utilisation de l'ATP dans l'assimilation photosynthétique de CO_2. - Biochim. biophys. Acta *234*: 344-352, 1971.

10863 - **MIKHALEVA, E.N.**: Osobennosti fotosinteza u gretskogo orekha v svyazi s protsessom adaptatsii. [Peculiarities of walnut photosynthesis in relation to processes of adaptation.] - In: Tezisy Dokladov Tret'ego Ural'skogo Soveshchaniya po Ekologii i Fiziologii Drevesnykh Rastenii. Pp. 86-87. Sverdlovsk 1971. [Ps; in R.]

10864 - **MIKHALEVA, E.N., SAZYKINA, N.A.**: Detstvie ponizhennoĭ temperatury na intensivnost' fotosinteza i aktivnost' fotokhimicheskikh reaktsiĭ izolirovannykh khloroplastov gorokha raznogo geograficheskogo proiskhozhdeniya. [Effect of low temperature on the photosynthetic rate and activity of photochemical reactions of isolated chloroplasts of pea of different geographical origin.] - In: Tezisy Dokladov Vtorogo Vsesoyuznogo Simpoziuma: " Fiziologicheskie Osnovy Ustoĭchivosti Rastenii k Zamorozkam i Ponizhennym Temperaturam". Pp. 23-25. Petrozavodsk 1971. [In R.]

10865 - **MIKULOVICH, T.P., KHOKHLOVA, V.A., KULAEVA, O.N., SVESHNIKOVA, I.N.**: Vliyanie 6-benzilaminopurina na izolirovannye semyadoli tykvy. [Effect of 6-benzylaminopurine on isolated gourd cotyledons.] - Fiziol. Rast. *18*: 98-106, 1971. [Chl; in R, ab: E.]

10866 - **MILLER, E.E., NORMAN, J.M.**: A sunfleck theory for plant canopies. I. Lengths of sunlit segments along a transect. - Agron. J. *63*: 735-738, 1971.

10867 - **MILLER, E.E., NORMAN, J.M.**: A sunfleck theory for plant canopies. II. Penumbra effect: intensity distribution along sunfleck segments. - Agron. J. *63*: 739-743, 1971.

10868 - **MILLER, M.M., NOBEL, P.S.**: Light-induced structural changes of pea chloroplasts *in vivo*. - Plant Physiol. *47* (Suppl.): 32, 1971.

10869 - **MILLER, R.B.**: Forest productivity in the temperate-humid zone of the Southern Hemisphere. - In: DUVIGNEAUD, P. (ed.): Productivity of Forest Ecosystems. Pp. 299-305. Unesco, Paris 1971. [Growth analysis.]

*10870 - **MILLER, R.L., WARD, C.H.**: Algal bioregenerative systems. - In: KAMMERMEYER, K. (ed.): Atmosphere in Space Cabins and Closed Environments. Pp. 186-222. Appleton-Century-Crofts, New York 1966. [Ps.]

10871 - **MILLERD, A., SIMON, M., STERN, H.**: Legumin synthesis in developing cotyledons of *Vicia faba* L. - Plant Physiol. *48*: 419-425, 1971. [Chl.]

*B10872 - **MILNER, C., HUGHES, R.E.** (ed.): Methods for the Measurement of the Primary Production of Grassland. - IBP Handbook No. 6. Blackwell Scientific Publications, Oxford - Edinburgh 1968.

*10873 - **MILTHORPE, F.L.**: From the qualitative to the quantitative, with special reference to the use of light by crops. - Aust. J. Sci. *32*: 345-349, 1970. [Ps.]

10874 - **MILWARD, A.F., SKERRETT, E.J.**: Carbon dioxide measurement: a coulometric method for use in photosynthesis experiments. - J. Phys. E: Sci. Instrum. *4*: 1016-1018, 1971.

10875 - **MILWARD, A.F., SKERRETT, E.J.**: Studies in phytotoxicity: The detection of incipient phytotoxic effects of sulphur preparations. - Pestic. Sci. *2*: 38-40, 1971. [Ps.]

*10876 - **MIRONYUK, V.I., EĬNOR, L.O.**: Vliyanie proizvodnykh fenola na kislorodnyĭ obmena *Dunaliella salina* TEOD. [Effect of phenol derivatives on oxygen exchange of *Dunaliella salina* TEOD.] - Gidrobiol. Zh. *6* (3): 91-95, 1970. [Chl, Car.]

10877 - **MIROSHNIKOVA, A.T., NIKOLAEVSKIĬ, V.S.**: Novye biofizicheskie metody diagnostiki gazoustoĭchivosti rasteniĭ. [New biophysical methods of diagnosis of gas resistance of plants.] - Uch. Zap. permsk. gos. Univ. A.M. Gor'kogo *256*: 197-210, 1971. [Chl; in R.]

10878 - MIROSHNIKOVA, A.T., NIKOLAEVSKIĬ, V.S., SHESTIMEROVA, G.I., SHISHOVA, L.V.: Vliyanie kislykh gazov na sposobnost' okisleniya kletochnogo soderzhimogo list'ev anodnym tokom. [Effect of acid gases on the capacity of oxidation of cell contents of leaves by an anode current.] - Uch. Zap. permsk. gos. Univ. A.M. Gor'kogo *277*: 157-159, 1971. [Ps; in R.]

10879 - MIROSLAVOV, E.A.: Stroenie khloroplastov trikhom epidermisa lista korovyaka. [Structure of trichome chloroplasts of leaf epidermis in *Verbascum thapsus* L.] - Tsitologiya *13*: 108-111, 1971. [In R, ab: E.]

10880 - MISRA, H.P., FRIDOVICH, I.: The generation of superoxide radical during the autoxidation of ferredoxins. - J. biol. Chem. *246*: 6886-6890, 1971.

10881 - MITROFANOV, B.A., OKANENKO, A.S., GULYAEV, B.I., MAKHOVSKAYA, M.A., LAVREN-TOVICH, D.I.: Znachenie otdel'nykh organov rasteniĭ pshenitsy v fotosinteze poseva. [Importance of individual organs of wheat plants in canopy photosynthesis.] - In: ZALENSKIĬ, O.V. (ed.): Fotosintez i ispol'zovanie SoinechnoĭEnergii. Pp. 82-86. Nauka, Leningrad 1971. [In R.]

10882 - MITSUI, A.: *Euglena* cytochromes. - In: COLOWICK, S.P., KAPLAN, N.O. (ed.): Methods in Enzymology. Vol. 23. Pp. 368-371. Academic Press, New York - London 1971.

10883 - MITSUI, A.: Purification and some chemical properties of crystalline *Euglena* ferredoxin. - Biochim. biophys. Acta *243*: 447-456, 1971.

10884 - MITSUI, A., ARNON, D.I.: Crystalline ferredoxin from a blue-green alga, *Nostoc* sp. - Physiol. Plant. *25*: 135-140, 1971.

10885 - MITSUI, A., SAN PIETRO, A.: Purification and effect on electron transport of a new factor (F_{333}) from *Porphyra tenera*. - Plant Physiol. *47* (Suppl.): 33, 1971.

10886 - MITTELHEUSER, C.J., van STEVENINCK, R.F.M.: Rapid action of abscisic acid on photosynthesis and stomatal resistance. - Planta *97*: 83-86, 1971.

10887 - MITTELHEUSER, C.J., van STEVENINCK, R.F.M.: The ultrastructure of wheat leaves. I. Changes due to natural senescence and the effects of kinetin and ABA on detached leaves incubated in the dark. - Protoplasma *73*: 239-252, 1971. [Chloroplast.]

10888 - MITTELHEUSER, C.J., van STEVENINCK, R.F.M.: The ultrastructure of wheat leaves. II. The effects of kinetin and ABA on detached leaves incubated in the light. - Protoplasma *73*: 253-262, 1971. [Chloroplast.]

*10889 - MIWA, Takuji, MIWA, Tomoko: The triplet energy transfer from anthracene to β-carotene. - Sci. Papers Inst. phys. chem. Res. *62*: 141-144, 1968.

*10890 - MIWA, Takuji, MIWA, Tomoko: [On the flash photolytic intermediate of β-carotene.] - Chiba Daigaku Kogakubu Kenkyu Hokoku *21* (39): 37-46, 1970. [In Jap., ab: E.]

*10891 - MIWA, Takuji, MIWA, Tomoko: [On the luminescence of β-carotene.] - Chiba Daigaku Kogakubu Kenkyu Kokoku *21* (40): 189-196, 1970. [In Jap., ab: E.]

10892 - MIYASAKA, T., MORIMOTO, Y.: [A method for measuring the vertical section geometry of the surface by electron microscopy.] - J. Electron Microscop. *20*: 325-326, 1971. [In Jap.]

10893 - MOCHALKIN, A.I., GOLOVAN', A.M., ALEKSEEV, S.I., SOKOLOV, M.S.: Sutochnaya dinamika fotoreemissii u list'ev soi i kukuruzy. [Diurnal dynamics of photoreemission in leaves of soybean and maize.] - Vestn. sel'skokhoz. Nauki *16* (8): 65, 1971. [In R.]

10894 - MOHANTY, P., MAR, T., GOVINDJEE: Action of hydroxylamine in the red alga *Porphyridium cruentum*. - Biochim. biophys. Acta *253*: 213-221, 1971. [Ps, Chl.]

10895 - MOHANTY, P., PAPAGEORGIOU, G., GOVINDJEE: Fluorescence induction in the red alga *Porphyridium cruentum*. - Photochem. Photobiol. *14*: 667-682, 1971.

*10896 - MOKRONOSOV, A.T.: Metodicheskie vozmozhnosti otsenki fotosinteticheskogo me-

tabolizma ugleroda v polevykh usloviyakh. [Methodological possibilities of
assessing photosynthetic carbon metabolism in field conditions.] - In: Tezisy
Dokladov Vsesoyuznogo Soveshchaniya po Unifikatsii Metodov i Priborov dlya
Massovykh Izmerenii Intensivnosti Fotosinteza. Pp. 77-81. Nauch.-issled.
Inst. Rastenievod. N.I. Vavilova, Leningrad - Pushkin 1970. [In R.]

10897 - MOKRONOSOV, A.T.: Fotosintez kartofelya. [Photosynthesis of potato.] - In:
POTAPOV, N.G. (ed.): Fiziologiya Sel'skokhozyaistvennykh Rastenii. Vol. XII.
Fiziologiya Kartofelya i Korneplodov. Pp. 99-128. Izd. mosk. Univ., Moskva
1971. [In R.]

10898 - MOKRONOSOV, A.T.: Peredvizhenie i ispol'zovanie produktov fotosinteza vo vto-
richnykh sinteticheskikh protsessakh. [Transport and use of photosynthates in
secondary synthetic processes.] - In: POTAPOV, N.G. (ed.): Fiziologiya Sel'-
skokhozyaistvennykh Rastenii. Vol. XII. Fiziologiya Kartofelya i Korneplodov.
Pp. 129-155. Izd. mosk. Univ., Moskva 1971. [In R.]

10899 - MOKRONOSOV, A.T., IVANOVA, N.A.: Osobennosti fotosinteticheskoi funktsii pri
chastichnoi defoliatsii rastenii. [Peculiarities of photosynthetic function
after partial defoliation of plants.] - Fiziol. Rast. 18: 668-676, 1971.
[In R, ab: E.]

*10900 - MOKRONOSOV, A.T., NEKRASOVA, G.F.: Prevrashchenie ekzogennogo alanina - $1C^{14}$
v list'yakh kartofelya. [Conversion of exogenous alanin - $1^{14}C$ in potato
leaves.] - Uch. Zap. ural'. gos. Univ. 113, Ser. biol. 8 (Voprosy Regulyatsii Fotosinteza, Sb. 2): 98-106, 1970.]In R.]

10901 - MOLCHANOV, M.I.: O lipoaminokislotnykh soedineniyakh plastid pri formirovanii
ul'trastruktury khloroplastov kukuruzy pod vozdeistviem sveta. [Lipoamino
acid compounds of plastids in formation of maize chloroplast ultrastructure
under the action of light.] - Dokl. Akad. Nauk SSSR 199: 1196-1199, 1971.
[In R.]

10902 - MOLDAU, H.: Model of plant productivity at limited water supply considering
adaptation. - Photosynthetica 5: 16-21, 1971.

10903 - MONSON, W.G., BURTON, G.W., WILKINSON, W.S., DUMFORD, S.W.: Effect of N fertilization and simazine on yield, protein, amino-acid content, and carotenoid
pigments of coastal bermudagrass. - Agron. J. 63: 928-930, 1971.

10904 - MONTEITH, J.L., BISCOE, P.V.: Meteorological measurements of photosynthesis
and transpiration. - Univ. Nottingham School Agr. Rep. 1970-71: 66-75, 1971.

10905 - MONTEITH, J.L., ELSTON, J.F.: Microclimatology and crop production. - In:
WAREING, P.F., COOPER, J.P. (ed.): Potential Crop Production. A Case Study.
Pp. 23-42. Heinemann Educational Books, London 1971. [Ps.]

10906 - MONTENY, B.A.: Caracteristiques morphologiques et échanges en gaz carbonique
des feuilles de Panicum maximum. - Pp. 1-28. ORSTOM, Centre d'Adiopodoumé,
Côte d'Ivoire 1971.

10907 - MONTENY, B.A.: Mesure de la productivité d'un couvert végétal par une méthode
gravimétrique. - Pp. 1-27. ORSTOM, Centre d'Adiopodoumé, Côte d'Ivoire 1971.

10908 - MONTOJOS, J.C., MAGALHAES, A.C.: Growth analysis of dry beans (Phaseolus
vulgaris L. var. Pintado) under varying conditions of solar radiation and
nitrogen application. - Plant Soil 35: 217-223, 1971.

10909 - MONTY, C.L.V.: An autoecological approach of intertidal and deep water stro-
matolites. - Ann. Soc. Geol. Belg. 94: 265-276, 1971. [Chl, Car, phycobilins.]

10910 - MOON, K.E., THOMPSON, E.O.P.: Separation and characterization of subunits of
ribulose diphosphate carboxylase. - Aust. J. biol. Sci. 24: 755-764, 1971.

10911 - MOONEY, H.A., DUNN, E.L., HARRISON, A.T., MORROW, P.A., BARTHOLOMEW, B.,
HAYS, R.L.: A mobile laboratory for gas exchange measurements. - Photosynthetica 5: 128-132, 1971.

10912 - MORARU, K.V., VIDRASHKO, G.E., BABINTSEVA, V.I.: Soderzhanie khlorofilla v
list'yakh i tekhnologo-biokhimicheskie osobennosti zerna pshenitsy. [Chloro-
phyll content in leaves and technological-biochemical features of wheat grain.]
- Izv. Akad. Nauk mold. SSR, Ser. biol. khim. Nauk 1971 (4): 8-13, 1971.

*10913 - MORELAND, D.E., BLACKMON, W.J., TODD, H.G., FARMER F.S.: Effects of diphenyl-
ether herbicides on reactions of mitochondria and chloroplasts. - Weed Sci.
18: 636-642, 1970.

10914 - MORELAND, D.E., BOOTS, M.R.: Effects of optically active 1-(α-methylbenzyl)
-3 - (3,4-dichlorophenyl)urea on reactions of mitochondria and chloroplasts. -
Plant Physiol. *47*: 53-58, 1971.

10915 - MORITA, S., MIYAZAKI, T.: Dichroism of bacteriochlorophyll in cells of the
photosynthetic bacterium *Rhodopseudomonas palustris*. - Biochim. biophys. Acta *245*: 151-159, 1971.

10916 - MORRIS, I., FARRELL, K.: Photosynthetic rates, gross patterns of carbon dioxide assimilation and activities of ribulose diphosphate carboxylase in marine algae grown at different temperatures. - Physiol. Plant. *25*: 372-377,
1971.

10917 - MORROW, P.A., SLATYER, R.O.: Leaf resistance measurements with diffusion
porometers: precautions in calibration and use. - Agr. Meteorol. *8*: 223-233,
1971.

10918 - MORROW, P.A., SLATYER, R.O.: Leaf temperature effects on measurements of
diffusive resistance to water vapor transfer. - Plant Physiol. *47*: 559-561,
1971.

10919 - MORTON, R.A.: Ubiquinones, plastoquinones and vitamins K. - Biol. Rev.
Cambridge phil. Soc. *46*: 47-96, 1971.

10920 - MOSS, D.N.: Carbon dioxide compensation in plants with C_4 characteristics.
- In: HATCH, M.D., OSMOND, C.B., SLATYER, R.O. (ed.): Photosynthesis and
Photorespiration. Pp. 120-123. Wiley-Interscience, New York - London - Syd-
ney - Toronto 1971.

10921 - MOSS, D.N., MUSGRAVE, R.B.: Photosynthesis and crop production. - Advances
Agron. *23*: 317-336, 1971.

10922 - MOSS, D.N., WILLMER, C.M., CROOKSTON, R.K.: CO_2 compensation concentration
in maize (*Zea mays* L.) genotypes. - Plant Physiol. *47*: 847-848, 1971.

10923 - MOSSER, J.L., BROCK, T.D.: Effect of wide temperature fluctuation on the blue-
green algae of Bead Geyser, Yellowstone National Park. - Limnol. Oceanogr.
16: 640-645, 1971. [Ps.]

10924 - MOURAVIEFF, I.: Action de colorants sur l'ouverture de l'appareil somatique
des feuilles entières en présence ou non de gaz carbonique. - Ann. Sci. nat.
- Bot. Biol. vég. *12*: 189-197, 1971.

10925 - MOUSSEAU, M.: La photosynthèse des formes d'ombre et de lumière de *Teucrium
scorodonia* L. *in situ* et au laboratoire. - Photosynthetica *5*: 241-248, 1971.

10926 - MÜHLETHALER, K.: The ultrastructure of plastids. - In: GIBBS, M. (ed.): Structure and Function of Chloroplasts. Pp. 7-34. Springer-Verlag, Berlin - Heidelberg - New York 1971.

10927 - MUKERJI, S.K., TING, I.P.: Phosphoenolpyruvate carboxylase isoenzymes: separation and properties of three forms from cotton leaf tissue. - Arch. Biochem. Biophys. *143*: 297-317, 1971.

10928 - MUKHAMADIEV, B.T., KVITKO, K.V., ZALENSKIĬ, O.V.: Mutanty khlorelly, ustoĭ-
chivye k ingibitoru fotofosforilirovaniya DCMU - 3(3,4)-dikhlorfenil-1,1-
dimetilmochevine. Soobshchenie II. Mutagennoe deĭstvie DCMU na razlichnye
shtammy. [Mutants of *Chlorella* resistant to the photophosphorylation inhibitor, 3(3,4)-dichlorophenyl-1,1-dimethylurea (DCMU). Report II. Mutagenic effect of DCMU on different strains.] - Genetika *7* (5): 36-41, 1971. [In R,
ab: E.]

10929 - MUKHAMADIEV, B.T., ZALENSKIĬ, O.V., KVITKO, K.V.: Mutanty khlorelly, ustoĭ-
chivye k ingibitoru fotofosforilirovaniya DCMU-3(3,4)-dikhlorfenil-1,1-dime-
tilmochevine. Soobshchenie I. Priroda vozniknoveniya mutantov. [Mutants of
Chlorella resistant to the inhibitor of photophosphorylation, 3(3,4)-dichlo-
rophenyl-1,1-dimethylurea. I. The origin of mutants.] - Genetika *7* (4): 38-
41, 1971. [In R, ab: E.]

10930 - **MUKHAMADIEV, B.T., ZALENSKIĬ, O.V., KVITKO, K.V.:** Mutanty khlorelly, ustoĭ-chivye k ingibitoram fotofosforilirovaniya. [*Chlorella* mutants resistant to inhibitors of photophosphorylation.] - Bot. Zh. *56*: 727-729, 1971.

10931 - **MUKHIN, E.N.:** Fermentativnye mekhanizmy vosstanovleniya piridinnukleotidov pri fotosinteze. [Enzymic mechanism of the reduction of pyridine nucleotides during photosynthesis.] - Usp. biol. Khim. *12*: 246-267, 1971. [In R.]

*10932 - **MUKOHATA, Y.:** Biophysical studies on subcellular particles. III. On the light scattering pattern of spinach chloroplast suspension. - Annu. Rep. biol. Works Fac. Sci. Osaka Univ. *14*: 121-134, 1966.

*10933 - **MUKOHATA, Y., MITSUDO, M., ISEMURA, T.:** Biophysical studies on subcellular particles. II. Reduction of the amount of accumulated protons and the light scattering response of spinach chloroplasts by alkaline solution. - Annu. Rep. biol. Works Fac. Sci. Osaka Univ. *14*: 107-119, 1966.

10934 - **MUKOHATA, Y., MITSUDO, M., KAKUMOTO, S., HIGASHIDA, M.:** Biophysical studies on subcellular particles V. Effects of temperature on the ferricyanide-Hill reaction, the light-induced pH shift and the light scattering response of isolated spinach chloroplasts. - Plant Cell Physiol. *12*: 869-880, 1971.

10935 - **MUKOHATA, Y., MITSUDO, M., NAKAE, T., MYOJO, K.:** Biophysical studies on subcellular particles IV. The light scattering response and photosynthetic activities of spinach chloroplasts treated with *n*-alkanes. - Plant Cell Physiol. *12*: 859-868, 1971.

10936 - **MULCHI, C.L.:** Photosynthetic and photorespiratory gas exchange kinetics in illuminated soybean leaves employing mass spectrometric techniques. - Diss. Abstr. Int. B *31*: 6224, 1971.

10937 - **MULCHI, C.L., VOLK, R.J., JACKSON, W.A.:** Oxygen exchange of illuminated leaves at carbon dioxide compensation. - In: HATCH, M.D., OSMOND, C.B., SLATYER, R.O. (ed.): Photosynthesis and Photorespiration. Pp. 35-50. Wiley-Interscience, New York - London - Sydney - Toronto 1971.

10938 - **MÜLLER, D., KNÜPP, H.:** Zur Messung der Primärproduktion und der biogenen Belüftung in Fliessgewässern. 1. Ein Laborvergleich der Messmethoden. - Int. Rev. ges. Hydrobiol. *56*: 49-67, 1971.

10939 - **MÜLLER, H.P.:** Untersuchungen über die Wirkung von Gibberellin A_3 auf das Längenwachstum und den Pigmentgehalt von Chlorophyllmutanten. - Z. Pflanzenphysiol. *64*: 34-40, 1971.

10940 - **MÜLLER, W.:** Helldunkelflaschen-Sauerstoffmethode. - Limnologica (Berlin) *4*: 281-290, 1966.

10941 - **MÜLLER, W., WARTENBERG, A.:** Zur Lokalisation der Photosynthesepigmente, insbesondere der Dunkelrot-Chlorophylle, in den Chloroplasten von *Mesotaenium caldariorum* (LAGERH.) HANSG. var. *caldariorum*. - Z. Pflanzenphysiol. *65*: 365-377, 1971.

10942 - **MUNTEANU, C., STIRBAN, M.:** Unele aspecte ale metabolismului semiparazitelor. Conţinutul în pigmenţi asimilatori. [Some aspects of semiparasite metabolism. Content of assimilatory pigments.] - Contribuţ. bot. Univ. "Babeş-Bolyai" (Cluj) *1971*: 369-375, 1971. [In Roum., ab: F.]

10943 - **MÜNTZ, K.:** Reversible Schäden im Pigmentsystem von *Chlorella pyrenoidosa* als Folge des Einfrierens der Zellen. - Photosynthetica *5*: 32-37, 1971.

10944 - **MURAKAMI, S., PACKER, L.:** The role of cations in the organization of chloroplast membranes. - Arch. Biochem. Biophys. *146*: 337-347, 1971. [Chl.]

10945 - **MURAMOTO, H., HESKETH, J.D., BAKER, D.N.:** Cold tolerance in a hexaploid cotton. - Crop Sci. *11*: 589-591, 1971. [Chl.]

10946 - **MURATA, N.:** Control of excitation transfer in photosynthesis. V. Correlation of membrane structure to regulation of excitation transfer between two pigment systems in isolated spinach chloroplasts. - Biochim. biophys. Acta *245*: 365-372, 1971.

10947 - **MURATA, N.:** Effects of monovalent cations on light energy distribution between two pigment systems of photosynthesis in isolated spinach chloroplasts. - Biochim. biophys. Acta *226*: 422-432, 1971.

10948 - MURATA, N.: Effects of monovalent cations on light energy distribution between two pigment systems of photosynthesis. - Carnegie Inst. Year Book 69: 699-704, 1971.

10949 - MURATA, N.: Comparison of chlorophyll fluorescence and light-scattering changes upon addition of cations in isolated chloroplasts. - Carnegie Inst. Year Book 70: 485-486, 1971.

10950 - MURATA, N., FORK, D.C.: Light-induced absorbance changes in Fraction 1 particles prepared from spinach chloroplasts by French-press treatment. - Biochim. biophys. Acta 245: 356-364, 1971.

10951 - MURATA, N., FORK, D.C.: Studies of light-induced changes of absorption in fraction 1 particles derived from spinach chloroplasts by French pressure cell treatment. - Carnegie Inst. Year Book 70: 461-468, 1971.

10952 - MURATA, N., FORK, D.C.: Oxidation-reduction reactions of system 1 particles prepared by French pressure cell treatment of spinach chloroplasts. - Carnegie Inst. Year Book 70: 468-472, 1971.

10953 - MURATA, T., MURATA, N.: Water-soluble chlorophyll proteins from Brassica nigra and Lepidum virginicum L. - Carnegie Inst. Year Book 70: 504-507, 1971.

10954 - MURATA, T., TODA, F., UCHINO, K., YAKUSHIJI, E.: Water-soluble chlorophyll protein of Brassica oleracea var. Botrys (cauliflower). - Biochim. biophys. Acta 245: 208-215, 1971.

10955 - MURRAY, A.E., KLEIN, A.O.: Relationship between photoconvertible and non-photoconvertible protochlorophyllides. - Plant Physiol. 48: 383-388, 1971.

10956 - MURRAY, D.R., BRADBEER, J.W.: Inhibition of photosynthetic CO_2 fixation in spinach chloroplasts by α-hydroxy 2-pyridinemethanesulphonate. - Phytochemistry 10: 1999-2003, 1971.

10957 - MURRAY, D.R., GIOVANELLI, J., SMILLIE, R.M.: Photometabolism of glycollate by Euglena gracilis. - Aust. J. biol. Sci. 24: 23-33, 1971.

*10958 - MURTY, K.S.: Effect of top-dressing nitrogen at heading time on carbon assimilation of rice plant during the ripening period. - Indian J. Plant Physiol. 12: 202-210, 1969.

*10959 - MURTY, K.S., NAYAK, S.K.: Radiotracer studies on photosynthetic efficiency of different rice varieties. - In: Proc. Dep. At. Energy Symp. Rad. Radioisotop. Soil Stud. Plant Nutr. Bangalore 1970. Pp. 187-197. Bombay 1971.

*10960 - MUTUSKIN, A.A., PSHENOVA, K.V.: Plastotsianin iz list'ev pshenitsy. [Plastocyanin from wheat leaves.] - In: Biologicheskaya Rol' Medi. Pp. 52-55. Nauka, Moskva 1970. [Isolation; in R.]

10961 - MUTUSKIN, A.A., PSHENOVA, K.V., ALEKHINA, S.K., KOLESNIKOV, P.A.: Kharakteristika plastotsianina list'ev pshenitsy. [Characteristics of plastocyanin in wheat leaves.] - Biokhimiya 36: 236-239, 1971. [In R, ab: E.]

10962 - MUZAFAROV, A.M., TAUBAEV, T.T., UMAROV, G.Ya., BERDYKULOV, Kh.A., RAKHIMOV, A., SADYKOV, M.S.: Vliyanie impul'snogo kontsentrirovannogo solnechnogo sveta na rost, razvitie, intensivnost' fotosinteza i nakoplenie biomassy khlorelly. [Effect of impulse concentrated solar radiation on growth, development, photosynthetic rate and biomass accumulation in Chlorella.] - In: Kul'tivirovanie Vodoroslei i Vysshikh Vodnykh Rastenii v Uzbekistane. Pp. 32-35. Fan, Tashkent 1971. [In R.]

10963 - MUZAFAROV, A.M., UMAROV, G.Ya., TAUBAEV, T.T., SADYKOV, M.S., BERDYKULOV, Kh.A., RAKHIMOV, A.: Posledeistvie impul'snogo kontsentrirovannogo solnechnogo sveta na rost, intensivnost' fotosinteza i produktivnost' khlorelly. [After-effect of impulse concentrated solar light on growth, photosynthetic rate and productivity in Chlorella.] - In: Svetoimpul'snaya Stimulyatsiya Rastenii. Pp. 247-258. Nauka, Moskva 1971. [In R.]

10964 - MYERS, J.: Enhancement studies in photosynthesis. - Annu. Rev. Plant Physiol. 22: 289-312, 1971.

10965 - MYERS, J., GRAHAM, J.-R.: The photosynthetic unit in Chlorella measured by repetitive short flashes. - Plant Physiol. 48: 282-286, 1971.

*10966 - MYGAL', O.K.: Vplyv deyakikh mikroelementiv na chysel'nist' saprofitiv u kul'-
turi *Microcystis aeruginosa* (KÜTZ.)ELENK. [Effect of microelements on frequency of saprophytes in culture of *Microcystis aeruginosa* (KÜTZ.)ELENK.] -
Ukr. bot. Zh. *27*: 658-661, 1970. [Ps; in Ukr., ab: E.]

10967 - NADLER, K.D., HERRON, H.A., GRANICK, S.: Chlorophyll and the development of
oxygen evolution. - Plant Physiol. *47* (Suppl.): 45, 1971.

10968 - NAGATOMI, S., KODAMA, S.: [Studies on the selection methods of sugarcane breeding: III. On het correlations between yield components, stalk forms, juice quality and leaf forms.] - Jap. J. trop. Agr. *15*: 6-10, 1971. [Ps; in Jap., ab: E.]

10969 - NAGY, A.H., PACSÉRY, M., FALUDI-DÁNIEL, Á.: Activity and compartmentation
of photosynthetic carboxylases in normal and chloroplast-mutant maize leaves. - Physiol. Plant. *24*: 301-305, 1971.

10970 - NAKAGAWA, H., KUSHIDA, T., OGURA, N., TAKEHANA, H.: [Studies on the color
of tomato. V. Thermal degradation of lycopene and β-carotene.] - Nippon Shokuhin Kogyo Gakkai-Shi *18*: 259-263, 1971. [In Jap., ab: E.]

10971 - NAKAMURA, S., KIMURA, T.: Studies on spinach ferredoxin-nicotinamide adenine
dinucleotide phosphate reductase. Kinetic studies on the interactions of the
reductase and ferredoxin and a possible regulation of enzyme activities by
ionic strength. - J. biol. Chem. *246*: 6235-6241, 1971.

10972 - NAKAMURA, S., KIMURA, T.: A possible regulation of activity of ferredoxin-
NADP reductase and ferredoxin system by ionic strength: catalytic significance of the one to one complex. - FEBS Lett. *15*: 352-354, 1971.

10973 - NAKANISHI, M., WARD, F.J.: Some sources of error in the ^{14}C method for estimating primary productivity and their relationship to light intensity during
incubation. - Jap. J. Limnol. *32*: 85-89, 1971.

*10974 - NAKAYAMA, O.: [Use of flagellate algae for the treatment of waste waters
with non-aerated photoheterotrophic process securing an algal protein.] - J.
Ferment. Technol. *48*: 416-424, 1970. [Dissolved O_2 determination; in Jap.,
ab: E.]

10975 - NAKHUTSRISHVILI, G.Sh.: O fotosinteze vysokogornykh rasteniĭ tsentral'nogo
Kavkaza zimoĭ. [Photosynthesis of the high-mountain plants in the Central
Caucasus.] - Soobshch. Akad. Nauk gruz. SSR *64*: 417-419, 1971. [In R, ab:
E, Georg.]

10976 - NALBANDYAN, R.M., MUTUSKIN, A.A., PSHENOVA, K.V.: Izmenenie parametrov signala E.P.R. plastotsianina pod vliyaniem nekotorykh denaturiruyushchikh agentov. [Alteration of the plastocyanin EPR signal parameters under the effect
of some denaturating agents.] - Dokl. Akad. Nauk SSR *201*: 1396-1398, 1971.
[In R.]

*10977 - NALEWAJKO, C., MARIN, L.: Extracellular production in relation to growth of
four planktonic algae and of phytoplankton populations from Lake Ontario. -
Can. J. Bot. *47*: 405-413, 1969. [Ps.]

10978 - NARASINGA RAO, C.: Radio tracer studies on translocation of photosynthates
in rice. - In: Proc. Dep. at. Energy Symp. Rad. Radioisotop. Soil Stud.
Plant Nutr. Bangalore 1970. Pp. 199-206. Bombay 1971.

10979 - NARBUT, S.I., VASIL'EVA, V.E., EVSTIGNEEVA, T.A.: Izuchenie spontannoĭ pigmentnoĭ mutatsii redisa. [Spontaneous pigment mutation of radish.] - Vestn.
leningrad. Univ. Biol. *1971*: 108-116, 1971. [In R, ab: E.]

B10980 - NASYROV, Yu. S.: Geneticheskie Aspekty Fotosinteza. [Genetic Aspects of Photosynthesis.] - Donish, Dushanbe 1971. [In R, ab: E.]

10981 - NASYROV, Yu.S., ABDURAKHMANOVA, Z.N., ERGASHEV, A., ALIEV, K.: O mekhanizme
deĭstviya vysokogornoĭ UF radiatsii na stanovlenie i funktsional'nuyu aktivnost' fotosinteticheskogo apparata. [Mechanism of action of alpine UV radiation on the formation and functional activity of the photosynthetic apparatus.]
- Dokl. Akad. Nauk tadzh. SSR *14* (9): 53-56, 1971. [In R, ab: Tadzh.]

10982 - NASYROV, Yu.S., ALIEV, K.A., ABDULLAEV, Kh.A., MUZAFFAROVA, S.: Yaderno-tsi-
toplazmaticheskiĭ kontrol' sinteza nukleinovykh kislot i belkov v protsesse

formirovaniya fotosinteticheskogo apparata. [Nuclear-cytoplasmatic control
of the synthesis of nucleic acids and proteins during the formation of photosynthetic apparatus.] - In: NASYROV, Yu.S. (ed.): Geneticheskie Aspekty
Fotosinteza. Pp. 5-23. Donish, Dushanbe 1971. [In R, ab: E.]

10983 - NASYROV, Yu.S., KAS'YANENKO, A.G., ABDURAKHMANOVA, Z.N.: Gennaya regulyatsiya
struktury i funktsii fotosinteticheskogo apparata. [Gene regulation of structure and function of photosynthetic apparatus.] - In: Biokhimiya i Biofizika
Fotosinteza. Pp. 121-128. Irkutsk 1971. [In R.]

10984 - NASYROV, Yu.S., MAKARLA, U.K., MAKHMADBEKOVA, L.M., PINKHASOV, Yu.I.: Metabolizm i transport produktov fotosinteza u khlopchatnika. [Metabolism and
photosynthate transport in cotton.] - Dokl. Akad. Nauk tadzh. SSR 14 (6):
70-73, 1971. [In R, ab: Tadzh.]

10985 - NASYROV, Yu.S., SHAKIROVA, S.V.: Metabolizm produktov fotosinteza u bledno-
zelenogo mutanta gorokha v zavisimosti ot intensivnosti sveta. [Photosynthate
metabolism in a light-green pea mutant in dependence on irradiance.] - In:
NASYROV, Yu.S. (ed.): Geneticheskie Aspekty Fotosinteza. Pp. 186-194. Donish,
Dushanbe 1971. [In R, ab: E.]

*10986 - NÁTR, L.: Gas exchange of barley leaves as influenced by mineral nutrient
deficiency. - Sci. Agr. bohemoslov. 2: 211-218, 1970.

10987 - NÁTR, L.: The effect of fusicoccin on gas exchange of detached barley leaves.
- Photosynthetica 5: 195-199, 1971.

10988 - NÁTR, L.: Vliv deficitu minerálních živin na intenzitu fotosyntézy. [The effect of mineral-nutrients deficiency on photosynthetic rate.] - Rostlinná
Výroba (Praha) 17: 149-155, 1971. [In Czech, ab: E,R.]

10989 - NÁTR, L.: Odrůdové rozdíly v působení deficitu minerálních živin na intenzitu
fotosyntézy a transpirace. [Varietal differences in the effect of deficiency
in mineral nutrients on photosynthetic and transpiration rates.] - Rostlinná
Výroba (Praha) 17: 411-418, 1971. [In Czech, ab: E, G, R.]

10990 - NÁTR, L., PURŠ, J., BEZDĚK, V.: Intenzita fotosyntézy po aplikaci minerálních
živin postřikem na list u jarního ječmene. [Photosynthetic rate after foliar
application of sprayed mineral nutrients in spring barley.] - Rostlinná Výroba (Praha) 17: 519-528, 1971. [In Czech, ab: E, R.]

10991 - NAZARENKO, A.V., SAMUILOV, V.D., SKULACHEV, V.P.: Fotoindutsirovannye izmeneniya pH v kletkakh i khromatoforakh *Rhodospirillum rubrum*. [Photoinduced
changes of pH in the cells and chromatophores of *Rhodospirillum rubrum*.] -
Biokhimiya 36: 780-782, 1971. [In R, ab: E.]

10992 - NAZAROVA, I.G., EVSTIGNEEV, V.B.: Spektral'nye svoĭstva i fotosensibilizi-
ruyushchaya sposobnost' vodorastvorimykh analogov khlorofilla v svyazannom
s vysokopolimernym substratom sostoyanii. [Spectral properties and photosensitizing effect of water soluble chlorophyll analogs bound with polymer
substrate.] - Mol. Biol. (Moskva) 5: 826-833, 1971. [In R, ab: E.]

10993 - NEALES, T.F., TREHARNE, K.J., WAREING, P.F.: A relationship between net
photosynthesis, diffusive resistance, and carboxylating enzyme activity in
bean leaves. - In: HATCH, M.D., OSMOND, C.B., SLATYER, R.O. (ed.): Photosynthesis and Photorespiration. Pp. 89-96. Wiley-Interscience, New York - Sydney-
Toronto 1971.

10994 - NEAMŢU, G.: Rolul pigmenţilor carotinoidici în organismele vegetale. [Role
of carotenoid pigments in plant organisms.] - Stud. Cercet. Biochim. 14: 215-
228, 1971. [In Roum.]

10995 - NEAMŢU, G., BODEA, C.: Über die Carotinoide aus högeren Saprophyten. - Rev.
roum. Biochim. 8: 129-133, 1971.

*10996 - NEČAS, J.: Možnosti využití růstově analytických charakteristik v analýze
struktury komplexního znaku"výnosový výkon bramborových odrůd". [Possibilities
of utilization of growth analysis chracteristics in analysing the structure
of the complex character "yielding capacity of potato varieties".] - Rostlinná Výroba (Praha) 12: 759-778, 1966. [Growth analysis; in Czech, ab: E,
G, R.]

*10997 - NECHITAÏLO, V.A., TEPLITSKAYA, E.V., KHARKEVICH, S.S.: Dinamika soderzhaniya
         karotina i askorbinovoĭ kisloty v perspektivnykh kormovykh rasteniyakh pri-
         rodnoĭ flory Kavkaza, introdutsirovannykh v usloviyakh Kieva. [Dynamics in
         content of carotene and ascorbic acid in perspective fodder plants of natural
         flora of Caucasus, introduced in Kiev conditions.] - Rast. Resursy 2: 494-
         504, 1966. [In R.]

 10998 - NEDRANKO, L.V.: Produktivnost' fotosinteza tomatnykh rastenii v zavisimosti
         ot rezhima azotno-fosfornogo pitaniya. [Productivity of photosynthesis in
         tomato plants as dependent on N a P nutrition.] - Tr. kishinev. sel'skokhoz.
         Inst. M.V. Frunze 85 (Mineral'noe Pitanie i Svet kak Faktory Fotosintetiches-
         koĭ Deyatel'nosti Sel'skokhozyaĭstvennykh Rastenii): 41-49, 77, 1971. [In R.]

*10999 - NEGISI, K.: [Respiration in non-photosynthetic organs of trees in relation
         to dry matter production of forests.] - J. Jap. Forestry Soc. 52: 331-345,
         1970. [In Jap.]

 11000 - NEKRASOVA, G.F.: Vliyanie CO_2 na fotosinteticheskiĭ metabolizm ugleroda.
         [Effect of CO_2 on photosynthetic carbon metabolism.] - In: ZALENSKIĬ, O.V.
         (ed.): Fotosintez i Ispol'zovanie Solnechnoĭ Energii. Pp. 164-169. Nauka,
         Leningrad 1971. [In R, ab: E.]

*11001 - NELSON, C.J., SMITH, D.: Growth of birdsfoot trefoil and alfalfa. IV. Car-
         bohydrate reserve levels and growth analysis under two temperature regimes.
         - Crop Sci. 9: 589-591, 1969. [Growth analysis.]

 11002 - NELSON, L.A.: Some varietal responses of barley measured by their carbon di-
         oxide exchange rates. - Diss. Abstr. int. B 32: 1319-B, 1971.

 11003 - NELSON, N., NELSON, H., NAIM, Y., NEUMANN, J.: Effect of pyridine on the
         light-induced pH rise and postillumination ATP synthesis in chloroplasts. -
         Arch. Biochem. Biophys. 145: 263-267, 1971.

 11004 - NEMKY, E.: Nyárklónok összehasonlító vizsgálata. [Comparison between cotton-
         wood populations.] - Erdészeti Faipari Egyetem Tud. Közl. (Sopron) 1970:
         115-125, 1971. [Ps; in Hung., ab: E, G, R.]

 11005 - NESTEROV, V.G., ISHIN, Yu.D.: Vliyanie zagryaznennosti assimilyatsionnogo
         apparata sosny na ego deyatel'nost'. [Effect of pollution of the pine assi-
         milation apparatus on its activity.] - Dokl. TSKhA (Moskva) 162: 337-340,
         1971. [In R.]

*11006 - NESTEROV, V.G., NIKOLAEV, L.V.: Primenenie metodov infrakrasnoĭ spektroskopii
         i betaspektroskopii pri opredelenii soderzhaniya vody i sukhogo veshchestva
         v zhivykh list'yakh. [Use of infra-red spectroscopy and beta-spectroscopy
         for determining the contents of water and dry matter in living leaves.] -
         Dokl. TSKhA (Moskva) 154: 359-363, 1969. [In R.]

 11007 - NEUMANN, J., ARNTZEN, C.J., DILLEY, R.A.: Two sites for adenosine triphos-
         phate formation in photosynthetic electron transport mediated by photosystem
         I. Evidence from digitonin subchloroplast particles. - Biochemistry 10: 866-
         873, 1971.

 11008 - NEUMANN, J., ARNTZEN, C.J., DILLEY, R.A.: Quantum requirement of photosystem
         I. Mediated ATP formation in chloroplast fragments. - FEBS Lett. 16: 164-166,
         1971.

 11009 - NEUMANN, J., LEVINE, R.P.: Reversible pH changes in cells of Chlamydomonas
         reinhardi resulting from CO_2 fixation in the light and its evolution in the
         dark. - Plant Physiol. 47: 700-704, 1971.

 11010 - NEUSCHELER-WIRTH, H.: Die lichtorientierte Chloroplastenbewegung von Mougeo-
         tia in Abhängigkeit von der Vorkultur, vom Licht-Dunkel-Wechsel und vom Zell-
         alter. - Z. Pflanzenphysiol. 65: 130-139, 1971.

 11011 - NEUSHUL, M.: Uniformity of thylakoid structure in a red, a brown, and two
         blue-green algae. - J. Ultrastruct. Res. 37: 532-543, 1971.

 11012 - NEUVILLE, D., DASTE, P.: Observations concernant la production de pigment
         bleu par la Diatomée Navicula ostrearia (GAILLON) BORY maintenue en culture
         unialgale sur un milieu synthétique. - Compt. rend. Acad. Sci. Paris, Sér.
         D 272: 2232-2234, 1971.

11013 - **NEWBOULD, P.J.**: Comparative production of ecosystems. - In: WAREING, P.F., COOPER, J.P. (ed.): Potnetial Crop Production. A Case Study. Pp. 228-238. Heinemann Educational Books, London 1971. [Growth analysis, Chl.]

11014 - **NEWCOMB, E.H., FREDERICK, S.E.**: Distribution and structure of plant microbodies (peroxisomes). - In: HATCH, M.D., OSMOND, C.B., SLATYER, R.O. (ed.): Photosynthesis and Photorespiration. Pp. 442-457. Wiley-Interscience, New York - London - Sydney - Toronto 1971.

11015 - **NGUYEN DUC, Anh Thu, VIEIRA DA SILVA, J.**: Evolution du point de compensation du CO_2 chez le Cotonnier sous l'action de traitements osmotiques. - Compt. rend. Acad. Sci. Paris, Sér. D *273*: 1291-1294, 1971. [Ps.]

*11016 - **NICHIPOROVICH, A.A.**: O metodakh otsenki fotosinteticheskoĭ funktsii rasteniĭ v svyazi s zadachami selektsii. [Methods of assessing plant photosynthetic function in relation to aims of breeding.] - In: Tezisy Dokladov Vsesoyuznogo Soveshchaniya po Unifikatsii Metodov i Priborov dlya Massovykh Izmereniĭ Intensivnosti Fotosinteza. Pp. 84-88. Nauch.-issled. Inst. Rastenievod. N.I. Vavilova, Leningrad - Pushkin 1970. [In R.]

11017 - **NICHIPOROVICH, A.A.**: Fotosintez i nekotorye printsipy primeneniya udobreniĭ kak sredstva optimizatsii fotosinteticheskoĭ deyatel'nosti i produktivnosti rasteniĭ. [Photosynthesis and principles of fertilizer application as means of optimizing photosynthetic activity and productivity of plants.] - Agrokhimiya *1971* (1): 3-13, 1971. [In R.]

11018 - **NICHIPOROVICH, A.A.**: K 200-letiyu otkrytiya fotosinteza. [To the 200th anniversary of the discovery of photosynthesis.] - Fiziol. Rast. *18*: 1077-1087, 1971. [In R, ab: E.]

11019 - **NICHIPOROVICH, A.A., ASROROV, K.A.**: O nekotorykh printsipakh optimizatsii fotosinteticheskoĭ deyatel'nosti rasteniĭ v posevakh. [Several principles of optimization of photosynthetic activity in plant stands.] - In: ZALENSKIĬ, O.V. (ed.): Fotosintez i Ispol'zovanie Solnechnoĭ Energii. Pp. 5-17. Nauka, Leningrad 1971. [In R, ab: E.]

11020 - **NICHOLS, M.A.**: Growth responses of radish to nitrogen and phosphorus fertilizers. - Hort. Res. *2*: 156-160, 1971. [Growth analysis.]

11021 - **NICOLSON, G.L.**: Structure of the photosynthetic apparatus in protein-embedded chloroplasts. - J. Cell Biol. *50*: 258-263, 1971.

11022 - **NIEDERMAN, R.A., GIBSON, K.D.**: The separation of chromatophores from the cell envelope in *Rhodopseudomonas spheroides*. - Prep. Biochem. *1*: 141-150, 1971.

*11023 - **NIFONTOVA, M.G.**: K voprosu o potentsial'nom fotosinteze nekotorykh rasteniĭ lesotundry. [Potential photosynthesis of some forest-tundra plants.] - Uch. Zap. ural'sk.gos. Univ. *113*, Ser. biol. 8 (Voprosy Regulyatsii Fotosinteza, Sb. 2): 180-191, 1970. [In R.]

11024 - **NIGON, V., HEIZMANN, P., VERDIER, G., FREYSSINET, G., SALVADOR, G., RICHARD, F., TRABUCHET, G., NICOLAS, P., INNOCENT, J.P.**: La formation des structures chloroplastiques et son déterminisme chez *Euglena gracilis*. - Groupe lyonnaise Rech. Biol. cell., Rapport annu. *1971*: 19-38, 1971.

11025 - **NIĬLISK, Kh.I.**: Uglovoe raspredelenie spektral'nykh intensivnosteĭ rasseyannoĭ radiatsii. [Angular distribution of spectral intensities of diffuse radiation.] - In: ZALENSKIĬ, O.V. (ed.): Fotosintez i Ispol'zovanie Solnechnoĭ Energii. Pp. 31-39. Nauka, Leningrad 1971. [In R, ab: E.]

*11026 - **NIKOLAEVSKIĬ, V.S.**: Bystryĭ metod opredeleniya ob"ema polosteĭ v gubchatoĭ parenkhime list'ev. [A rapid method of determining cavities in spongy parenchyma of leaves.] - Uch. Zap. perm. Ord. trud. kras. Znameni gos. Univ. A.M. Gor'kogo *222*: 159-164, 1969. [In R.]

11027 - **NIKOLAEVSKIĬ, V.S., FIRGER, V.V., BELOKRYLOVA, L.M.**: Svetovoĭ metabolizm ugleroda-14 u dekorativnykh rasteniĭ i ego rol' v gazoustoĭchivosti. [Light metabolism of carbon-14 in ornamental plants and its role in gas stability.] - Uch. Zap. perm. gos. Univ. *256*: 55-92, 1971. [Ps; in R.]

*11028 - **NILOV, G.I., BLAGONRAVOVA, L.N.**: Fitotoksicheskoe deĭstvie nekotorykh khlororganicheskikh pestitsidov na rasteniya. [Phytotoxic effect of certain chlororganic pesticides on plants.] - Tr. gos. nikit. bot. Sada *46*: 114-122, 1970. [Chl; in R, ab: E.]

11029 - **NILOVSKAYA, N.T., BOKOVAYA, M.M.:** K voprosu ob izuchenii gazoobmena i urozhaĭnosti posevov morkovi pri razlichnykh svetovykh rezhimakh. [Gas-exchange and productivity of carrots under different light conditions.] - In: ZALENSKIĬ, O.V. (ed.): Fotosintez i Ispol'zovanie Solnechnoĭ Energii. Pp. 75-79. Nauka, Leningrad 1971. [In R, ab: E.]

*11030 - **NILOVSKAYA, N.T., LAPTEV, V.V.:** Metodika issledovaniya dinamiki fotosinteza i dykhaniya posevov sel'skokhozyaĭstvennykh rasteniĭ v fitotronakh. [Method of studying photosynthesis and respiration dynamics of crop stands in phytotrons.] - In: Tezisy Dokladov Vsesoyuznogo Soveshchaniya po Unifikatsii Metodov i Priborov dlya Massovykh Izmereniĭ Intensivnosti Fotosinteza. Pp. 82-83. Nauch.-issled. Inst. Rastenievod. N.I. Vavilova, Leningrad - Pushkin 1970. [In R.]

*11031 - **NISHIMURA, M.:** Kinetics of light-induced electron transfer, internal and external H$^+$ changes and phosphorylation in *Rhodospirillum rubrum* and *Chromatium*. - Fed. Proc. *27*: 343, 1968.

*11032 - **NISHIMURA, M.:** Carotenoid absorption shift as indicator of membrane phenomena and its relationship to energy transfer processes in chromatophores of photosynthetic bacteria. - Fed. Proc. *28*: 884, 1969.

11033 - **NISHIMURA, M.:** Fluorochrome labeling of chromatophores of photosynthetic bacteria with auramine O. - In: CHANCE, B., LEE, C.-P., BLASIE, J.K.: Probes of Structure and Function of Macromolecules and Membranes. Vol. 1. Pp. 227-233. Academic Press, New York - London 1971.

*11034 - **NISHIMURA, M., KADOTA, K.:** Light-induced electron transfer, internal and external H$^+$ changes and phosphorylation in *Rhodospirillum rubrum*. - Fed. Proc. *26*: 610, 1967.

*11035 - **NISHIMURA, M., PRESSMAN, B.C.:** Effects of ionophorous antibiotics on the light induced internal and external hydrogen ion changes and phosphorylation in bacterial chromatophores. - Biochemistry *8*: 1360-1370, 1969.

11036 - **NISHIZAKI, Y., JAGENDORF, A.T.:** Kinetics of acid efflux from chloroplasts following the acid-base transition. - Biochim. biophys. Acta *226*: 172-186, 1971.

*11037 - **van NOORT, D., WALLACE, A.:** Role of iron in chlorophyll synthesis. - In: WALLACE, A. (ed.): Current Topics in Plant Nutrition. Pp. 27-28. Los Angeles, Calif. 1966.

11038 - **NORMAN, J.M., MILLER, E.E., TANNER, C.B.:** Light intensity and sunfleck-size distributions in plant canopies. - Agron. J. *63*: 743-748, 1971. [Canopy irradiance model.]

11039 - **NORRIS, J.R., UPHAUS, R.A., CRESPI, H.L., KATZ, J.J.:** Electron spin resonance of chlorophyll and the origin of signal I in photosynthesis. - Proc. nat. Acad. Sci. USA *68*: 625-628, 1971.

11040 - **NOSTICZIUS, Á.:** Is it justified to relate the photosynthetic activity to the soluble protein? - Acta agron. Acad. Sci. hung. *20*: 446-447, 1971.

11041 - **NULTSCH, W.:** Phototactic and photokinetic action spectra of the diatom *Nitzschia communis*. - Photochem. Photobiol. *14*: 705-712, 1971.

11042 - **NUR, I.M.:** Different methods for determining leaf area of some oil crops. - J. agr. Sci. *77*: 19-24, 1971.

11043 - **NYUPPIEVA, K.A.:** Vliyanie zamorozkov na sostoyanie pigmentov v list'yakh razlichnykh po ustoĭchivosti vidov kartofelya. [Effect of frost on leaf pigments of potato species with different resistance.] - Uch. Zap. petrozavodsk. gos. Univ. *17*:(6 - Voprosy Zimostoĭkosti Rasteniĭ v Usloviyakh Karelii): 78-91, 1969 (1971).

*11044 - **OAKS, A., BIDWELL, R.G.S.:** Compartmentation of intermediary metabolites. - Annu. Rev. Plant Physiol. *21*: 43-66, 1970. [Ps.]

*11045 - **OBYDENNYĬ, P.T.:** Usovershenstvovanie peredvizhnoĭ ustanovki dlya avtomaticheskoĭ registratsii fotosinteza i dykhaniya rasteniĭ. [Improvement of a portable device for automatic recording of plant photosynthesis and respiration.] - In: Tezisy Dokladov Vsesoyuznogo Soveshchaniya po Unifikatsii Metodov i Priborov dlya Massovykh Izmereniĭ Intensivnosti Fotosinteza. Pp. 89-94. Nauch.-issled. Inst. Rastenievod. N.I. Vavilova, Leningrad - Pushkin 1970. [In R.]

11046 - O'CARRA, P., KILLILEA, S.D.: Subunit structures of *C*-phycocyanin and *C*-phycoerythrin. - Biochem. biophys. Res. Commun. *45*: 1192-1197, 1971.

11047 - OCHIAI, H., SHIBATA, H., SUEKANE, T.: Chloroplast development in 4-thiouridine treated radish cotyledons. - Agr. biol. Chem. *35*: 1259-1266, 1971.

11048 - OCHIAI, H., SHIBATA, H., SUEKANE, T., KONO, Y.: [Effect of 4-thiouridine on chloroplast development in a higher plant.] - Amino Acid nucl. Acid *24*: 1-13, 1971. [Chl; in Jap., ab: E.]

11049 - OELZE, J., KAMEN, M.D.: Adenosine triphosphate cellular levels in *Rhodospirillum rubrum* during transition from aerobic to anaerobic metabolism. - Biochim. biophys. Acta *234*: 137-143, 1971.

11050 - OELZE, J., WEAVER, P.: The adjustment of photosynthetically grown cells of *Rhodospirillum rubrum* to aerobic light conditions. - Arch. Mikrobiol. *79*: 108-121, 1971. [Chl.]

11051 - OELZE-KAROW, H., BUTLER, W.L.: The development of photophosphorylation and photosynthesis in greening bean leaves. - Plant Physiol. *48*: 621-625, 1971.

11052 - OGASAWARA, N., MIYACHI, S.: Effects of dark preincubation and chloramphenicol on blue light-induced CO_2 incorporation in *Chlorella* cells. - Plant Cell Physiol. *12*: 675-682, 1971.

11053 - OGAWA, T.: [Molecular extinction coefficient of chlorophyll in the reaction center.] - Chemistry and Life [Kagaku to Seibutsu] *9*: 135-140, 1971. [In Jap.]

11054 - OGAWA, T., VERNON, L.P.: Increased content of cytochromes 554 and 562 in *Anabaena variabilis* cells grown in the presence of diphenylamine. - Biochim. biophys. Acta *226*: 88-97, 1971.

11055 - OGREN, W.L., BOWES, G.: Ribulose diphosphate carboxylase regulates soybean photorespiration. - Nature-new Biol. *230*: 159-160, 1971.

*11056 - OGURA, N.: [Stability of chlorophyll pigment.] - Tech. Bull. Fac. Hort., Chiba Univ. [Chiba Daigaku Engeigakubu Gakujutsu Hokoku] *14*: 49-55, 1966. [In Jap., ab: E.]

11057 - OHAD, I., WALLACH, D., EYTAN, G., BAR-NUN, S.: Development of photophosphorylation and proton pump activity during formation of chloroplast membranes in etiolated *Chlamydomonas reinhardi* Y-1 greening in the presence of inhibitors of protein synthesis. - In: QUAGLIARIELLO, E., PAPA, S., ROSSI, C.S. (ed.): Energy Transduction in Respiration and Photosynthesis. Pp. 393-394. Adriatica Editrice, Bari 1971.

11058 - Ó hEOCHA, C.: Pigments of the red algae. - Oceanogr. mar. Biol. annu. Rev. *9*: 61-82, 1971.

11059 - OHKI, R., KUNIEDA, R., TAKAMIYA, A.: Effects of various cations on separation of the two photochemical systems by digitonin treatment. - Biochim. biophys. Acta *226*: 144-153, 1971.

11060 - OKABE, K.: [Use of radioisotopes for studies on photosynthesis.] - Nucl. Eng. [Genshiryoku Kogyo] *17* (6): 68-72, 1971. [In Jap.]

11061 - OKADA, M., TAKAMIYA, A.: Light-induced absorption spectrum changes of carotenoid in chromatophores of photosynthetic bacterium, *Rhodopseudomonas spheroides*. II. Rapid and slow absorption changes of caroteinoid in chromatophores. - Plant Cell Physiol. *12*: 683-694, 1971.

11062 - OKALI, D.U.U.: Rates of dry-matter production in some tropical forest-tree seedlings. - Ann. Bot. *35*: 87-97, 1971. [Growth analysis.]

11063 - OKANENKO, A.S., POCHINOK, Kh.N., GOLIK, K.N., SMELYANSKAYA, E.P.: Fotosintez i produktivnost' v svyazi s vodnym rezhimom rasteniĭ. [Photosynthesis and

productivity in relation to plant water relations.] - In: Fotosintez, Rost
i UstoTchivost' RasteniT. Pp. 5-28. Naukova Dumka, Kiev 1971. [In R.]

11064 - OKANENKO, A.S., POCHINOK, Kh.N., MITROFANOV, B.A., GULYAEV, B.I.: Ispol'zovanie solnechnoT energii posevami sel'skokhozyaTstvennykh kul'tur. [Use of
solar energy by crop canopies.] - Fiziol. Biokhim. kul't. Rast. *3*: 241-251,
1971. [In R, ab: E.]

11065 - OKANENKO, A.S., POCHINOK, Kh.N., MITROFANOV, B.A., GULYAEV, B.I., LAVRENTO-
VICH, D.I., MAKHOVSKAYA, M.A., GOLIK, K.N., POGOL'SKAYA, V.I.: Intensivnost'
i produktivnost' fotosinteza i ispol'zovanie solnechnoT radiatsii posevami
sel'skokhozyaTstvennykh rasteniT. [Rate and productivity of photosynthesis
and the use of solar radiation by stands of agricultural crops.] - In: Fotosintez, Rost i UstoTchivost' RasteniT. Pp. 28-85. Naukova Dumka, Kiev 1971.
[In R.]

11066 - OKAYAMA, S., EPEL, B.L., ERIXON, K., LOZIER, R., BUTLER, W.L.: The effects
of lipase on spinach and *Chlamydomonas* chloroplasts. - Biochim. biophys. Acta *253*: 476-482, 1971.

*11067 - OKAZAWA, Y.: [Occurrence of natural cytokinin in potato tuber.] - Proc. Crop
Sci. Soc. Jap. *38*: 25-30, 1969. [Chl; in Jap., ab: E.]

11068 - OKU, T., KAWAHARA, H., TOMITA, G.: The Hill reaction and oxygen uptake in
isolated pine chloroplasts. - Plant Cell Physiol. *12*: 559-566, 1971.

11069 - OKU, T., TOMITA, G.: Effect of polyethylene glycol on heat inactivation of
the Hill reaction. - Biochem. biophys. Res. Commun. *44*: 958-962, 1971.

11070 - OKU, T., TOMITA, G.: The preparation of photoactive chloroplasts from pine
leaves. - Photosynthetica *5*: 28-31, 1971.

11071 - OKU, T., TOMITA, G.: Protochlorophyllide holochrome 2. Effect of ultraviolet
irradiation on the phototransformation. - Photosynthetica *5*: 133-138, 1971.

11072 - OKUDA, J., INOUE, T., MIWA, I.: Rapid polarographic microdetermination of
dissolved oxygen in water with flavin enzyme. - Analyst *96*: 858-864, 1971.

11073 - OKUNTSOV, M.M., SIMONOVA, E.I., VOZILOVA, L.D., IL'INA, L.P.: SeTstvie sveta
i temperaturnykh usloviT na zaklyuchitel'nuyu stadiyu biosinteza khlorofilla.
[Effect of light and temperature on the final phase of chlorophyll biosynthesis.] - In: Biokhimiya i Biofizika Fotosinteza. Pp. 59-65. Irkutsk 1971.
[In R.]

11074 - OKUNTSOV, M.M., VERKHOTUROVA, G.S.: Vliyanie sveta na postuplenie radioaktiv-
nogo ugleroda iz ekzogennykh rastvorov glyukozy v di- i trikarbonovye kisloty
list'ev fasoli. [Effect of light on labeling of di- and tricarboxylic acids
by radioactive carbon from exogenous solution in bean leaves.] - Nauch. Dokl.
vyssh. Shkoly, biol. Nauki *14* (10): 76-81, 1971. [In R.]

*11075 - CKUNTSOV, M.M., VOZILOVA, L.D.: Osobennosti biosinteza predshestvennikov
Khlorofilla v zelenoT khvoe prorostkov sosny. [Features of biosynthesis of
chlorophyll precursors in green needles of pine seedlings.] - In: Materialy
NauchnoT Konferentsii Molodykh Uchenykh Vuzov g. Tomska. Vol. 1. Pp. 197-
198. Izd. tomskogo Univ., Tomsk 1968. [In R.]

*11076 - OKUNTSOV, M.M., VOZILOVA, L.D.: O vozmozhnosti fotokhimicheskogo biosinteza
protokhlorofillida zelenoT khvoeT prorostkov sosny. [Possibility of photochemical biosynthesis of protochlorophyllide by green needles of pine seedlings.] - In: Inform. Byull. Sibir'. Otdel. Akad. Nauk SSSR. No. 4. Pp. 74-
75. Irkutsk 1969. [In R.]

*11077 - OKUNTSOV, M.M., VOZILOVA, L.D.: TemnovoT biosintez khlorofillida *A* prorost-
Kami sosny, vyrashchennymi na svetu. [Dark biosynthesis of chlorophyllide *a*
in pine seedlings grown in light.] - In: Inform. Byull. Sibir'. Otdel. Akad.
Nauk SSSR. No. 5. P. 50. Irkutsk 1969. [In R.]

11078 - OKUNTSOV, M.M., VOZILOVA, L.D.: Usloviya biosinteza protokhlorofillida v
khvoe prorostkov sosny. [Conditions of protochlorophyllide biosynthesis in
needles of pine seedlings.] - Nauch. Dokl. vyssh. Shkoly, biol. Nauki *14*
(8): 83-88, 1971. [In R.]

*11079 - OLD, S.M.: Microclimates, fire, and plant production in an Illinois prairie.
- Ecol. Monogr. *39*: 355-384, 1969. [Chl.]

*11080 - **OLECH, K.**: Wpływ niektórych herbicydów na fotosyntezę oraz oddychanie roślin uprawnych i chwastów. I. Atrazinu (2-chloro-4-etyloamino-6-izopropyloamino-S-triazyny). [Effect of some herbicides on photosynthesis and respiration of crop plants and weeds. I. Atrazine (2-chloro-4-ethylamino-6-isopropylamino-S-triazine).] - Ann. Univ. M. Curie-Skłodowska, Sect. E *21*: 289-308, 1966. [In Pol., ab: E, R.]

*11081 - **OLECH, K.**: Wpływ niektórych hercibydów na fotosyntezę i oddychanie roślin uprawnych i chwastów. II. Działanie Atrazinu (2-chloro-4-etyloamino-6-izopropyloamino-S-triazyny) w zależności od warunków zewnętrznych. [Effect of some herbicides on photosynthesis and respiration of crop plants and weeds. II. Correlation between the action of Atrazine (2-chloro-4-ethylamino-6-isopropyl-amino-S-triazine).] - Ann. Univ. M. Curie-Skłodowska, Sect. E *22*: 149-166, 1967. [In Pol., ab: E, R.]

*11082 - **OLECH, K.**: Wpływ niektórych herbicydów na fotosyntezę i oddychanie roślin uprawnych i chwastów. III. Działanie Afalonu (N-metylo-N-metoksy-N-3,4-dwu-chlorofenylomocznika) i Tenoranu (N,N-dwumetylo-N-4(4-chlorofenoksy)-fenylo-mocznika). [Effect of some herbicides on photosynthesis and respiration of crop plants and weeds. III. Action of Linuron (N-(3,4-dichlorophenyl)-N-methylurea) and Tenoran (N'-4(4-chlorophenoxy)phenyl-N,N-dimethylurea.] - Ann. Univ. M. Curie-Skłodowska, Sect. E *22*: 167-183, 1967. [In Pol., ab: E, R.]

*11083 - **OLECH, K.**: Wpływ niektórych herbicydów na fotosyntezę oraz oddychanie roślin uprawnych i chwastów: IV. Działanie Burexu (1-fenylo-4-amino-5-chloro-6-ketopirydazyny. [Effect of some herbicides on photosynthesis and respiration of crop plants and weeds. IV. Action of Burex (1-phenyl-4-amino-5-chloro-6-ketopyridazine.] - Ann. Univ. M. Curie-Skłodowska, Sect. E *22*: 185-195, 1967. [In Pol., ab: E, R.]

*11084 - **OLECH, K.**: Ocena przydatności Afalonu (Linuronu) jako preparatu chwastobój-czego w uprawach marchwi i pietruszki na podstawie działania tego herbicydu na fotosyntezę roślin. [Evaluation of applicability of Afalon (Linuron) as herbicide in cultures of carrot and parsley based on the action of this herbicide on plant photosynthesis.] - Ann. Univ. M. Curie-Skłodowska, Sect. E *23*: 193-200, 1968. [In Pol., ab: E, R.]

11085 - **OLIMID, V.**: Intensitatea fotosintezei la orz - Cenad 396 sub influenţa sarurilor minerale aplicate la semanat şi în cursul perioadei de vegetaţie. [Effect of mineral salts applied during sowing and during vegetation period on photosynthetic rate of barley cv. 396 Cenad.] - In: Anale - Univ. Craiova, Ser. a III-a Biol. - Ştiinte agr. 2 (12): 185-191, 1971. [In Roum., ab: E, F.]

11086 - **OLSON, J.M.**: Bacteriochlorophyll-protein of green photosynthetic bacteria. - In: COLOWICK, S.P., KAPLAN, N.O. (ed.): Methods in Enzymology. Vol. 23. Pp. 636-644. Academic Press, New York - London 1971.

11087 - **OLSON, J.S.**: Primary productivity: temperate forests, especially American deciduous types. - In: DUVIGNEAUD, P. (ed.): Productivity of Forest Ecosystems. Pp. 235-258. Unesco, Paris 1971.

11088 - **OMRAN, R.G., BENEDICT, C.R., POWELL, R.D.**: Effects of chilling on protein synthesis and CO_2 fixation in cotton leaves. - Crop Sci. *11*: 554-556, 1971.

11089 - **ONDOK, J.P.**: Calculation of mean leaf area ratio in growth analysis. - Photosynthetica *5*: 269-271, 1971.

11090 - **ONDOK, J.P.**: Horizontal structure of some macrophyte stands and its production aspects. - Hidrobiologia (Bucureşti) *12*: 47-55, 1971.

11091 - **ONDOK, J.P.**: Indirect estimation of primary values used in growth analysis. - In: ŠESTÁK, Z., ČATSKÝ, J., JARVIS, P.G. (ed.): Plant Photosynthetic Production. Manual of Methods. Pp. 392-411. Dr. W. Junk N.V. Publ., The Hague 1971.

11092 - **ONDOK, J.P., KVĚT, J.**: Integral and differential formulae in growth analysis. - Photosynthetica *5*: 358-363, 1971.

11093 - **O'NEAL, D., GIBBS, M., PEAVEY, D.**: Photosynthetic carbon metabolism of isolated corn chloroplasts. - In: HATCH, M.D., OSMOND, C.B., SLATYER, R.O. (ed.): Photosynthesis and Photorespiration. Pp. 240-245. Wiley-Interscience, New York - London - Sydney - Toronto 1971.

11094 - van OORSCHOT, J.L.P., de ROOY, M.: The selectivity of some herbicides in bulb growing. - Acta hort. *23*: 305-311, 1971. [Ps, growth analysis.]

11095 - OPARINA, N.V.: Osobennosti stroeniya khloroplastov u *Zygnema* sp. (*Chlorophyceae*). [Characteristics of the chloroplast structure in *Zygnema* sp. (*Chlorophyceae*).] - Tsitol. Genet. *5*: 467-468, 1971. [In R, ab: E.]

11096 - ÖQUIST, G.: Changes in pigment composition and photosynthesis induced by iron-deficiency in the blue-green alga *Anacystis nidulans*. - Physiol. Plant. *25*: 188-191, 1971.

*11097 - ORLOVA, I.F.: Vliyanie mikroelementov na urozhaĭ i kachestvo gorokha. [Effects of trace nutrients on the yield and quality of peas.] - Nauch. Tr., orlovskaya obl. sel'skokhoz. opyt. Sta. *5*: 83-96, 1970. [Chl; in R.]

*11098 - ORLOVA, I.F., GULYAEVA, V.E.: Deĭstvie molibdena na urozhaĭ i khimicheskiĭ sostav krasnogo klevera. [Effect of molybdenum on the yield and chemical composition of red clover.] - Nauch. Tr., orlovskaya obl. sel'skokhoz. opyt. Sta. *5*: 105-110, 1970. [Chl, Car; in R.]

11099 - ORR, A.R., KESSLER, J.E., TePASKE, E.R.: Diuron (DCMU) induced growth, morphological and photosynthetic changes in *Eudorina elegans*. - Amer. J. Bot. *58*: 459-460, 1971.

11100 - ORSENIGO, M., MARZIANI, G.: Chloroplast structural changes during temperature-induced bleaching and re-greening in a "golden leaf" mutant of maize. - Caryologia *24*: 347-364, 1971.

*11101 - OSIPOVA, O.P.: Osnovnye trebovaniya pri vydelenii khloroplastov. [Principal requirements for the isolation of chloroplasts.] - In: KIRICHENKO, E.B. (ed.): Metody Vydeleniya Khloroplastov. Pp. 7-17. Pushchino-na-Oke 1970. [In R, ab: E.]

11102 - OSIPOVA, O.P., KHEĬN, Kh.Ya., NICHIPOROVICH, A.A.: Aktivnost' fotosinteticheskogo apparata rasteniĭ, vyrosshikh pri raznoĭ intensivnosti sveta. [The activity of the photosynthetic apparatus in plants grown at different light intensity.] - Fiziol. Rast. *18*: 257-263, 1971. [In R, ab: E.]

11103 - OSMAN, A.M.: Dry-matter production of a wheat crop in ralation to light interception and photosynthetic capacity of leaves. - Ann. Bot. *35*: 1017-1035, 1971.

11104 - OSMAN, A.M.: Root respiration of wheat plants as influenced by age, temperature, and irradiation of shoots. - Photosynthetica *5*: 107-112, 1971.

11105 - OSMAN, A.M., MILTHORPE, F.L.: Photosynthetis of wheat leaves in relation to age, illuminance and nutrient supply I. Techniques. - Photosynthetica *5*: 55-60, 1971.

11106 - OSMAN, A.M., MILTHORPE, F.L.: Photosynthesis of wheat leaves in relation to age, illuminance and nutrient supply II. Results. - Photosynthetica *5*: 61-70, 1971.

11107 - OSMOND, C.B.: Respiration in the light. - Pac. Sci. Congr. Proc. *1*: 71, 1971.

11108 - OSMOND, C.B.: The absence of photorespiration in C_4 plants: real or apparent? - In: HATCH, M.D., OSMOND, C.B., SLATYER, R.O (ed.): Photosynthesis and Photorespiration. Pp. 472-482. Wiley-Interscience, New York - London - Sydney - Toronto 1971.

11109 - OSMOND, C.B.: Photorespiration: assessment. - In: HATCH, M.D., OSMOND, C.B., SLATYER, R.O. (ed.): Photosynthesis and Photorespiration. Pp. 544-546. Wiley-Interscience, New York - London - Sydney - Toronto 1971.

11110 - OSMOND, C.B., HARRIS, B.: Photorespiration during C_4 photosynthesis. - Biochim. biophys. Acta *234*: 270-282, 1971.

11111 - OSNITSKAYA, L.K., CHUDINA, V.I.: Sposobnost' fotosinteziruyushchikh bakteriĭ k avtotrofnomu i geterotrofnomu razvitiyu. [Capacity of photosynthesizing bacteria for autotrophic and heterotrophic development.] - In: Biokhimiya i Biofizika Fotosinteza. Pp. 75-79. Irkutsk 1971. [In R.]

11112 - ØSTGÅRD, O.: Klorofyllinnhald i grassortar. [Chlorophyll content in varieties of forage grasses.] - Forskning Forsøk Landbruket *22*: 419-429, 1971. [In Norweg.]

11113 - ØSTGÅRD, O., EAGLES, C.F.: Variation in growth and development in natural
populations of *Dactylis glomerata* from Norway and Portugal. II. Leaf development and tillering. - J. appl. Ecol. *8*: 383-391, 1971.

11114 - OSTROVSKAYA, L.K.: Fragmentirovanie khloroplastov rasteniĭ kak metod issledovaniya organizatsii fotosinteticheskikh edinits. [Fragmentation of plant
chloroplasts as a method for studying organization of photosynthetic units.]
- Fiziol. Biokhim. kul't. Rast. *3*: 252-263, 1971. [In R, ab: E.]

11115 - OTTEN, H.A.: Absorption changes in the reaction center of photosynthetic bacteria and π-electron calculations on bacteriochlorophyll, its mono cation
and anion. - Photochem. Photobiol. *14*: 589-596, 1971.

*11116 - OVSYANNIKOV, A.S.: Produktivnost' fotosinteza list'ev v raznych chastyakh
krony yabloni. [Productivity of photosynthesis of leaves in different parts
of crown of apple tree.] - Sadovodstvo *107* (12): 30-31, 1969. [In R.]

11117 - OVSYANNIKOV, A.S.: Sravnitel'naya produktivnost' fotosinteza u sortov yabloni
v period formirovaniya urozhaya. [Comparative productivity of photosynthesis
in apple varieties during yield formation.] - In: Sbornik Dokladov Pervoĭ
Vsesoyuznoĭ Konferentsii Molodykh Uchenykh po Sadovodstvu. Vol. II. Pp. 179-
184. Michurinsk 1971. [In R.]

*11118 - OVSYANNIKOV, A.S., LEBEDEV, V.M., OVSYANNIKOVA, R.S., LOBANOVA, E.V.: Meto-
dika opredeleniya kaloriĭnosti razlichnykh organov i tkaneĭ u plodovykh rasteniĭ. [Determination of caloric values of different organs and tissues of
fruit plants.] - Sbor. nauch. Rabot vsesoyuz. nauch.-issled. Inst. Sadovod.
I.V. Michurina *14*: 200-206, 1970. [In R.]

11119 - OVSYANNIKOV, A.S., OVSYANNIKOVA, R.S.: Metodika otsenki produktivnosti fotosinteza list'ev kryzhovnika v period formirovaniya urozhaya. [Methods of
assessment of the photosynthetic productivity of gooseberry leaves during
yield formation.] - Sbor. nauch. Rabot vsesoyuz. nauch.-issled. Inst. Sadovod. I.V. Michurina *16*: 131-140, 1971. [In R.]

11120 - OWEN, W.J., ROGERS, L.J.: Inhibition of photosynthetic electron transport
by 1,1,1-trichloro-2,2-bis-(p-chlorophenyl)ethane (DDT) at a site in the intermediate electron-transport chain. - Biochem. J. *125*: 43P-44P, 1971.

11121 - OWEN, W.J., ROGERS, L.J., HAYES, J.D.: Inhibition of the Hill reaction and
photophosphorylation by 1,1,1,-trichloro-2,2-bis-(p-chlorophenyl)ethane (DDT).
- Biochem. J. *121*: 6P-7P, 1971.

11122 - OZOLS, A., PĒTERSONS, E.: Fotosintēzes pētījumu nozīme un uzdevumi lauksaim-
niecības kultūru produktivitātes palielināšanā Latvijas PSR. [The role of
photosynthesis studies and the tasks of Latvian SSR in the raising of agricultural productivity.] - Izv. Akad. Nauk latv. SSR *1971* (9): 3-5, 1971.
[In Lithu.]

11123 - PAASCHE, E.: Effect of ammonia and nitrate on growth, photosynthesis, and
ribulosediphosphate carboxylase content of *Dunaliella tertiolecta*. - Physiol.
Plant. *25*: 294-299, 1971.

11124 - PADAN, E., RABOY, B., SHILO, M.: Endogenous dark respiration of the blue-
green alga, *Plectonema boryanum*. - J. Bacteriol. *106*: 45-50, 1971. [Ps.]

11125 - PADAN, E., RIMON, A., GINZBERG, D., SHILO, M.: A thermosensitive cyanophage
(LPP1-G) attacking the blue-green alga *Plectonema boryanum*. - Virology *45*:
773-776, 1971. [Ps.]

11126 - PAE, A.: Fotosinteticheskaya deyatel'nost' posevov kormovoĭ bryukvy "Kuuziku". [Photosynthetic activity of the Swedish turnip "Kuuziku".] - Eesti NSV
Tead. Akad. Toim., Biol. *20*: 311-320, 1971. [In R, ab: Est., G.]

11127 - PAINE, R.T.: The measurement and application of the calorie to ecological
problems. - Annu. Rev. Ecol. Systematics *2*: 145-164, 1971.

11128 - PALLAS, J.E. Jr., WILKINSON, S.R., MONSON, W.G., BURTON, G.W.: Thermo and
photoperiod effects on growth and chlorophyll concentration of *Cynodon dac-
tylon* cultivars. - Plant Physiol. *47* (Suppl.): 30, 1971.

11129 - PALMER, G., DUNHAM, W.R., FEE, J.A., SANDS, R.H., IIZUKA, T., YONETANI, T.:
The magnetic susceptibility of spinach ferredoxin from 77-250 °K: a mesure-

ment of the antiferromagnetic coupling between the two iron atoms. - Biochim.
biophys. Acta *245*: 201-207, 1971.

11130 - PAN, D., WAYGOOD, E.R.: A fundamental thermostable cyanide-sensitive phosphoenolpyruvate acid carboxylase in photosynthetic and other organisms. - Can.
J. Bot. *49*: 631-643, 1971.

11131 - PAPAGEORGIOU, G., GOVINDJEE: pH control of the chlorophyll *a* fluorescence
in algae. - Biochim. biophys. Acta *234*: 428-432, 1971.

11132 - PARISH, R.W.: The isolation of peroxisomes, mitochondria and chloroplasts
from leaves of spinach beet (*Beta vulgaris* L. ssp. *vulgaris*). - Europe. J.
Biochem. *22*: 423-429, 1971.

11133 - PARK, R.B.: Chloroplasts (and grana): photosynthetic electron transport in
aldehyde-fixed material. - In: COLOWICK, S.P., KAPLAN, N.O. (ed.): Methods
in Enzymology. Vol. 23. Pp. 248-250. Academic Press, New York - London 1971.

11134 - PARK, R.B.: The architecture of photosynthesis. - In: HARRIS, P.J. (ed.):
Biological Ultrastructure: the Origin of Cell Organelles. Pp. 25-40. Oregon
State Univ. Press, Corvallis 1971.

11135 - PARK, R.B., SANE, P.V.: Distribution of function and structure in chloroplast
lamellae. - Annu. Rev. Plant Physiol. *22*: 395-430, 1971.

11136 - PARK, R.B., STEINBACK, K.E., SANE, P.V.: Distribution of variable fluorescence
among subchloroplast fractions. - Biochim. biophys. Acta *253*: 204-207, 1971.

*11137 - PARKASH, V.: Chloroplast: its genetical and biochemical autonomy. - J. Post-
Graduate School *7*: 75-80, 1969.

11138 - PARKINSON, K.J.: Carbon dioxide infra-red gas analysis - Effects of water
vapour. - J. exp. Bot. *22*: 169-176, 1971.

11139 - PARLANGE, J.-Y., WAGGONER, P.E., HEICHEL, G.H.: Boundary layer resistance
and temperature distribution on still and flapping leaves. I. Theory and laboratory experiments. - Plant Physiol. *48*: 437-442, 1971.

11140 - PARSHINA, Z.S., BEDENKO, V.P., INTYKBAEVA, B.B., MAICHEKINA, R.M.: Opticheskie svoĭstva list'ev. [Optical properties of leaves.] - In: Ekologo-fiziologicheskie Issledovaniya Gornykh Rasteniĭ. Pp. 64-81. Nauka kaz. SSR, Alma-
Ata 1971. [In R.]

*11141 - PARSHINA, Z.S., PARSHIN, N.G.: Deĭstvie korotkovolnovoĭ radiatsii na rost i
razvitie rasteniĭ. [Effect of short-wave radiation on growth and development
of plants.] - Izv. Akad. Nauk kaz. SSR, Ser. biol. *1969* (6): 8-13, 1969.
[Chl, Car; in R, ab: Kazakh.]

11142 - PARSHINA, Z.S., SHOKOVA, R.I., BELOSLYUDOVA, L.F.: Pigmenty plastid list'ev.
[Plastid pigments of leaves.] - In: Ekologo-fiziologicheskie Issledovaniya
Gornykh Rasteniĭ. Pp. 59-64. Nauka kaz. SSR, Alma-Ata 1971. [In R.]

11143 - PASHCHENKO, V.N., MUREĬ, I.A., NICHIPOROVICH, A.A.: Issledovanie fiziologicheskikh osobennosteĭ tomatov v zavisimosti ot intensivnosti sveta, kontsentratsii elementov mineral'nogo pitaniya i tsenoticheskogo vzaimodeĭstviya
rasteniĭ. [Physiological features of tomato as related to irradiance, concentration of mineral elements, and plant interactions in the canopy.] -
Fiziol. Rast. *18*: 1134-1140, 1971. [Ps; in R, ab: E.]

11144 - PASHCHENKO, V.P.: Deĭstvie azotnogo i fosfornogo pitaniya rasteniĭ na vosstanovitel'nuyu funktsiyu fotosinteticheskogo apparata. [Effect of nitrogen
and phosphorus nutrition of plants on the reductive function of the photosynthetic apparatus.] - Fiziol. Rast. *18*: 695-700, 1971. [In R, ab: E.]

11145 - PATALAS, K.: Crustacean plankton communities in forty-five lakes in the Experimental Lakes Area, northwestern Ontario. - J. Fish. Res. Board Can. *28*:
231-244, 1971. [Chl.]

11146 - PATARAVA, B.D., DUMBADZE, V.Z.: Vliyanie doz azotnykh udobreniĭ na nekotorye
fiziologicheskie protsessy dvukhletnikh seyantsev chaĭnogo rasteniya. [Ef-
fect of the doses of nitrogen fertilizers on some physiological processes in
two-years old tea seedlings.] - Subtrop. Kul't. *1971* (4): 26-29, 1971. [Chl,
Car; in R.]

*11147 - PATHAK, P.S.: Growth of *Tribulus terrestris* LINN. under reduced light intensities. - Trop. Ecol. *10*: 240-255, 1969. [Growth analysis.]

*11148 - PATHAK, P.S.: Growth and dry matter production of *Tribulus terrestris* LINN.
at different temperatures. - Proc. nat. Acad. Sci. India, Sect. B *40*: 203-
211, 1970. [Growth analysis.]

11149 - PATIL, B.A., JOSHI, G.V.: Photosynthetic studies in *Ulva lactuca*. Ethanol
insoluble fraction. - Bot. Mar. *14*: 22-23, 1971.

11150 - PATIL, B.C., SALUNKHE, D.K., SINGH, B.: Metabolism of solanine and chlorophyll in potato tubers as affected by light and specific chemicals. - J.
Food Sci. *36*: 474-476, 1971.

11151 - PATIL, B.C., SINGH, B., SALUNKHE, D.K.: Formation of chlorophyll and solanine
in Irish potato (*Solanum tuberosum* l.) tubers and their control by gamma radiation and CO_2 enriched packaging. - Lebensm.-Wiss. Technol *4*: 123-125, 1971.

11152 - PATTERSON, B.D., ATKINS, C.A., GRAHAM, D., WILLS, R.B.H.: Carbonic anhydrase:
a new method of detection on polyacrylamide gels using low-temperature fluorescence. - Anal. Biochem. *44*: 388-391, 1971.

11153 - PATTERSON, B.D., SMILLIE, R.M.: Developmental changes of ribosomal ribonucleic
acid and fraction I protein in wheat leaves. - Plant Physiol. *47*: 196-198,
1971.

11154 - PAULE, M.R.: The effect of temperature on the kinetics of adenosine diphos-
phoglucose pyrophosphorylase from *Rhodospirillum rubrum*. - Biochemistry *10*:
4509-4517, 1971.

11155 - PAULECH, C., HERRERA, S.: The influence of the fungus *Capnodium citri* BERK.
et DESM. on the photosynthesis and transpiration of sour orange leaves. -
Biológia (Bratislava) *26*: 501-505, 1971.

*11156 - PEARCE, R.B., FISSEL, G., CARLSON, G.E.: Carbon uptake and distribution be-
fore and after defoliation of alfalfa. - Crop Sci. *9*: 756-759, 1969. [Ps.]

11157 - PEARCY, R.W., BJÖRKMAN, O.: Biochemical characteristics. - Carnegie Inst.
Year Book *69*: 632-640, 1971. [Ps.]

11158 - PEARCY, R.W., BJÖRKMAN, O., HARRISON, A.T., MOONEY, H.A.: Photosynthetic
performance of two desert species with C_4 photosynthesis in Death Valley,
California. - Carnegie Inst. Year Book *70*: 540-550, 1971.

11159 - PECORA, R.A.: A comparative study of selected xanthophycean algae in pure
culture. - Diss. Abstr. int. B *32*: 1401-B, 1971. [Chl, Car.]

11160 - PEISKER, M., APEL, P.: Untersuchungen zum Einfluss von Sauerstoff auf den
CO_2-Gaswechsel assimilierender Blätter. - Biochem. Physiol. Pflanzen *162*:
165-176, 1971.

11161 - PENKA, M., ČERMÁK, J., ŠTĚPÁNEK, V.: Contribution to studies of the photosynthetic rate in oak (*Quercus robur* L.). - Acta Univ. Agr. (Brno) Ser. C
(Fac. Silvic.) *40*: 283-301, 1971.

11162 - PENTH, B., WEIGL, J.: Anionen-Influx, ATP-Spiegel und CO_2-Fixierung in *Lim-
nophila gratioloides* und *Chara foetida*. - Planta *96*: 212-223, 1971.

11163 - PERRIER, A.: Leaf temperature measurement. - In: ŠESTÁK, Z., ČATSKÝ, J.,
JARVIS, P.G. (ed.): Plant Photosynthetic Production. Manual of Methods. Pp.
632-671. Dr. W. Junk N.V. Publ., The Hague 1971.

*11164 - PERRY, T.O., BALDWIN, G.W.: Winter breakdown of the photosynthetic apparatus
of evergreen species. - Forest Sci. *12*: 298-300, 1966. [Chl.]

11165 - PETERING, D., FEE, J.A., PALMER, G.: The oxygen sensitivity of spinach ferredoxin and other iron-sulfur proteins. The formation of protein-bound sulfur-
zero. - J. biol. Chem. *246*: 643-653, 1971.

11166 - PETERS, G.A.: Morphological and spectral aspects of pigment development and
related studies in the non-sulfur, purple photosynthetic bacterium, *Rhodopseudomonas spheroides*. - Diss. Abstr. int. B *31*: 7142-B, 1971.

11167 - PETERSON, L.W., KLEINKOPF, G.E., HUFFAKER, R.C.: Evidence for lack of turnover of ribulose-1,5-diP carboxylase in barley leaves. - Plant Physiol. *47* (Suppl.): 10, 1971.

11168 - PETKOVA, R.A.: Light-induced changes of the redox potential in chloroplast suspensions. - Dokl. bolg. Akad. Nauk *24*: 103-106, 1971.

11169 - PETRUKHIN, Yu.A.: Produkty fotosinteza i ikh posleduyushchaya metabolizatsiya na svetu i v temnote. [Photosynthates and their further metabolization in light and darkness.] - Uch. Zap. perm. gos. ped. Univ. *98*: 75-78, 1971. [In R.]

11170 - PFAU, J., WERTHMÜLLER, K., SENGER, H.: Permanent automatic synchronization of micro algae achieved by photoelectrically controlled dilution. - Arch. Mikrobiol. *75*: 338-345, 1971. [Ps, Chl.]

11171 - PFENNIG, N., TRÜPER, H.G.: Type and neotype strains of the species of phototrophic bacteria maintained in pure culture. - Int. J. syst. Bacteriol. *21*: 19-24, 1971.

*11172 - PFLÜGER, R.: Der biochemische Weg zu höheren Erträgen unter Berücksichtigung der Auswirkung auf die Düngungsmethoden. - In: Role of Fertilization in the Intensification of Agricultural Production. Pp. 435-448. Int. Potash Inst., Bern 1970. [Ps.]

11173 - PFLUGER, U.N., BACHOFEN, R.: Der Einfluss von anorganischem Pyrophosphat auf die CO_2-Fixierung in photosynthetischen Bakterien. - Arch. Mikrobiol. *77*: 36-46, 1971.

*11174 - PHAN, C.-T.: Activité photosynthétique des fruits. - Fruits *25*: 109-111, 1970.

11175 - PHILIP, T., FRANCIS, F.J.: Isolation and chemical properties of capsanthin and derivatives. - J. Food Sci. *36*: 823-827, 1971.

11176 - PHILIPSON, K.D., TSAI, S.C., SAUER, K.: Circular dichroism of chlorophyll and related molecules calculated using a point monopole model for the electronic transitions. - J. phys. Chem. *75*: 1440-1445, 1971.

11177 - PHILLIPS, E.T.: Evolution of a light scattering photometer. - BioScience *21*: 865-867, 1971.

11178 - PHILLIPS, P.J., McWILLIAM, J.R.: Thermal responses of the primary carboxylating enzymes from C_3 and C_4 plants adapted to contrasting temperature environments. - In: HATCH, M.D., OSMOND, C.B., SLATYER, R.O (ed.): Photosynthesis and Photorespiration. Pp. 97-104. Wiley-Interscience, New York - London - Sydney - Toronto 1971.

11179 - PICKETT, J.M.: Enhancement models at high light intensities. - Plant Physiol. *47*: 222-225, 1971.

11180 - PICKETT, J.M.: Effect of flashes of red or blue light on the composition of starved *Chlorella pyrenoidosa*. - Plant Physiol. *47*: 226-229, 1971.

11181 - PICKETT, J.M.: Photosynthetic enhancement at high light intensities. - Plant Physiol. *48*: 265-266, 1971.

11182 - PIERSON, B.K., CASTENHOLZ, R.W.: Bacteriochlorophylls in gliding filamentous prokaryotes from hot springs. - Nature - new Biol. *233*: 25-27, 1971.

*11183 - PIH, T.-S., TOOP, E.W.: Effect of gibberellic acid on growth of *Antirrhinum majus* "Utah White" in a carbon dioxide-enriched environment. - Can. J. Plant Sci. *48*: 617-620, 1968. [Chl.]

*11184 - PIPINIS, I.A., KIRVYALITE, D.Yu.: Biokhimicheskaya kharakteristika korneĭ i zelenoĭ massy tarana dubil'nogo. (2.Biokhimicheskaya kharakteristika travyanoĭ muki.) [Biochemical characteristics of roots and green material of knotweed. (2. Biochemical characteristics of grass flour.)] - Lietuvos TSR Mokslы Akad. darbai, Ser. C *3* (47): 45-53, 1968. [Car; in R, ab: E, Lithu.]

*11185 - PIPINIS, I.A., SMALYUKAS, D.Yu.: Biokhimicheskaya kharakteristika korneĭ i zelenoĭ massy tarana dubil'nogo. (3. Vliyanie Mg, B, Zn, Cu na urozhaĭ i ego kachestvo.)[Biochemical characteristics of roots and green material of knotweed. (3. The effect of Mg, B, Zn, Cu on the yield and quality of knotweed.).] - Lietuvos TSR Mokslы Akad. darbai, Ser. C *2* (49): 47-53, 1969. [In R, ab: E. Lithu.]

11186 - PIPINIS, I.A., SMALYUKAS, D.Yu.: Biokhimicheskaya kharakteristika korneĭ i
zelenoĭ massy tarana dubil'nogo. (6. Vliyanie nekotorykh gerbitsidov na urozhaĭ i ego kachestvo.). [Biochemical characteristics of roots and green
material of knotweed. (6. Influence of some herbicides on the yield and its
quality.).] - Lietuvos TSR Mokslų Akad. darbai, Ser. C 2 (55): 39-49, 1971.
[Chl; in R, ab: E, Lithu.]

*11187 - PÎRJOL, L., MILICĂ, C.I.: Cercetări morfo-fiziologice la fasole. [Morpho-
physiological investigations in beans.] - An. Inst. Cercet. Cereale Plante
teh., Ser. C 35: 523-535, 1969. [Chl; in Roum., ab: E, R.]

*11188 - PÎRJOL, L., MILICĂ, C.I.: Dinamica proceselor fiziologice şi biochimice la
grîul de toamnă în cursul dezvoltării ontogenetice. [Dynamics of physiological and biochemical processes in winter wheat during the ontogenetic development.] - An. Inst. Cercet. Cereale Plante teh., Ser. C 36: 103-115, 1970.
[Chl; in Roum., ab: E, R.]

*11189 - PÎRJOL, L., MILICĂ, C.I., DRĂGHICI, L.: Dinamica unor procese fiziologice
şi biochimice în cursul vegetaţiei la cîteva soiuri şi linii de orz do toamnă. [Dynamics of physiological and biochemical processes during growth period
in some winter barley varieties and lines.] - An. Inst. Cercet. Cereale Plante teh., Ser. C 36: 131-143, 1970. [Chl; in Roum., ab: E, R.]

11190 - PÎRJOL, L., MILICĂ, C.I., PICU, I.: Cercetări privind influenţa regimului de
nutriţie asupra unor indici morfofiziologici şi biochimici la soiul de grîu
capitole în condiţii de irigare. [Effect of nutrition regime on some mor-
phophysiological and biochemical indices in wheat cv. Capitole under irrigation.] - An. Inst. Cercet. Cereale Plante Teh., Ser.C 37: 39-53, 1971.
[Ps, Chl; in Roum., ab: E, R.]

*11191 - PÎRJOL, L., MILICĂ, C.I., VARGA, P.: Modificări fiziologice şi biochimice
în cursul vegetaţiei unele soiuri de lucernă. [Physiological and biochemical
changes during the growth of some alfalfa varieties.] - An. Inst. Cercet.
Cereale Plante teh., Ser. C 36: 353-376, 1970. [Chl, Car; in Roum., ab: E, R.]

*11192 - PÎRJOL, L., MILICĂ, C.I., VARGA, P., CIURDĂRESCU, G.: Cercetări fiziologice
asupra rezistenţei la secetă a lucernei. [Physiological investigations on
alfalfa resistance to drought.] - An. Inst. Cercet. Cereale Plante teh.,
Ser. C 35: 615-627, 1969. [Chl, Car; in Roum., ab: E, R.]

11193 - PÎRJOL, L., MILICĂ, C.I., VRÎNCEANU, V.: Rezistenţa la secetă a florii-soa-
relui în diferite faze de vegetaţie. [Sunflower drought resistance at various
growth stages.] - An. Inst. Cercet. Cereale Plante teh., Ser. C 37: 191-208,
1971. [Ps, Chl, Car; in Roum., ab: E. R.]

*11194 - PÎRJOL, L., PICU, I., MILICĂ, C.I.: Cercetări morfo-fiziologice la cîteva
soiuri de grîu în cultura irigată. [Morpho-physiological investigations in
some wheat varieties under irrigation.] - An. Inst. Cercet. Cereale Plante
teh., Ser. C 35: 119-136, 1969. [Chl; in Roum., ab: E, R.]

11195 - PÎRJOL, L., PICU, I., MILICĂ, C.I.: Influenţa îngrăşămintelor chimice asupra principaleror însuşiri morfologice şi fiziologice la grîu în condiţii de
irigare. [Effect of chemical fertilizers on main morphological and physiological features in irrigated wheat.] - An. Inst. Cercet. Cereale Plante teh.,
Ser. C 37: 9-26, 1971. [Chl; in Roum., ab: E, R.]

*11196 - PÎRJOL, L., PINTILIE, C., MILICĂ, C.I.: Influenţa unor măsuri agrotehnice
asupra principalelor caracteristici morfologice şi fiziologice la grîu. [Ef-
fect of some agrotechnical measures on the main morphological and physiological wheat characters.] - An. Inst. Cercet. Cereale Plante teh., Ser. B
35: 419-430, 1969. [Chl; in Roum., ab: E, R.]

11197 - PISAREV, B.A., GANZIN, G.A.: Prorashchivanie kartofelya pri estestvennom i
iskusstvennom osveshchenii. [Germination of potato under natural and artificial illumination.] - Vestn. sel'skokhoz. Nauki 16 (11): 75-78, 1971. [Chl;
in R, ab: E, F, G.]

11198 - PISEK, A.: Zur Geschichte der experimentellen Ökologie (besonders des in
Innsbruck hierzu geleisteten Beitrages). - Ber. deut. bot. Ges. 84: 365-379,
1971. [Ps.]

*11199 - PISKORNIK, Z.: Wpływ przemysłowych zanieczyszczeń powietrza na fotosyntezę
w drzewach liściastych. [Effect of industrial polluted air on photosynthesis
of deciduous trees.] - Biul. Zakł. Badań nauk. gornosląskiego Okręgu Przem.
Pol. Akad. Nauk 12: 155-178, 1969. [In Pol., ab: E, R.]

11200 - PISKUNKOVA, N.F., PIMENOVA, M.N., BAKLASHOVA, T.G.: Nekotorye dannye o roli
fotosinteza v ispol'zovanii atsetata i piruvata Scenedesmus quadricauda.
[Role of photosynthesis in the utilization of acetate and pyruvate by Scenedesmus quadricauda.] - Mikrobiologiya 40: 386-388, 1971. [In R, ab: E.]

11201 - PLANCHON, C.: Relations entre l'assimilation chlorophyllienne nette, la teneur en chlorophylle a et la température chez le Blé tendre (Triticum aestivum L.). - Compt. rend. Acad. Sci. Paris, Sér. D 272: 68-71, 1971.

11202 - PLATONENKOVA, L.S., USPENSKAYA, V.E.: Konyugirovannye pteridiny fotosinteziruyushchikh bakterii Rhodospirillum rubrum. [Conjugated pteridines of photosynthesizing bacteria Rhodospirillum rubrum.] - Mikrobiologiya 40: 988-992,
1971. [In R, ab: E.]

11203 - PLAUT, Z.: Inhibition of photosynthetic carbon dioxide fixation in isolated
spinach chloroplasts exposed to reduced osmotic potentials. - Plant Physiol.
48: 591-595, 1971.

11204 - PLOAIE, P.G.: The fine structure of chloroplasts and pyrenoids of Euglena
gracilis strain Z. - Rev. roum. Biol., Sér. Bot. 16: 179-183, 1971.

11205 - PŁOSZYŃSKI, M., ŽURAWSKI, H.: The phytotoxic action of triazine herbicides
on flax, beets and buckwheat seedlings, and some physiological changes connected with it. - Acta agrobot. 24: 205-215, 1971. [Production.]

*11206 - PLUSCEC, J., BOGORAD, L.: A dipyrrylmethane intermediate in the enzymatic
synthesis of uroporphyrinogen. - Biochemistry 9: 4736-4743, 1970.

*11207 - POCHINOK, Kh.N.: Pribor dlya gazometricheskogo opredeleniya intensivnosti fo-
tosinteza v polevykh usloviyakh. [Apparatus for gasometric determination of
photosynthetic rate in field conditions.] - In: Tezisy Dokladov Vsesoyuznogo
Soveshchaniya po Unifikatsii Metodov i Priborov dlya Massovykh Izmerenii
Intensivnosti Fotosinteza. Pp. 101-104. Nauch.-issled. Inst. Rastenievod.
N.I. Vavilova, Leningrad - Pushkin 1970. [In R.]

11208 - POCHINOK, Kh.N.: Vliyanie gerbitsidov i antitranspirantov na intensivnost'
fotosinteza i transpiratsii rastenii. [Effect of herbicides and antitrans-
pirants on photosynthesis and transpiration rates.] - Fiziol. Biokhim. kul't.
Rast. 3: 538-544, 1971. [In R, ab: E.]

11209 - PODEŠVA, J., BARÁK, K., JURČÍK, F.: Vliv různého poměru a hladiny NPK-živin
u vojtěšky seté na intenzitu fotosyntézy, obsah chlorofylu a jejich vztahy
k některým morfogenním a výnosovým ukazatelům. I. Vztah intenzity fotosyntézy k obsahu chlorofylu, k pokryvnosti listoví a k tvorbě susiny nadzemní
biomasy. [Effect of different ratios and levels of NPK on photosynthetic
rate, chlorophyll content and their relations to morphogenic and yield indices of alfalfa. I. Relations between photosynthetic rate, and chlorophyll
formation, leaf area index and formation of aboveground biomass dry matter.]
- Acta Univ. Agr., Fac. Agron. (Brno) A 19: 235-242, 1971. [In Czech.]

11210 - PODEŠVA, J., BARÁK, K., JURČÍK, F.: Vliv různého poměru a hladiny NPK-živin
u vojtěšky seté na intenzitu fotosyntézy, obsah chlorofylu a jejich vztahy
k některým morfogenním a výnosovým ukazatelům. II. Vztah intenzity fotosyntézy k tvorbě sušiny podzemní biomasy, k tloušťce kořenového krčku a k některým morfogenním a kvalitativním ukazatelům sklizně; závěry a souhrn. [Ef-
fect of different ratios and levels of NPK on photosynthetic rate, chlorophyll content and their relations to morphogenic and yield indices of alfalfa. II. Relations between photosynthetic rate and underground biomass dry
matter formation, root neck thickness, morphogenic and qualitative yield
indices.] - Acta Univ. Agr., Fac. Agron. (Brno) A 19: 243-250, 1971. [In
Czech, ab: E, F, R.]

11211 - POE, M., PHILLIPS, W.D., GLICKSON, J.D., McDONALD, C.C., SAN PIETRO, A.:
Proton magnetic resonance studies of the ferredoxins from spinach and parsley. - Proc. nat. Acad. Sci. USA 68: 68-71, 1971.

11212 - POGOSYAN, S.A., MELKONYAN, M.V.: O dinamike intensivnosti fotosinteza vinograda po yarusam list'ev. [Dynamics of photosynthetic rate in grapevine leaves of different insertion levels.] - Vinodelie Vinogradarstvo SSSR *1971*
(3): 34-36, 1971. [In R.]

11213 - POINCELOT, R.P.: Differences in lipid composition between intact and membrane-
stripped spinach chloroplasts. - Biochim. biophys. Acta *239*: 57-60, 1971.

11214 - POLYA, G.M., JAGENDORF, A.T.: Wheat leaf RNA polymerases. I. Partial purification and characterization of nuclear, chloroplast and soluble DNA-dependent
enzymes. - Arch. Biochem. Biophys. *146*: 635-648, 1971.

11215 - POLYA, G.M., JAGENDORF, A.T.: Wheat leaf RNA polymerases. II. Kinetic characterization and template specificities of nuclear, chloroplast, and soluble
enzymes. - Arch. Biochem. Biophys. *146*: 649-657, 1971.

11216 - POLYAKUV, I.M., PRIMAK, N.N.: Svyaz' izmeneniT ul'tratonkoT struktury khloroplastov s razlichnoT okraskoT list'ev v ontogeneze *Nicotiana tabacum* L. [Re-
lationship between chloroplast ultrastructure and leaf colour during ontogenesis of *Nicotiana tabacum* L.] - Tsitol. Genet. *5*: 359-363, 1971. [In R.]

11217 - POLYAKOV, P.V.: Izmenenie immuniteta u podsolnechnika pod vliyaniem mikroelementov. [Change in sunflower immunity under the influence of microelements.]
- Sel'.-khoz. Biol. *6*: 471-472, 1971. [Ps; In R.]

11218 - PONOMAREVA, S.A., PROTSENKO, D.F., SIVTSEV, M.V.: Vliyanie khloristogo natriya na plastidnye pigmenty list'ev tomatov. [Effect of sodium chloride on
plastid pigments of tomato leaves.] - Fiziol. Rast. *18*: 404-408, 1971. [In
R, ab: E.]

11219 - POPESCU, F.: Cercetări comparative asupra fotosintezei la sorg (*Sorghum vulgare*) şi soia (*Glycine hispida*) sub influenţa ingraşamintelor minerale şi a
densiţatilor diferite. [Comparative research on photosynthesis of sorgum
(*Sorthum vulgare*) and soybean (*Glycine hispida*) under the influence of mineral fertilizers and different plant densities.] - Ann. Univ. Craiova, Ser.
3, *3*: 111-118, 1971. [In Roum., ab: E, F.]

11220 - POPESCU, I.: Cloroplastele. [Chloroplasts.] - Natura (Bucureşti) *23* (2):
11-19, 1971. [In Roum.]

11221 - POPOV, K., DILOVA, S.: Vliyanie na zat"mnyavaneto v"rkhu s"stoyanieto na
plastidnite pigmenti v echemicheni lista. [Influence of darkening on the state
of plastid pigments in barley leaves.] - Izv. Inst. Fiziol. Rast. "M. Popov"
b"lg. Akad. Nauk *17*: 245-256, 1971. [In Bulg., ab: E, R.]

11222 - POPOV, K., STANEV, V.: Produktivnost na fotosintezata pri sl"nchogledov po-
sev s razlichna g"stota. [Productivity of photosynthesis in sunflower sown
at a different density.] - Izv. Inst. Fiziol. Rast. "M. Popov" b"lg. Akad.
Nauk *17*: 257-271, 1971. [In Bulg., ab: E, R.]

*11223 - POPOV, V.F.: SvetovoT rezhim v krone derev'ev grushi v zavisimosti ot pod-
voya. [Light regime in pear tree crown in dependence on the rootstock.] -
Tr. kishinev. sel'skokhoz. Inst. *69*: 72-101, 1970. [In R.]

*11224 - POPOV, S.I., SAĬCHENKO, Ya.N., KOLOV, O.V., GAP, L.P.: Trekhkanal'nyT kon-
duktometricheskiT pribor dlya opredeleniya intensivnosti fotosinteza i dykhaniya rasteniT v polevykh usloviyakh. [Three-channel conductimetric apparatus for determination of photosynthetic and respiration rates of plants in
field conditions.] - In: Tezisy Dokladov Vsesoyuznogo Soveshchaniya po Unifikatsii Metodov i Priborov dlya Massovykh IzmereniT Intensivnosti Fotosinteza. Pp. 95-100. Nauch.-Issled. Inst. Rastenievod. N.I. Vavilova, Leningrad - Pushkin 1970. [In R.]

11225 - POPOVA, I.A., RYZHKOVA, E.F., SAPOZHNIKOV, D.I.: Nekotorye osobennosti reaktsii dezepoksidatsii violaksantina. [Violaxanthin de-epoxidation reaction.]
- Dokl. Akad. Nauk SSSR *201*: 494-496, 1971. [In R.]

11226 - POPOVA, N.A.: Dinamika sakharov i khlorofilla v list'yakh vinogradnogo rasteniya. [Dynamics of sugars and chlorophyll in grape leaves.] - Tr. kishinev.
sel'.-khoz. Inst. *85*: 65-70, 78, 1971. [In R.]

*11227 - POPOVA, O.F.: Stanovlenie reaktsii violaksantinovogo tsikla v zeleneyushchikh prorostkakh kukuruzy. [Development of the violaxanthin cycle in greening seedlings of maize.] - In: Tezisy II Biokhimicheskogo S"ezda. Sektsiya 19. Pp. 43-44. Tashkent 1969. [In R.]

11228 - POPOVA, O.F.: Prevrashcheniya ksantofillov v zelensyushchikh prorostkakh kukuruzy. [Transformations of xanthophylls in greening maize seedlings.] - Fiziol. Rast. *18*: 677-682, 1971. [In R, ab: E.]

11229 - POPOVICI, N., GOGEA, V.: L'influence de la caféine sur l'intensité de la photosynthèse et sur la teneur en chlorophylle chez le maïs dans la période postgerminatoire. - An. şti. Univ. "Al. I. Cuza" Iaşi, Sec. 2a, *17*: 263-272, 1971.

11230 - POPOVICI, N., ŞTEFAN, N.: Influenţa cafeinei asupra unor procese fiziologice la fasole in perioada postgerminatorie. [Effect of caffeine on physiological processes in bean seedlings during postgermination period.] - An. şti. Univ. "Al. I. Cuza", Sec. 2a, *17*: 273-285, 1971. [Ps, Chl; in Roum., ab: F.]

11231 - PORATH, D., BEN-SHAUL, Y.: Structural and physiological changes during "heat-bleaching" in *Spirodela oligorrhiza*. - Israel J. Bot. *20*: 157-168, 1971. [Ps, Chl.]

*11232 - POROKHNEVICH, N.V.: Vliyanie tsinka na rost i razvitie l'na i fotosintezi-ruyushchego apparata. [Effect of zinc on the growth, development and photo-synthetic apparatus in flax.] - Botanika (Issledovaniya) (Minsk) *10*: 143-151, 1968. [In R.]

11233 - POROKHNEVICH, N.V.: Izmenenie soderzhaniya karotinoidov v list'yakh i khloro-plastakh l'na pri vklyuchenii v pitanie rastenii mikroelementov v razlichnykh sochetaniyakh. [Change in carotene content in leaves and chloroplasts of flax following application of microelements in different combination.] - Vestn. beloruss. gos. Univ. V.I. Lenina, Ser. 2, *1971* [1 (7)]: 49-53, 1971. [In R.]

11234 - POROKHNEVICH, N.V.: Vliyanie tsinka na fotosinteticheskii apparat rastenii l'na v prisutstvii medi v pitatel'nom rastvore s razlichnym soderzhaniem kisloroda. [Effect of zinc on photosynthetic apparatus of flax in presence of copper in the nutrient solution with different oxygen content.] - Fiziol. Rast. *18*: 690-694, 1971. [In R.]

*11235 - POROKHNEVICH, N.V., KALISHEVICH, S.V.: O vliyanii medi na fotosinteziruyush-chii apparat rastenii. [Effect of copper on photosynthetic apparatus.] - Botanika (Issledovaniya) (Minsk) *11*: 39-49, 1969. [In R.]

*11236 - POROKHNEVICH, N.V., KALISHEVICH, S.V., ANTIPOVA, A.I., TOLLAR, G.V.: Deist-vie tsinka i medi pri razdel'nom i sovmestnom vnesenii na formirovanie fo-tosinteticheskogo apparata i urozhainost' l'na v usloviyakh polevogo opyta. [Effect of zinc and copper applied separately or in combination on the for-mation of photosynthetic apparatus and productivity of flax in field con-ditions.] - Vestn. beloruss. gos. Univ. V.I. Lenina, Ser. 2, *1969* (3): 55-59, 1969. [In R.]

*11237 - POROKHNEVICH, N.V., PELAGEICHIK, T.Ya.: Vliyanie tsinkovogo i mednogo udo-brenii v posledeistvii na morfologicheskie i anatomo-fiziologicheskie poka-zateli i urozhainost' l'na. [After-effect of zinc and copper fertilizers on morphological, anatomical and physiological parameters and productivity of flax.] - Agrokhimiya *1970* (4): 75-80, 1970. [Chl; in R.]

11238 - POSKUTA, J., BARANKIEWICZ, T., ANTOSZEWSKI, R.: Gaseous exchange rates of strawberry plants as influenced by irradiance and oxygen concentration. - Photosynthetica *5*: 395-398, 1971.

11239 - POSSINGHAM, J.V.: Some effects of mineral nutrient deficiencies on the chlo-roplasts of higher plants. - In: SAMISH, R.M. (ed.): Recent Advances in Plant Nutrition. Vol. 1. Pp. 155-165. Gordon and Breach Sci. Publ., New York - London - Paris 1971.

11240 - POSTIUS, S., JACOBI, G.: Dark starvation and chloroplast function. I. The decrease of enzyme activities corelated with NADP reduction and their re-generation by light. - Plant *99*: 222-229, 1971.

11241 - POYARKOVA, N.M., DROZDOVA, I.S., VOSKRESENSKAYA, N.P.: Vliyanie sinego sveta na aktivnost' karboksilazy ribulezodifosfata v list'yakh *Vicia faba*. [Effect

of the blue light on the activity of ribulose diphosphate carboxylase in the
leaves of *Vicia faba.*] - Fiziol. Rast. *18:* 683-689, 1971. [In R, ab: E.]

11242 - POZSÁR, B.I.: The determination of the effect of soluble protein level on
the intensity of photosynthetic carbon dioxide fixation. - Acta agron. Acad.
Sci. hung. *20:* 197-203, 1971.

11243 - PREISS, J., GREENBERG, E.: Spinach leaf D-fructose 1,6-diphosphate 1-phos-
phohydrolase (FDPase) EC 3.1.3.11. - In: COLOWICK, S.P., KAPLAN, N.O. (ed.):
Methods in Enzymology. Vol. 23. Pp. 691-696. Academic Press, New York -
London 1971.

11244 - PREMUZIC, E.: Chemistry of natural products derived from marine sources. -
Fortschr. Chem. org. Naturstoffe *29:* 417-488, 1971. [Car.]

*11245 - PRICE, L.W.: Control units for use with electrochemical electrodes. - Bio-
med. Eng. *1970:* 193-197, 1970. [Clark O_2 electrode.]

11246 - PRIOUL, J.-L.: Réactions des feuilles de *Lolium multiflorum* à l'éclairement
pendant la croissance et variation des résistances aux échanges gazeux pho-
tosynthétiques. - Photosynthetica *5:* 364-375, 1971.

11247 - PROKHARCHYK, R.A., MASHTAKOV, S.M.: Paraŭnal'naya kharaktarystyka dzeyannya
khimichnykh regulyataraŭ rostu na reaktsyyu Khila khlaraplastaŭ raslin lu-
binu. [Comparative characteristics of the effect of chemical growth regulators on the Hill reaction of lupine chloroplasts.] - Vestsy Akad. Navuk belarus. SSR, Ser. biyal. Navuk *1971* (2): 50-55, 140, 1971. [In Belorus.,
ab: R.]

11248 - PROKOF'EV, A.A., PINKHASOV, Yu.I., EGAMBERDIEV, A.M.: Postuplenie assimilyatov iz list'ev glavnogo steblya v plodovye organy khlopchatnika.[Trans-
fer of photosynthates from the leaves of main stem into the fruit organs of
cotton.] - Fiziol. Rast. *18:* 29-36, 1971. [In R, ab: E.]

11249 - PROKOF'EV, A.A., PINKHASOV, Yu.I., EGAMBERDIEV, A.M.: Vliyanie vodnogo de-
fitsita na postuplenie assimilyatov iz list'ev v plodovye organy khlopchatnika. [Effect of water deficit on transfer of photosynthates from leaves
into fruit organs of cotton.] - Fiziol. Rast. *18:* 563-567, 1971. [In R,
ab: E.]

11250 - PROTSENKO, D.F., KOLOSHA, O.I., MISHUSTINA, P.S., SHMAT'KO, I.G., OSTAPLYUK,
E.D., BELETSKAYA, E.K.: Faktory, opredelyayushchie ustoĭchivost' rasteniĭ.
[Factors determining plant resistance.] - In: Fotosintez, Rost i UstoĭchI-
vost' Rasteniĭ. Pp. 201-324. Naukova Dumka, Kiev 1971. [Ps, Chl, Car; in R.]

11251 - PROTSENKO, D.F., MUSIENKO, N.N., SLAVNYĬ, P.S.: Genotipicheskie razlichiya
termoustoĭchivosti zelenykh pigmentov ozimoĭ pshenitsy. [Genotypic differences in thermal resistance of green pigments in winter wheat.] - Fiziol.
Biokhim. kul't. Rast. *3:* 482-486, 1971. [In R, ab: E.]

11252 - PUCKRIDGE, D.W.: Photosynthesis of wheat under field conditions, III. Seasonal trends in carbon dioxide uptake of crop communities. - Aust. J. agr.
Res. *22:* 1-9, 1971.

11253 - PUCKRIDGE, D.W., RATKOWSKY, D.A.: Photosynthesis of wheat under field conditions. IV. The influence of density and leaf area index on the response
to radiation. - Aust. J. agr. Res. *22:* 11-20, 1971.

11254 - PUECH, A.A., REBEIZ, C.A., CRANE, J.C.: The effect of 2-chloroethyl phos-
phonic-acid on pigment changes in the mission fig fruit. - Plant Physiol.
47 (Suppl.): 15, 1971.

*11255 - PUGH, P.R.: Liquid scintillation counting of ^{14}C-diatom material on filter
papers for use in productivity studies. - Limnol. Oceanogr. *15:* 652-655,
1970.

11256 - PUNNETT, T.: Environmental control of photosynthetic enhancement. - Science
171: 284-286, 1971.

*11257 - PYARNIK, T.R., KEERBERG, O.F.: Uchet fotodykhaniya pri opredelenii fotosin-
teza radioizotopnym metodom. [Photosynthesis determination by the radioiso-
tope method with respect to photorespiration.] - In: Tezisy Dokladov Vsesoyuznogo Soveshchaniya po Unifikatsii Metodov i Priborov dlya Massovykh Iz-

merenïT Intensivnosti Fotosinteza. Pp. 105-108. Nauch.-issled. Inst. Rastenievod. N.I. Vavilova, Leningrad - Pushkin 1970. [In R.]

11258 - PYARNIK, T., VIÏL', Yu., VYARK, E., KEERBERG, O., KEERBERG, Kh.: O primenenii diskov iz list'ev dlya izucheniya gazoobmena fotosinteza. [Use of leaf
discs in measurements of photosynthetic gas exchange.] - Izv. Akad. Nauk
est. SSR, Biol. *20*: 232-241, 1971. [In R, ab: E, Eston.]

11259 - PYARNIK, T., VYARK, E., KEERBERG, O., KEERBERG, Kh.: O primenenii diskov
iz list'ev dlya izucheniya produktov fotosinteticheskoT assimilyatsii CO_2.
[Use of leaf discs for studying products of photosynthetic CO_2 assimilation.]
- Izv. Akad. Nauk est. SSR, Biol. *21*: 52-60, 1971. [In R, ab: E.]

11260 - PYLIOTIS, N.A., WOO, K.C., DOWNTON, W.J.S.: Thylakoid aggregation correla-
ted with chlorophyll *a*/chlorophyll *b* ratio in some C_4 species. - In: HATCH,
M.D., OSMOND, C.B., SLATYER, R.O. (ed.): Photosynthesis and Photorespiration.
Pp. 406-412. Wiley-Interscience, New York - London - Sydney - Toronto 1971.

11261 - PYRINA, I.L., ELIZAROVA, V.A.: Spektrofotometricheskoe opredelenie khlorofillov v kul'turakh nekotorykh vodorosleT. [Spectrophotometric determination
of chlorophylls in cultures of some algae.] - Tr. Inst. Biol. vnutr. Vod
Akad. Nauk SSSR *21* (24) - Biologiya i Produktivnost' Presnovodnykh Organizmov: 56-65, 296, 1971. [In R.]

*11262 - PYT'EVA, N.F., RUBIN, A.B.: Matematicheskoe modelirovanie protsessov trans-
porta elektronov pri fotosinteze, soprovozhdayushchikhaya khemilyuminests-
entsieT. [Mathematical modelling of electron transport during photosynthesis accompanied by chemiluminescence.] - Biofizika *15*: 47-52, 1970. [In R,
ab: E.]

B11263 - QUAGLIARIELLO, E., PAPA, S., ROSSI, C.S. (ed.): Energy Transduction in Respiration and Photosynthesis. - Adriatica Editrice, Bari 1971.

11264 - QUINLAN, K.P.: Proton movement accompanying the light-induced electron
transfer in the chlorophyll-quinone systems. - Photochem. Photobiol. *13*:
113-121, 1971.

11265 - RAAFAT, A., HÖFNER, W.: Effect of age on the fixation of $^{14}CO_2$ in sugars,
organic acids and amino acids of bean leaves. - Phytochemistry *10*: 2373-
2381, 1971.

11266 - RAAFAT, A., HÖFNER, W., LINSER, H.: $^{14}CO_2$ assimilation during photosynthesis of ageing bean seedlings. - Z. Pflanzenphysiol. *64*: 22-33, 1971.

*11267 - RAAFAT, A., STUR, J., SIPOS, M., MARKE, N.: Some aspects of oxidation re-
duction changes in chloropasts during the process of aging. - Acta biochim.
biophys. Acad. Sci. hung. *3*: 453, 1968.

11268 - RABINOWITCH, E.: An unfolding discovery. - Proc. nat. Acad. Sci. USA *68*:
2875-2876, 1971. [Ps.]

11269 - RADMER, R., CHENIAE, G.M.: Photoactivation of the manganese catalyst of O_2
evolution II. A two-quantum mechanism. - Biochim. biophys. Acta *253*: 182-
186, 1971.

11270 - RADUNZ, A.: Phosphatidylglycerin-Antiserum und seine Reaktionen mit Chloroplasten. - Z. Naturforsch. *26 b*: 916-919, 1971.

11271 - RADUNZ, A., SCHMID, G.H., MENKE, W.: Antibodies to chlorophyll and their
reactions with chloroplast preparations. - Z. Naturforsch. *26 b*: 435-446,
1971.

11272 - RAEBURN, S., RABINOWITZ, J.C.: Pyruvate: ferredoxin oxidoreductase I. The
pyruvate-CO_2 exchange reaction. - Arch. Biochem. Biophys. *146*: 9-20, 1971.

11273 - RAEBURN, S., RABINOWITZ, J.C.: Pyruvate: ferredoxin oxidoreductase II. Characteristics of the forward and reverse reactions and properties of the enzyme. - Arch. Biochem. Biophys. *146*: 21-33, 1971.

*11274 - RAGGI, V.: Influenza della luce sulle alterazioni prodotte da *Uromyces pha-
seoli* (PERS) WINT. su fagiolo principalmente a carico degli aminoacidi.
[Effect of light on the changes induced by *Uromyces phaseoli* (PERS) WINT.
in bean, mainly in the contents of amino acids.] - Riv. Patol. veg. (Pavia),
Ser. IV, *2*: 263-302, 1966. [Ps; in Ital., ab: E.]

*11275 – RAGGI, V., D'ARMINI, M.: Influenza dei trattamenti anticrittogamici sull'-
         attività fotosintetica e respiratoria della vite. [Effect of antimicrobial
         agents on the activities of photosynthesis and respiration in grapevine.]
         – Riv. Patol. veg. (Pavia), Ser. IV, 6: 91-106, 1970. [In Ital., ab: E.]

11276 – RAGHAVAN, V., DeMAGGIO, A.E.: Enhancement of protein synthesis in isolated
        chloroplasts by irradiation of fern gametophytes with blue light. – Plant
        Physiol. 48: 82-85, 1971.

11277 – RAINNIE, D.J., BRAGG, P.D.: Loss of silver ions from oxygen electrode. –
        Anal. Biochem. 44: 392-396, 1971.

11278 – RAJAN, A.K., BETTERIDGE, B., BLACKMAN, G.E.: Changes in the growth of Sal-
        vinia natans induced by cycles of light and darkness of widely different
        duration. – Ann. Bot. 35: 597-604, 1971.

11279 – RAJAN, A.K., BETTERIDGE, B., BLACKMAN, G.E.: Interrelationships between the
        nature of the light source, ambient air temperature, and the vegetative
        growth of different species within growth cabinets. – Ann. Bot. 35: 323-
        343, 1971. [Growth analysis.]

11280 – RAJU, P.V.: The effekt of in situ application of growth hormones and fer-
        tilizers on photosynthetic ^{14}C incorporation in some marine algae. – Bot.
        mar. 14: 129-131, 1971.

11281 – RAKHI, M.: Ob anatomicheskikh parametrakh lista v svyazi s diffuzionnymi
        soprotivleniyami. [The characteristics of leaf anatomy and diffusion resis-
        tances.] – Izv. Akad. Nauk est. SSR, Biol. 20: 84-94, 1971. [In R, ab: E,
        Eston.]

11282 – RAKHIMOV, G.T.: Ob aktivnosti khloroplastov pustynnykh rastenĭ. [Chloro-
        plast activity in desert plants.] – Uzb. biol. Zh. 15 (4): 20-22, 1971.
        [In R.]

11283 – RAKHMANINA, K.P., AKHMEDOV, N.: Radiatsionnyĭ balans i summarnoe isparenie
        posevov khlopchatnika i bobovo-zlakovykh trav. [Radiation balance and total
        evaporation of cotton stands and of mixed stands of beans and cereals.] –
        In: ZALENSKIĬ, O.V. (ed.): Fotosintez i Ispol'zovanie Solnechnoĭ Energii.
        Pp. 58-62. Nauka, Leningrad 1971. [In R, ab: E.]

11284 – RAMAN, R., TOLLIN, G.: Chlorophyll one-electron photochemistry: flash pho-
        tolysis studies of the chlorophyll-quinone reaction in pyridine. – Photo-
        chem. Photobiol. 13: 135-145, 1971.

11285 – RAMANA, K.V.R., SINGH, N.: β-carotene in leaf proteins. – Current Sci. 40:
        293-294, 1971.

11286 – RAMANA, K.V.R., SINGH, N.: Studies on carotene and xanthophyll pigments in
        leaf protein and their stability during storage. – Indian J. exp. Biol. 9:
        478-480, 1971.

11287 – RANDALL, D.D., TOLBERT, N.E.: Two phosphatases associated with photosyn-
        thesis and the glycolate pathway. – In: HATCH, M.D., OSMOND, C.B., SLATYER,
        R.O. (ed.): Photosynthesis and Photorespiration. Pp. 259-266. Wiley-Inter-
        science, New York – London – Sydney – Toronto 1971.

11288 – RANDALL, D.D., TOLBERT, N.E.: 3-phosphoglycerate phosphatase in plants. I.
        Isolation and chracterization from sugarcane leaves. – J. biol. Chem. 246:
        5510-5517, 1971.

11289 – RANDALL, D.D., TOLBERT, N.E.: 3-phosphoglycerate phosphatase in plants. III.
        Activity associated with starch particles. – Plant Physiol. 48: 488-492,
        1971.

11290 – RANDALL, D.D., TOLBERT, N.E., GREMEL, D.: 3-phosphoglycerate phosphatase
        in plants. II. Distribution, physiological considerations, and comparison
        with P-glycolate phosphatase. – Plant Physiol. 48: 480-487, 1971.

11291 – RAO, D.M., REDDY, T.P., REDDY, G.M.: Induction of chlorophyll mutations by
        deuterium oxide (D_2O) in Oryza sativa. – Mutat. Res. 12: 109-111, 1971.

11292 – RAO, K.K., CAMMACK, R., HALL, D.O., JOHNSON, C.E.: Mössbauer effect in
        Scenedesmus and spinach ferredoxins. The mechanism of electron transfer in

plant-type iron-sulphur proteins. - Biochem. J. *122*: 257-265, 1971.

11293 - RAPER, C.D. jr.: Factors affecting the development of flue-cured tobacco grown in artificial environments. III. Morphological behavior of leaves in simulated temperature, light-duration, and nutrition progressions during growth. - Agron. J. *63*: 848-852, 1971. [Growth analysis.]

11294 - RAU, I., GRIMME, L.H.: Zum Einfluss verschieden substituierter s-Triazine auf Stoffwechselreaktionen der Grünalge *Ankistrodesmus braunii*. - Z. Naturforsch. *26 b*: 919-921, 1971. [Ps.]

11295 - RAUNER, Yu.L.: Izuchenie fizicheskikh faktorov mikroklimata rastitel'nogo pokrova. [Physical factors of microclimate within a plant canopy.] - In: ZALENSKII, O.V. (ed.): Fotosintez i Ispol'zovanie Solnechnoĭ Energii. Pp. 50-57. Nauka, Leningrad 1971. [In R, ab: E.]

*11296 - RAVEN, J.A.: Ion transport in *Hydrodictyon africanum*. - J. gen. Physiol. *50*: 1607-1625, 1967. [Ps.]

*11297 - RAVEN, J.A.: The linkage of light-stimulated Cl influx to K and Na influxes in *Hydrodictyon africanum*. - J. exp. Bot. *19*: 233-253, 1968. [Ps.]

11298 - RAVEN, J.A.: Cyclic and non-cyclic photophosphorylation as energy sources for active K influx in *Hydrodictyon africanum*. - J. exp. Bot. *22*: 420-433, 1971.

11299 - RAVEN, J.A.: Energy metabolism in green cells. - Trans. bot. Soc. Edinburgh *41*: 219-225, 1971.

11300 - RAVEN, J.A.: Inhibitor effects on photosynthesis, respiration and active ion transport in *Hydrodictyon africanum*. - J. Membrane Biol. *6*: 89-107, 1971.

11301 - RAVEN, J.A.: Ouabain-insensitive K influx in *Hydrodictyon africanum*. - Planta *97*: 28-38, 1971. [Ps.]

11302 - RAVEN, J.A.: The effects of visible light on the influx and efflux of solutes in plant cells. - Chemistry Industry *1971*: 859-865, 1971. [Ps.]

11303 - RAWSON, H.M., EVANS, L.T.: The contribution of stem reserves to grain development in a range of wheat cultivars of different height. - Aust. J. agr. Res. *22*: 851-863, 1971. [Ps.]

11304 - RAYMOND, J.C.: The characterization of *Ectothiorhodospira halophila* and its photosynthetic membranes. - Diss. Abstr. Int. B *31*: 6160-B, 1971.

11305 - RAYMUNDO, L.C.: The biosynthesis of carotenoids in the tomato fruit. - Diss. Abstr. Int. B *32*: 5025-B - 5026-B, 1971.

11306 - RAZORENOVA, T.A., NILOVSKAYA, N.T.: Vidovaya i vozrastnaya spetsifika temnovogo dykhaniya posevov nekotorykh ovoshchnykh rasteniĭ. [Species and ontogenetic features of dark respiration in stands of some vegetables.] - Fiziol. Rast. *18*: 501-506, 1971. [Growth analysis; in R, ab: E.]

*11307 - REBEIZ, C.A., ABOU-HAIDAR, M.: Biosynthèse du magnesium protoporphyrine monestère à partir de l'acide δ-aminolevulinique par un système extracellulaire. - In: Proc. First Science Meeting of the Lebanese Ass. for the Advancement of Science. Pp. 21-22. Beirut, Lebanon 1969.

*11308 - REBEIZ, C.A., CASTELFRANCO, P.: Protochlorophyllide biosynthesis in a cell-free system from higher plants. - Plant Physiol. *46* (Suppl.): 42, 1970.

11309 - REBEIZ, C.A., CASTELFRANCO, P.A.: Protochlorophyll biosynthesis in a cell-free system from higher plants. - Plant Physiol. *47*: 24-32, 1971.

11310 - REBEIZ, C.A., CASTELFRANCO, P.A.: Chlorophyll biosynthesis in a cell-free system from higher plants. - Plant Physiol. *47*: 33-37, 1971.

11311 - REBEIZ, C.A., NISHIJIMA, C., CRANE, J.C.: The biosynthesis and accumulation of microgram quantities of protochlorophyll, chlorophyll *a* and chlorophyll *b* in a cell-free system from higher plants. - Plant Physiol. *47* (Suppl.): 45, 1971.

*11312 - REBEIZ, C.A., YAGHI, M.: Biosynthèse de la protochlorophyllide et de la protochlorophyllide ester durant la biosynthèse étioplastique dans les cotylédons

de concombre. - In: Proc. First Science Meeting of the Lebanese Ass. for the
Advancement of Science. Pp. 22-23. Beirut, Lebanon 1969.

*11313 - REBEIZ, C.A., YAGHY, M., ABOU-HAIDAR, M.: Etioplast-pigment accumulation
in germinating cucumber cotyledons. - In: Abstracts XI International Botanical Congress. P. 178. Seattle 1969.

11314 - REDMANN, R.E.: Carbon dioxide exchange by native Great Plants grasses. -
Can. J. Bot. 49: 1341-1345, 1971.

11315 - REED, D.W., MAYNE, B.C.: The subcellular localization of the pteridines in
strain R-26 of Rhodopseudomonas spheroides. - Biochim. biophys. Acta 226:
477-480, 1971.

11316 - REIFSNYDER, W.E., FURNIVAL, G.M., HOROWITZ, J.L.: Spatial and temporal distribution of solar radiation beneath forest canopies. - Agr. Meteorol. 9:
21-37, 1971/72.

11317 - REIN, R., NIR, S., STAMATIADOU, M.N.: Photochemical survival principle in
molecular evolution. - J. theor. Biol. 33: 309-318, 1971. [Ps.]

11318 - REÏNGARD, T.A., VOLOVIK, O.I., ZAÏTSEVA, N.A., POLISHCHUK, A.I., SHIRYAEV,
A.I., OSTROVSKAYA, L.K.: Tsiklicheskoe fotofosforilirovanie fragmentov
khloroplastov. [Cyclic photophosphorylation in chloroplast fragments.] -
In: Biokhimiya i Biofizika Fotosinteza. Pp. 170-175. Irkutsk 1971. [In R.]

*11319 - REISS, E.: Chloroplastenzahlen in Epidermiszellen und Schliesszellen bei
Oenotheren. - Biol. Zentralbl. 85: 735-758, 1966.

11320 - REISS-HUSSON, F., De KLERK, H., JOLCHINE, G., JAUNEAU, E., KAMEN, M.D.:
Some effects of iron deficiency on Rhodopseudomonas spheroides strain Y.
- Biochim. biophys. Acta 234: 73-82, 1971. [Chl.]

11321 - REMY, R.: Resolution of chloroplast lamellar proteins by electrophoresis
in polyacrylamide gels. Different patterns obtained with fractions enriched
in either chlorophyll a or chlorophyll b.- FEBS Lett. 13: 313-317, 1971.

11322 - RENGER, G.: The watersplitting system of photosynthesis. II. The accelera-
tion of the deactivation reactions in the watersplitting system by certain
chemicals. - Z. Naturforsch. 26 b: 149-153, 1971.

11323 - RENNERT, A., KNYPL, J.S.: Hamowanie syntezy chlorofilu i RNA w izolowanych
liścieniach ogórka przez N-hydroksymocznik. [Inhibition of chlorophyll and
RNA synthesis by N-hydroxyurea in detached cucumber cotyledons.] - Acta Soc.
Bot. Pol. 40: 669-679, 1971. [In Pol., ab: E.]

11324 - van RENSEN, J.J.S.: Action of some herbicides in photosynthesis of Scenedesmus, as studied by their effects on oxygen evolution and cyclic photophosphorylation. - Med. Landbouwhogeschool Wageningen 71-9: 1-80, 1971.

11325 - REPKA, Ï., SARICH, M., MAREK, Ï., ZIMA, M.: Vliyanie nedostatka makroelementov na strukturu khloroplastov i produktivnost' fotosinteza u rasteniï kukuruzy. [Effect of macronutrient deficiency on chloroplast structure and
photosynthetic productivity in maize plants.] - Fiziol. Rast. 18: 1107-1112,
1971. [In R, ab: E.]

11326 - REPKA, J., KOSTREJ, A.: Analýza produkčného procesu kultúrnych rastlín v
podmienkach južného Slovenska. III. Štúdium optimálnej štruktúry porastu.
[Analysis of the production process of cultivated plants in southern Slo-
vakia. III. The optimum structure of plant stand.] - Pol'nohospodárstvo 17:
923-938, 1971. [Growth analysis; in Slovak, ab: E, R.]

*11327 - RÉTALJINÉ GYÖRGYFY, K., JAKABFI FRIGYESNÉ: Adatok kétféle módon tartós tott
zöld lucerna karotinoid festékeinek összehasonlításáról. [Levels of carotenoid
pigments in alfalfa preserved by two methods.] - Agrokémia Talajtan 18: 117-
132, 1969. [In Hung., ab: E, F, R.]

11328 - REUNOV, A.V.: K voprosu o vozrastnykh osobennostyakh paramagnetizma list'ev
rasteniï. [Ontogenetic features of paramagnetism of plant leaves.] - Fiziol.
Rast. 18: 432-433, 1971. [Chl; in R.]

11329 - REUVENI, R., COHEN, Y., ROTEM, J.: Sporulation of Erysiphe cichoracearum as
influenced by conditions favoring photosynthesis in the host. - Israel J.
Bot. 20: 78-83, 1971.

11330 - **REVELANTE, N., KVEDER, S.:** Hydrographic and biotical conditions in North-Adriatic. XI. Some relations between phytoplankton abundance, primary productivity and plant pigments in Rovinj area. - Rapp. Commun. int. Mer Médit. *20*: 331-334, 1971.

11331 - **REYNOLDS, S.G.:** A note on the estimation of leaf areas of cocoa (*Theobroma cacao* L.) from three leaf parameters. - Trop. Agr. (Trinidad) *48*: 177-179, 1971.

11332 - **REYSS, A., BOURDU, R.:** Influence des héméropériodes très courtes sur la croissance de *Lolium multiflorum*, sa composition pigmentaire et l'ultrastructure chloroplastique. - Planta *97*: 230-244, 1971.

11333 - **RHODES, I.:** The relationship between productivity and some components of canopy structure in ryegrass (*Lolium* spp.). II. Yield, canopy structure and light interception. - J. agr. Sci. *77*: 283-292, 1971.

11334 - **RIBÉREAU-GAYON, G., PREISS, J.:** ADP-glucose pyrophosphorylase from spinach leaf. - In: COLOWICK, S.P., KAPLAN, N.O. (ed.): Methods in Enzymology. Vol. 23. Pp. 618-624. Academic Press, New York - London 1971.

11335 - **RICCI, E.:** Determination of carbon-12, carbon-13 isotopic abundances and nitrogen/carbon ratios in biological substances by proton-reaction analysis. - Anal. Chem. *43*: 1866-1871, 1971.

11336 - **RICHMOND, A.E., SACHS, B., OSBORNE, D.J.:** Chloroplasts, kinetin and protein synthesis. - Physiol. Plant. *24*: 176-180, 1971.

11337 - **RICKETTS, T.R.:** The structures of siphonein and siphonaxanthin from *Codium fragile*. - Phytochemistry *10*: 155-160, 1971.

11338 - **RICKETTS, T.R.:** Identification of xanthophylls KI and KIS of the *Prasinophyceae* as siphonein and siphonaxanthin. - Phytochemistry *10*: 161-164, 1971.

*11339 - **ŘÍMAN, L., MIŠTINOVÁ, A.:** Polyploidy pri viacročných krmovinách VI. Identifikácia polyploidov ďateliny lúčnej (*Trifolium pratense* L.) podľa chloroplastov v prieduchoch. [Polyploids in perennial forage crops. VI. Identification of polyploids in clover (*Trifolium pratense* L.) by chloroplasts in stomata.] - Ved. Práce výsk. Ústavu rast. Výroby Piešťanoch *7*: 59-74, 1969. [In Slovak, ab: E, R.]

11340 - **ROBERTS, B.R., TOWNSEND, A.M., DOCHINGER, L.S.:** Photosynthetic response to SO_2 fumigation in red maple. - Plant Physiol. *47* (Suppl.): 30, 1971.

*11341 - **ROCHA, V., MUKERJI, S.K., TING, I.P.:** Chloroplast-malic dehydrogenase: A new malic dehydrogenase isozyme from spinach. - Biochem. biophys. Res. Commun. *31*: 890-894, 1968.

11342 - **RÖCHLING, H., BÜCHEL, K.H.:** Hemmstoffe der Photosynthese, IX Notiz über herbizide Imidazo-pyridine und Imidazo-chinoxaline. - Chem. Ber. *104*: 344-347, 1971.

11343 - **RODIONOV, V.S.:** Posledeīstvie ponizhennoī polozhitel'noī temperatury na potentsial'nuyu intensivnost' fotosinteza nekotorykh vidov *Lycopersicon* TOURN. i *Beta* L. [After-effect of decreased temperature above zero on the potential photosynthetic rate of some species of *Lycopersicon* TOURN. and *Beta* L.] - Bot. Zh.*56* : 1019-1024, 1971. [In R, ab: E.]

11344 - **RODSKJER, N.:** A pyranometer with dome of RG 8 for use in plant communities. - Arch. Meteorol. Geophys. Bioklimatol., Ser. B *19*: 307-320, 1971.

11345 - **RODSKJER, N., KORNHER, A.:** Über die Bestimmung der Strahlungsenergie im Wellenlängenbereich von 0.3-0.7 μ in Pflanzenbeständen. - Agr. Meteorol. *8*: 139-150, 1971.

11346 - **RODZILLER, I.D., ZOTOV, V.M.:** Rol' biologicheskogo doochistnogo pruda v obogashchenii vody kislorodom. [Role of biological purification ponds in enriching the oxygen content of water.] - Gigiena Sanitariya (Moskva) *36* (10): 110-112, 1971. [Ps; in R.]

11347 - **ROGACHENKO, A.D.:** Radiatsionnye faktory i produktivnost' fotosinteza posevov kukuruzy v usloviyakh orosheniya. [Radiation factors and the productivity of photosynthesis in irrigated maize crops.] - In: ZALENSKIĬ, O.V. (ed.):

Fotosintez i Ispol'zovanie SolnechnoT Energii. Pp. 62-66. Nauka, Leningrad
1971. [In R, ab: E.]

11348 - ROMANENKO, V.I.: Produtsirovanie organicheskogo veshchestva fitoplanktonom
v Rybinskom vodokhranilishche. [Production of phytoplankton organic matter
in the Rybinsk reservoir.] - Gidrobiol. Zh. 7 (4): 5-10, 1971. [Ps; in R,
ab: E.]

11349 - ROMANENKO, V.I.: Opredelenie fotosinteza fitoplanktona vo vnutrennikh vodo-
emakh. [Determination of phytoplankton photosynthesis in the inland reservoirs.]
- Tr. Inst. Biol. vnutr. Vod Akad. Nauk SSSR 21 (24) - Biologiya i Produktivnost' Presnovodnykh Organizmov: 234-240, 1971. [In R.]

11350 - ROMANENKO, V.I., KUZNETSOV, S.I., DAUKSHTA, A.S.: Microbiologicheskie protse-
ssy v ozerakh Latvii. [Microbiological processes in Latvian lakes.] - Tr.
Inst. Biol. vnutr. Vod Akad. Nauk SSSR 21 (24)- Biologiya i Produktivnost'
Presnovodnykh Organizmov: 31-42, 295, 1971. [Ps; in R.]

11351 - ROMASHKO, Ya.D.: O nekotorykh aspektakh fiziologii lista kak organa fotosin-
teza. [Some aspects of physiology of the leaf as a photosynthetic organ.] -
In: Fotosintez, Rost i UstoTchivost' RasteniT. Pp. 85-111. Naukova Dumka,
Kiev 1971. [In R.]

11352 - RÖMER, W.: Untersuchungen über die Auslastung des Photosyntheseapparates
bei Gerste (Hordeum distichon L.) und Weissem Senf (Sinapis alba L.) in Abhängigkeit von den Umweltbedingungen. - Arch. Bodenfruchtbark. Pflanzenpro-
dukt. 15: 415-423, 1971.

11353 - RÖMKENS, M.J.M., MILLER, R.D.: Predicting root size and frequency from one-
dimensional consolidation data - a mathematical model. - Plant Soil 35: 237-
248, 1971.

11354 - RONSAL', G.A., LAGODA, P.P., RONSAL', L.N., PRISHCHEPA, A.G.: Izmenenie nekotorykh pokazateleT fotosinteticheskoT deyatel'nosti rasteniT pod vliyaniem
fosfornykh udobreniT v usloviyakh yuga Ukrainy. [Changes in some indices of
photosynthetic activity as affected by phosphorus fertilizers under conditions
of South Ukraine.] - Tr. kishinev. sel'skokhoz. Inst. M.V. Frunze 85 (Mineral'noe Pitanie i Svet kak Faktory FotosinteticheskoT Deyatel'nosti Sel'-
skokhozyaTstvennykh RasteniT): 70-74, 78-79, 1971. [Chl; in R.]

11355 - ROOK, D.A., SWEET, G.B.: Photosynthesis and photosynthate distribution in
Douglas-fir strobili grafted to young seedlings. - Can. J. Bot. 49: 13-17,
1971.

11356 - ROSINSKI, J.: Chloroplast development in leaves of Phaseolus vulgaris: factors controlling fine structural changes. - Diss. Abstr. int. B 31: 5182-B,
1971.

11357 - ROSS, J., MÄGI, H.: A calculation method for determining the leaf area and
its vertical distribution in the barley crop. - In: Estonian Contributions
to the International Biological Programme. Progress Report III. Pp. 101-112.
Acad. Sci. est. SSR, Est. rep. Comm. IBP. Tartu 1971.

11358 - RÖSSLER, F.: Messung der Dunsttrübung mittels Sonnenphotometer. - Arch. Meteorol. Geophys. Bioklimatol., Ser. B 19: 157-164, 1971.

11359 - ROTFARB, R.M., PAROMCHIK, I.I.: K voprosu ob ingibirovanii biosinteza khlorofilla amitrolom, simazinom i khloramfenikolom. [Inhibition of chlorophyll
biosynthesis by amitrole, simazine, and chloramphenicol.] - In: PlastidnyT
Apparat i Zhiznedeyatel'nost' RasteniT. Pp. 62-65. Nauka i Tekhnika, Minsk
1971. [In R.]

*11360 - ROTINI, O.T., NAVARI-IZZO, F., COVI, A.: La presenza dell'ALA-deidratasi
nelle piante superiori. [Presence of δ-aminolaevulinic acid dehydratase in
higher plants.] - Agrochimica 14: 277-284, 1970. [In Ital., ab: E, F, G, Span.]

11361 - ROTTENBERG, H., GRUNWALD, T., AVRON, M.: Direct determination of Δ pH in chloroplasts, and its relation to the mechanism of photoinduced reactions. - FEBS
Lett. 13: 41-44, 1971.

11362 - ROY, H., MOUDRIANAKIS, E.N.: Interactions between ADP and the coupling factor of photophosphorylation. - Proc. nat. Acad. Sci. USA 68: 464-468, 1971.

11363 - ROY, H., MOUDRIANAKIS, E.N.: Synthesis and discharge of the coupling factor. Adenosine diphosphate complex in spinach chloroplast lamellae. - Proc. nat. Acad. Sci. USA *68*: 2720-2724, 1971.

11364 - ROZHKO, I.I.,GRODZINSKIĬ, D.M.: K voprosu o regulyatornom deĭstvii CO_2 na fotosinteticheskiĭ metabolizm. [Role of CO_2 in the control of photosynthetic metabolism.] - In: ZALENSKIĬ, O.V. (ed.): Fotosintez i ispol'zovanie Solnechnoĭ Energii. Pp. 160-164. Nauka, Leningrad 1971. [In R, ab: E.]

11365 - RUBIN, A.B.: Regulation of primary stages of photosynthesis. - In: BRODA, E., LOCKER, A., SPRINGER-LEDERER, H. (ed.): Proceedings of the First European Biophysics Congress. Vol. 4. Pp. 53-54. Verlag Wiener med. Akad., Wien 1971.

11366 - RUBIN, A.B., FORK, D.C.: The effect of CCCP on the electron transport reactions in chloroplasts. - Carnegie Inst. Year Book *70*: 482-484, 1971.

11367 - RUBIN, B., ESHEL, Y.: Phytotoxicity of fluometuron and its derivatives to cotton and weeds. - Weed Sci. *19*: 592-594, 1971. [Ps.]

11368 - RUBIN, L.B., DUBROVIN, V.N.: Intensivnost' sveta kak regulyator fotoindutsirovannykh prevrashcheniĭ tsitokhromov pri fotosinteze u *Ectothiorhoodospira shaposhnikovii.* [Illuminance as the regulating parameter of the light-induced changes in cytochromes during photosynthesis in *Ectothiorhodospira shaposhnikovii.*] - Mol. Biol. (Moskva) *5*: 518-526, 1971. [In R, ab: E.]

11369 - RUBIN, L.B., SHVINKA, Yu.E., DUBROVIN, V.N., ADAMOVA, N.P.: Vliyanie donorov elektronov na okislitel'no-vosstanovitel'noe sostoyanie elektronno-transportnoĭ tsepi *Ectothiorhodospira shaposhnikovii.* [Effect of electron donors on the oxidative-reductive state of the electron-transport chain in *Ectothiorhodospira shaposhnikovii.* - Mikrobiologiya *40*: 949-955, 1971. [In R, ab: E.]

11370 - RUBY, R.H.: Delayed fluorescence from *Chlorella pyrenoidosa*: Effect of background illumination. - Photochem. Photobiol. *13*: 97-111, 1971.

11371 - RUBY, R.H., BROWN, J.S.: Delayed fluorescence emission by fractions of *Scenedesmus* mutants. - Carnegie Inst. Year Book *69*: 706-709, 1971.

11372 - RUDOĬ, A.B., VEZITSKIĬ, A.Yu., SHLYK, A.A.: Temnovye prevrashcheniya pigmentov postetiolirovannykh list'ev posle razlichnykh periodov osveshcheniya. [Transformations of pigments in post-etiolated leaves in the dark after varous periods of illumination.] - Dokl. Akad. Nauk SSSR *201*: 230-233, 1971. [In R.]

11373 - RUNGE, M.: Determination of energy values. - In: ELLENBERG, H. (ed.): Integrated Experimental Ecology. Pp. 75-80. Springer-Verlag, Berlin - Heidelberg - New York 1971.

11374 - RUNGE, M.: IV. Standortslehre (Okologische Geobotanik). - Fortschr. Bot. *33*: 359-366, 1971. [Ps.]

11375 - RURAINSKI, H.J., RANDLES, J., HOCH, G.E.: The relationship between P 700 and NADP reduction in chloroplasts. - FEBS Lett. *13*: 98-100, 1971.

*11376 - RUSKOV, P., MALCHEV, E.: Izmenenie na khlorofilite pri sushene na zelen fasul. [Change in chlorophylls during drying of green beans.] - Nauch. Trudy vissh. Inst. Khranit. vkusova Prom. (Plovdiv) *16* (1): 65-69, 1969. [In Bulg., ab: F, R.]

*11377 - RUSKOV, P., MALCHEV, E.: Izmenenie na khlorofila na zelen piper pri mlechnokisela fermentatsiya. [Change in chlorophylls in green pepper during lactic acid fermentation.] - Nauch. Trudy vissh. Inst. Khranit. vkusova Prom. (Plovdiv) *16* (1): 101-108, 1969. [In Bulg., ab: G, R.]

11378 - RUTGER, J.N.: Effect of plant density on yield of inbred lines and single crosses of maize (*Zea mays* L.). - Crop Sci. *11*: 475-476, 1971. [Growth analysis.]

11379 - RUTGER, J.N., FRANCIS, C.A., GROGAN, C.O.: Diallel analysis of ear leaf characteristics in maize (*Zea mays* L.). - Crop Sci. *11*: 194-195, 1971.

*11380 - RUTNER, A.C., LANE, M.D.: Nonidentical subunits of ribulose diphosphate carboxylase. - Biochem. biophys. Res. Commun. *28*: 531-537, 1967.

11381 - RYAN, F.J.: Studies on chlorophyll. Investigations on the preparation of water-soluble chlorophyll derivatives. Studies on the oxidation and reduction properties of chlorophyll analogues. - Diss. Abstr. int. B *31*: 3898-B, 1971.

11382 - RYBAKOVA, M.I., DENISOVA, R.R., POPOV, B.A.: Struktura khloroplastov list'-ev ozimoĭ pshenitsy i rzhi posle zimovki. [Structure of leaf chloroplasts in winter wheat and rye after wintering.] - Fiziol. Rast. *18*: 1260-1263, 1971. [In R, ab: E.]

*11383 - RYBIN, I.A., EFIMOV, A.K.: Bioelektricheskie potentsialy perekhodnykh sosto-yaniĭ v razlichnykh usloviyakh temperatury i soveshchennosti rasteniya. [Bio-electric potentials of transient states at different temperatures and irra-diances of the plant.] - Uch. Zap. ural'sk. gos. Univ. *113*, Ser. biol. *8* (Voprosy Regulyatsii Fotosinteza, Sb. 2): 45-56, 1970. [In R.]

11384 - RYBIN, I.A., EFIMOV, A.K.: Ob izmeneniyakh bioelektricheskogo potentsiala list'ev konskikh bobov v svetloe vremya sutok. [Changes in the bioelectric potential of broad bean leaves during day-light hours.] - Ekologiya *2* (5): 91-93, 1971. [In R.]

*11385 - RYBIN, I.A., EFIMOV, A.K., MAKAROV, N.M.: Bioelektricheskaya reaktsiya lista rasteniya v protsesse formirovaniya fotosinteticheskogo apparata. [Bioelec-tric reaction of a plant leaf during formation of photosynthetic apparatus.] - Uch. Zap. ural'sk. gos. Univ. *113*, Ser. biol. *8* (Voprosy Regulyatsii Foto-sinteza, Sb. 2): 57-64, 1970. [Ps, Chl; in R.]

11386 - RYRIE, I.J., JAGENDORF, A.T.: Inhibition of photophosphorylation in spinach chloroplasts by inorganic sulfate. - J. biol. Chem. *246*: 582-588, 1971.

11387 - RYRIE, I.J., JAGENDORF, A.T.: An energy-linked conformational change in the coupling factor protein in chloroplasts. Studies with hydrogen exchange. - J. biol. Chem. *246*: 3771-3774, 1971.

11388 - SAAKOV, V.S.: Deĭstvie ATF, ingibitorov i razobshchiteleĭ fotofosforiiirovan-iya na prevrashchenie ksantofillov v liste. [Effect of ATP, inhibitors and photophosphorylation uncouplers on conversion of xanthophylls in the leaf.]- Dokl. Akad. Nauk SSSR *198*: 966-969, 1971. [In R.]

11389 - SAAKOV, V.S.: O svyazi reaktsii diepoksidatsii ksantofillov s elektrontrans-portnoĭ cep'yu fotosinteza. [Relation between xanthophyll deepoxidation and the electron-transport chain of photosynthesis.] - Dokl. Akad. Nauk SSSR *201*: 1257-1260, 1971. [In R.]

11390 - SAAKOV, V.S.: O vzaimootnosheniyakh mezhdu svetoindutsiruemymi prevrashchen-iyamiksantofillow i elektrontransportnoĭ tsep'yu fotosinteza. [Correlation between light-induced xanthophyll conversions and the electron transport chain of photosynthesis.] - Fiziol. Rast. *18*: 1088-1097, 1971. [In R, ab: E.]

11391 - SAAKOV, V.S., NAZAROVA, G.D., MYL'NIKOVA, E.V., ALEKSEEVA, N.R.: Deĭstvie ingibitorov fotosinteza na pigmentnuyu sistemu. [Action of inhibitors of pho-tosynthesis on the pigment system.] - In: Biokhimiya i Biofizika Fotosinteza. Pp. 28-36. Irkutsk 1971. [In R.]

11392 - SAHA, S., OUITRAKUL, R., IZAWA, S., GOOD, N.E.: Electron transport and photo-phosphorylation in chloroplasts as a function of the electron acceptor. - J. biol. Chem. *246*: 3204-3209, 1971.

11393 - SAHU, S.C.: Quantum accumulation and activation kinetics of the photosynthe-tic oxygen evolution. - Diss. Abstr. int. B *32*: 1392-B - 1393-B, 1971.

11394 - SAIDOV, A.S., POLEVOĬ, V.V.: Vliyanie krasnogo i dal'nego krasnogo sveta na sintez pigmentov i pogloshchenie kisloroda u etiolirovannykh list'ev masha. [Influence of red and far-red light on the synthesis of pigments and absorp-tion of oxygen in etiolated leaves of *Phaseolus aureus*.] - Vestn. leningrad. Univ., Biol. *1971* (4): 140-142, 1971. [In R, ab: E.]

11395 - SAIER, M.H. Jr., FEUCHT, B.U., ROSEMAN, S.: Phosphoenolpyruvate-dependent fructose phosphorylation in photosynthetic bacteria. - J. biol. Chem. *246*: 7819-7821, 1971.

11396 - SAKAMOTO, M.: Chemical factors involved in the control of phytoplankton pro-duction in the Experimental Lakes Area, northwestern Ontario. - J. Fish. Res. Board Can. *28*: 203-213, 1971. [Ps, Chl.]

11397 - SĂLĂGEANU, N.: 200 de ani de la descoperirea fotosintezei (1771-1971). [200 years from the discovery of photosynthesis (1771-1971).] - Natura (Bucureşti) 23: 3-13, 1971. [In Roum.]

*11398 - SĂLĂGEANU, N., OLIMID, V.: Contribuţii la cunoaşterea nevoii de elemente minerale a plantelor. [Contribution to the diagnosis of mineral-nutrient deficiency in plants.] - Stud. Cercet. Biol., Ser. bot. 21: 285-293, 1969. [Ps; in Roum., ab: F.]

11399 - SALAZAR, A.G., PAULSEN, G.M.: Some physiological responses of Sorghum bicolor to benzene hexachloride. - J. agr. Food Chem. 19: 1005-1007, 1971. [Ps.]

11400 - SALCHEVA, G., GRAMATIKOVA, C.: Aftereffect of low temperatures on the cyclic photophosphorylation of chloroplasts, isolated from hardened and non-hardenec wheat plants. - Dokl. Akad. sel'skokhoz. Nauk Bolg. 4 (1): 55-59, 1971.

11401 - SALCHEVA, G., GRAMATIKOVA, C., ZAFIROV, I.: Aftereffect of lowered temperature on the cyclic photophosphorylation of chloroplasts isolated from Vicia faba, Zea mays and Phaseolus vulgaris. - Dokl. Akad. sel'skokhoz. Nauk Bolg. 4 (2): 183-188, 1971.

11402 - SALEMA, R.: Ultrastructural changes of β-irradiated chloroplasts. - An. Fac. Ciênc. Univ. Porto 54: 219-225, 1971.

11403 - SALEMA, R.: The production of thylakoids in the roots of a Triticale. - An. Fac. Ciênc. Univ. Porto 54: 227-235, 1971.

11404 - SALIN, M.L., HOMANN, P.H.: Changes of photorespiratory activity with leaf age. - Plant Physiol. 48: 193-196, 1971.

11405 - SALVADOR, G., LEFORT-TRAN, M., NIGON, V., JOURDAN, F.: Structure et évolution du corps prolamellaire dans les proplastes d'Euglena gracilis. - Exp. Cell Res. 64: 457-462, 1971.

11406 - SAMBORSCHI, E.: Studiul chimic al fracţiunii carotenoidice din Euphorbia salicifolia HOST. [Chemical study of carotenoïd fractions of Euphorbia salicifolia HOST. plants.] - Farmacia (Bucureşti) 19: 749-753, 1971. [In Roum., ab: E, F, G, R.]

11407 - SAMEJIMA, M., MIYACHI, S.: Light-enhanced carbon dioxide fixation in maize leaves. - In: HATCH, M.D., OSMOND, C.B., SLATYER, R.O. (ed.): Photosynthesis and Photorespiration. Pp. 211-217. Wiley-Interscience, New York - London - Sydney - Toronto 1971.

11408 - SAMISH, Y.B.: Kinetics of the substrate for the evolution of CO_2 in light by photosynthesizing organs. - J. theor. Biol. 33: 557-564, 1971. [Methods.]

11409 - SAMISH, Y.B.: The rate of photorespiration as measured by means of oxygen uptake and its respiratory quotient. - Plant Physiol. 48: 345-348, 1971.

11410 - SAMISH, Y.B., PALLAS, J.E. Jr.: CO_2 compensation point as influenced by previous photosynthesis. - Plant Physiol. 47 (Suppl.): 11, 1971.

11411 - SAMUEL, S., SHAH, N.M., FOGG, G.E.: Liberation of extracellular products of photosynthesis by tropical phytoplankton. - J. mar. biol. Ass. U.K. 51: 793-798, 1971.

11412 - SANE, P.V., PARK, R.B.: Action spectra of Photosystem I and Photosystem II in spinach chloroplast grana and stroma làmellae. - Biochim. biophys. Acta 253: 208-212, 1971.

11413 - SANE, P.V., PARK, R.B.: Cofactor effects on enhancement in NADP reduction by chloroplasts. - Biochem. biophys. Res. Commun. 44: 491-496, 1971.

11414 - SANGER, J.E.: Identification and quantitative measurement of plant pigments in soil humus layers. - Ecology 52: 959-963, 1971.

11415 - SANGER, J.E.: Quantitative investigations of leaf pigments from their interception in buds through autumn coloration to decomposition in falling leaves. - Ecology 52: 1075-1089, 1971.

11416 - SANKHLA, N.: Inhibition of abscisic acid induced chlorophyll degradation by potassium. - Current Sci. 40: 302-303, 1971.

11417 - SANTARIUS, K.A.: The effect of freezing on thylakoid membranes in the presence of organic acids. - Plant Physiol. 48: 156-162, 1971.

11418 - SAPOZHNIKOV, D.I., KOROLEVA, O.Ya.: Zavisimost' reaktsii dezepoksidatsii violaksantina ot partsial'nogo davleniya kisloroda. [Dependence of the deepoxidation reaction of violaxanthin on partial pressure of oxygen.] - Fiziol.
Rast. 18: 1273-1275, 1971. [In R.]

11419 - SAPOZHNIKOV, D.I., POPOVA, I.A., MASLOVA, T.G., KOROLEVA, O.Ya.: Osushchest-
vlenie reaktsii dezepoksidatsii violaksantina v raznykh uchastkakh krasnogo
sveta. [Deepoxidation of violaxanthin in different regions of red light.]
- Dokl. Akad. Nauk SSSR 198: 1465-1467, 1971. [In R.]

11420 - SAPOZHNIKOV, D.I., POPOVA, I.A., RYZHOVA, E.F.: K voprosu o sushchestvovanii
svetovogo poroga reaktsii dezepoksidatsii violaksantina. [Existence of light
threshold of the deepoxydation reaction of violaxanthin.] - Pp. 1-4. Nauka,
Leningrad 1971. [In R.]

11421 - SARGENT, D.F., TAYLOR, C.P.S.: Proof of two distinct enhancement effects by
blue light on oxygen uptake in Chlorella. - Nature 232: 649-650, 1971.

11422 - SASAMORI, T., EMORI, Y.: Laboratory measurement of diffuse spectral reflectance and transmittance of some natural objects in near-infrared. - Arch.
Meteorol. Geophys. Bioklimatol., Ser. B 19: 133-148, 1971.

11423 - SASTRY, Y.S.R., LAKSHMINARAYANA, G.: Chlorophyll-sensitized peroxidation of
saturated fatty acid esters. - J. amer. Oil chem. Soc. 48: 452-454, 1971.

11424 - SATO, V.L., LEVINE, R.P.: Studies of a mutant strain of Chlamydomonas reinhardi with impaired photosynthetic phosphorylation. - Plant Physiol. 47
(Suppl.): 9, 1971.

11425 - SATO, V.L., LEVINE, R.P., NEUMANN, J.: Photosynthetic phosphorylation in Chlamydomonas reinhardi. Effects of a mutation altering an ATP-synthesizing enzyme. - Biochim. biophys. Acta 253: 437-448, 1971.

11426 - SATOH, K.: Mechanism of photoinactivation in photosynthetic systems. IV.
Light-induced changes in the fluorescence transient. - Plant Cell Physiol.
12: 13-27, 1971.

11427 - SATOH, M., HAZAMA, K.: [Studies on photosynthesis and translocation of photosynthate in mulberry tree. I. Photosynthetic rate of remained leaves after
short pruning.] - Proc. Crop. Sci. Soc. Jap. 40: 7-11, 1971. [In Jap., ab: E.]

11428 - SATOH, M., OHYAMA, K.: [Studies on photosynthesis and translocation of photosynthate in mulberry tree. II. The effect of shoot pruning and lateral buds
removal on the photosynthetic rate of remained leaves.] - Proc. Crop Sci. Soc.
Jap. 40: 525-529, 1971. [In Jap., ab: E.]

11429 - SATOO, T.: Primary production relations of coniferous forests in Japan. - In:
DUVIGNEAUD, P. (ed.): Productivity of Forest Ecosystems. Pp. 191-205. Unesco,
Paris 1971.

11430 - SATPATHY, D., ARNASON, T.J.: Temperature sensitivity of some chlorophyll mutants in barley. - Rad. Bot. 11: 397-399, 1971.

11431 - SAUNDERS, G.W., STORCH, T.A.: Coupled oscillatory control mechanism in a
planktonic system. - Nature 230: 58-60, 1971. [Ps.]

11432 - SAVIN, V.N., STEPANENKO, O.G.: Vliyanie γ-izlucheniya Co⁶⁰ na produktivnost'
podsolnechnika pri ostrom obluchenii vegetiruyushchikh rastenii. [Effect of
γ-radiation from ⁶⁰Co on the productivity of sunflower after acute irradiation of growing plants.] - In: Primenenie Izotopov i Yadernykh Izluchenii v
Sel'skom Khozyaistve. Pp. 45-50. Atomizdat, Moskva 1971. [Chl; in R.]

11433 - SAWHNEY, R., CUMMING, B.G.: Inhibition of flower development in Chenopodium
rubrum by a photosynthetic inhibitor. - Can. J. Bot. 49: 2233-2237, 1971.

11434 - SCAWEN, M.D., HEWITT, E.J.: Plastocyanin from Cucurbita pepo L. - Biochem.
J. - Proc. biochem. Soc. 124: 32 P, 1971.

11435 - SCHACTER, B., ELEY, J.H., GIBBS, M.: Involvement of photosynthetic carbon
reduction cycle intermediates in CO₂ fixation and O₂ evolution by isolated
chloroplasts. - Plant Physiol. 48: 707-711, 1971.

11436 - SCHACTER, B.Z., GIBBS, M., CHAMPIGNY, M.-L.: Effect of antimycin A on photosynthesis of intact spinach chloroplasts. - Plant Physiol. 48: 443-446, 1971.

11437 - SCHAEDLE, M., FOOTE, K.C.: Seasonal changes in the photosynthetic capacity
of *Populus tremuloides* bark. - Forest Sci. *17*: 308-313, 1971.

11438 - SCHÄFER, K.: Wurzeltemperatur und Gaswechsel des Sprosses von *Lolium perenne*
L. und *Dactylis glomerata* L. - Wirtschaftseigene Futter *1971* (3): 212-217,
1971. [Ps.]

11439 - SCHÄFER, K.: Zur quantitativen Bestimmung der Atmung von Gräserwurzeln. -
Acker- Pflanzenzüchtung *133*: 123-136, 1971. [Ps.]

11440 - SCHÄFER, K.: Zur Temperaturabhängigkeit der apparenten Photosynthese einiger
Sorten von *Dactylis glomerata* L. bei verschiedener Beleuchtungsstärke. - Z.
Acker- Pflanzenbau *134*: 113-124, 1971.

11441 - SCHÄFER, K.: Zur Frage der Durchflussmenge bei CO_2-Gaswechselmessungen mit
"Sprossküvetten". - Z. Acker- Pflanzenbau *134*: 200-206, 1971.

11442 - SCHANTZ, R., SALAÜN, J.-P., SCHANTZ, M.-L., DURANTON, H.: Variations de la
teneur en glucides libres chez *Euglena gracilis* durant la formation de structures plastidiales. - Compt. rend. Acad. Sci. Paris, Sér. D *273*: 1795-1798,
1971.

11443 - SCHARLEMANN, W., CZYGAN, F.C.: Zusammensetzung und Stoffwechsel der Tetrater-
pene und Chlorophylle in Kalluskulturen von *Ruta graveolens* L. - Herba hung.
10 (2-3): 43-48, 1971.

11444 - SCHICK, H.-J.: Substrate and light dependent fixation of molecular nitrogen
in *Rhodospirillum rubrum*. - Arch. Mikrobiol. *75*: 89-101, 1971. [Ps.]

11445 - SCHICK, H.-J.: Interrelationship of nitrogen fixation, hydrogen evolution
and photoreduction in *Rhodospirillum rubrum*. - Arch. Mikrobiol. *75*: 102-109,
1971.

11446 - SCHICK, H.-J.: Regulation of photoreduction in *Rhodospirillum rubrum* by ammonia. - Arch. Mikrobiol. *75*: 110-120, 1971.

11447 - SCHIFF, J.A.: Developmental interactions among cellular compartments in *Eug-
lena*. - In: BOARDMAN, N.K., LINNANE, A.W., SMILLIE, R.M. (ed.): Autonomy and
Biogenesis of Mitochondria and Chloroplasts. - Pp. 98-118. North-Holland Publ.
Comp., Amsterdam - London 1971. [Ps, Chl, Car.]

11448 - SCHIFF, J.A.: The informational and nutritional requirements of cellular organelles. - Stadler Symp. (Missouri) *3*: 89-113, 1971. [Ps.]

*11449 - SCHILLING, G.: Neue Ergebnisse und Probleme der Pflanzenernährung in der
Deutschen Demokratischen Republik. - Sitzungsber. deut. Akad. Landwirtschaftswiss. Berlin *18* (1): 1-25, 1969. [Ps and yield formation.]

11450 - SCHILLING, G.: Der Produktionsprozess der höheren grünen Pflanze. - Wiss.
Beitr. Martin-Luther-Univ. Halle-Wittenberg *1971* (5, N 1): 7-34, 1971.

11451 - SCHINDLER, D.W.: Light, temperature, and oxygen regimes of selected lakes in
the Experimental Lakes Area. northwestern Ontario. - J. Fish. Res. Board Can.
28: 157-169, 1971. [Productivity.]

11452 - SCHINDLER, D.W.: A hypothesis to explain differences and similarities among
lakes in the Experimental Lakes Area, northwestern Ontario. - J. Fish. Res.
Board Can. *28*: 295-301, 1971. [Chl.]

11453 - SCHINDLER, D.W., HOLMGREN, S.K.: Primary production and phytoplankton in the
Experimental Lakes Area, northwestern Ontario, and other low-carbonate waters, and a liquid scintillation method for determining ^{14}C activity in photosynthesis. - J. Fish. Res. Board Can. *28*: 189-201, 1971.

11454 - SCHMALZ, H., GRUMMT, C., JERUZEL, G., STOLBERG, S.: Die Reaktion von chloro-
phyllhaltigen bzw. chlorophyllfreien Weizenkeimpflanzen auf eine Chlorcholinchlorid (CCC)-Applikation. - Arch. Züchtungsforsch. *1*: 163-166, 1971. [Chl.]

11455 - SCHMID, G.H.: Origin and properties of mutant plants: Yellow tobacco. - In:
COLOWICK, S.P., KAPLAN, N.O. (ed.): Methods in Enzymology. Vol 23. Pp. 171-
194. Academic Press, New York - London 1971. [Ps, Chl, Car.]

11456 - SCHMID, G.H., GAFFRON, H.: Fluctuating photosynthetic units in higher plants
and fairly constant units in algae. - Photochem. Photobiol. *14*: 451-464, 1971.

11457 - SCHMIDT, G.L., KAMEN, M.D.: Control of chlorophyll synthesis in *Chromatium vinosum*. - Arch. Mikrobiol. *76*: 51-64, 1971.

11458 - SCHMIDT, G.L., KAMEN, M.D.: Redox properties of the "P-836" pigment complex of *Chromatium*. - Biochim. biophys. Acta *234*: 70-72, 1971.

*11459 - SCHMIDT, G.W.: Zum Problem der Bestimmung der Kohlensäure in kalkarmen tropischen Gewässern. - Amazoniana *1*: 323-326, 1968.

11460 - SCHMIDT, K.: Carotenoids of purple nonsulfur bacteria. Composition and biosynthesis of the carotenoids of some strains of *Rhodopseudomonas acidophila, Rhodospirillum tenue,* and *Rhodocyclus purpureus*. - Arch. Mikrobiol. *77*: 231-238, 1971.

11461 - SCHMIDT, S., REICH, R., WITT, H.T.: Electrochromism of chlorophylls and carotenoids in multilayers and in chloroplasts. - Naturwissenschaften *58*: 414, 1971.

11462 - SCHNEIDER, H.A.W.: Light mediated increase in activity of porphobilinogen deaminase/uroporphyrinogen III cosynthetase and δ-aminolevulinate dehydratase in tissue cultures of tobacco. - Phytochemistry *10*: 319-321, 1971.

11463 - SCHNEIDER, H.A.W.: Porphyrinsynthesis in isolated particles from tissue cultures of tobacco. - Z. Naturforsch. *26 b*: 908-912, 1971.

11464 - SCHNEIDER, M.J., STIMSON, W.R.: Further evidence for photosynthetic involvment in a high energy reaction (HER) response. - Plant Physiol. *47* (Suppl.): 2, 1971.

11465 - SCHNEIDER, M.J., STIMSON, W.R.: Contributions of photosynthesis and phytochrome to the formation of anthocyanin in turnip seedlings. - Plant Physiol. *48*: 312-315, 1971.

11466 - SCHNEIDER, V.: Die Wirkung der unspezifisch mit Proteinen reagierenden Tannine auf photosynthetische Reaktionen der Chloroplastenlamellarsysteme. - Z. Pflanzenphysiol. *64*: 1-14, 1971.

11467 - SCHNEIDER, V.: Der Enfluss von Tannin auf die reversible, lichtabhängige Volumenänderung von isolierten Thylakoiden. - Z. Pflanzenphysiol. *64*: 15-21, 1971.

11468 - SCHÖCH, E.: Malat und Aspartat als Hauptprodukte der $^{14}CO_2$-Kurzzeit-Fixierung nun auch bei einer Composite. - Z. Pflanzenphysiol. *64*: 367-368, 1971.

11469 - SCHÖCH, E., KRAMER, D.: Korrelation von Merkmalen der C_4-Photosynthese bei Vertretern verschiedener Ordnungen der Angiospermen. - Planta *101*: 51-66, 1971.

11470 - SCHÖN, G.: Der Einfluss der Reservestoffe auf den ATP-Spiegel in Zellen von *Rhodospirillum rubrum* beim Übergang von aerober zu anaerober Dunkelkultur. - Arch. Mikrobiol. *68*: 40-50, 1969.

11471 - SCHÖN, G.: Der Einfluss der Kulturbedingungen auf den Nicotinamid-Adenin-Dinucleotid(phosphat)-Gehalt in Zellen von *Rhodospirillum rubrum*. - Arch. Mikrobiol. *79*: 147-163, 1971.

11472 - SCHÖNBOHM, E.: Über die Lokalisierung des Photoreceptors für den tonischen Blaulicht-Effekt bei der Verlagerung des *Mougeotia*-Chloroplasten im Starklicht. - Z. Pflanzenphysiol. *65*: 453-457, 1971.

11473 - SCHÖNBOHM, E.: Untersuchungen zum Photoreceptorproblem beim tonischen Blaulicht-Effekt der Starklichtbewegung des *Mougeotia*-Chloroplasten. - Z. Pflanzenphysiol. *66*: 20-33, 1971.

11474 - SCHRAM, B.L., KROES, H.H.: Structure of phycocyanobilin. - Europe. J. Biochem. *19*: 581-594, 1971.

11475 - SCHREIBER, U. BAUER, R., FRANCK, U.F.: Chlorophyllfluorescenz-Induktion an *Scenedesmus* bei Sauerstoffmangel. - Z. Naturforsch. *26 b*: 1195-1196, 1971.

11476 - SCHREUDER, H.T., SWANK, W.T.: A comparison of several statistical models in forest biomass and surface area estimation. - In: Forest Biomass Studies. Misc. Publ. Vol. 132. Pp. 125-136. Life Sci. and Agr. Exp. Sta., Univ. Maine, Orono 1971.

*11477 - SCHRÖDER, J., DREWS, G.: Quantitative Bestimmung der Fettsäuren von *Rhodospi-rillum rubrum* und *Rhodopseudomonas capsulata* während der Thylakoidmorphoge-nese. - Arch. Mikrobiol. *64*: 59-70, 1968. [Chl.]

*11478 - SCHRÖDER, J., DREWS, G.: Fettsäuregehalte in Lichtkulturen von *Rhodospirillum rubrum* während der Thylakoidmorphogenese. - Arch. Mikrobiol. *69*: 20-33, 1969. [Chl.]

11479 - SCHULDINER, S., AVRON, M.: Anion permeability of chloroplasts. - Europe. J. Biochem. *19*: 227-231, 1971.

11480 - SCHULDINER, S., AVRON, M.: On the mechanism of the energy-dependent quenching of atebrin fluorescence in isolated chloroplasts. - FEBS Lett. *14*: 233-236, 1971.

11481 - SCHULTZ, A.J. Jr.: The development and organization of photosynthetic pigment systems. - Diss. Abstr. Int. B *32*: 1393-B, 1971.

11482 - SCHULTZ, R.C., GATHERUM, G.E.: Photosynthesis and distribution of assimilate of Scotch pine seedlings in relation to soil moisture and provenance. - Bot. Gaz. *132*: 91-96, 1971.

11483 - SCHULZE, E.D., KOCH, W.: Measurement of primary production with cuvettes. - In: DUVIGNEAUD, P. (ed.): Productivity of Forest Ecosystems. Pp. 141-157. Unesco, Paris 1971.

11484 - SCHURER, K.: Direct reading optical leaf area planimeter. - Acta bot. neerl. *20*: 132-140, 1971.

11485 - SCHURR, A., HERSCOVICI, A., SHAVIT, N.: Cannabidiol (hashish component): a new inhibitor of the photosynthetic apparatus. - Israel J. Chem. *9*: 36BC, 1971.

11486 - SCHWARTZ, M.: The relation of ion transport to phosphorylation. - Annu. Rev. Plant Physiol. *22*: 469-484, 1971.

11487 - SCHWARTZBACH, S.D., SCHIFF, J.A.: Synthetic events during the lag period of chloroplast development in *Euglena gracilis* var. *bacillaris*. - Plant Physiol. *47* (Suppl.): 45, 1971.

11488 - SCHWARZ, W.: Das Photosynthesevermögen einiger Immergrüner während des Win-ters und seine Reaktivierungsgeschwindigkeit nach scharfen Frösten. - Ber. deut. bot. Ges. *84*: 585-594, 1971.

11489 - SCHWARZ, Z.: Das Verhalten der NAD- und der NADP-abhängigen Glycerinaldehyd-Phosphat-Dehydrogenase im Verlaufe der Entwicklung des Photosyntheseapparates bei *Phaseolus vulgaris* L. - Biol. Rundschau *9*: 333-335, 1971.

11490 - SCHWENKER, U.: Einfluss des Stoffwechsels auf die Pigmentzusammensetzung in alternden Kulturen von *Euglena gracilis*. - Planta *101*: 101-116, 1971.

11491 - SCOTT, N.S., MUNNS, R., GRAHAM, D., SMILLIE, R.M.: Origin and synthesis of chloroplast ribosomal RNA and photoregulation during chloroplast biogenesis. - In: BOARDMAN, N.K., LINNANE, A.W., SMILLIE, R.M. (ed.): Autonomy and Bio-genesis of Mitochondria and Chloroplasts. Pp. 383-392. North Holland Publ. Comp., Amsterdam - London 1971.

11492 - SECHENSKA, M., TOMOVA, N.: Khloroplasti - struktura i funktsii. [Chloroplasts - structure and function.] - Priroda (Sofia) *20* (2): 46-50, 1971. [In Bulg.]

11493 - SECKBACH, J.: Pigmentation of algae under pressure. - Limnol. Oceanogr. *16*: 567-572, 1971.

11494 - SECKBACH, J.: Size, composition and fine structure of *Cyanidium caldarium* cultured under pure CO_2. - Israel J. Bot. *20*: 302-310, 1971. [Chloroplast.]

11495 - SECKBACH, J., GROSS, H., NATHAN, M.B.: Growth and photosynthesis of *Cyani-dium caldarium* cultured under pure CO_2. - Israel J. Bot. *20*: 84-90, 1971.

11496 - SECKBACH, J., NATHAN, M.B.: Method for the separation of algal pigments by thin layer chromatography: CO_2 effect on pigmentation. - Lab. Pract. *20*: 933-934, 1971.

11497 - SELLY, G.R., MEYER, T.H.: The photosensitized oxidation of β-carotene. - Pho-tochem. Photobiol. *13*: 27-32, 1971.

11498 - SEGEN, B.J., GIBSON, K.D.: Deficiencies of chromatophore proteins in some
mutants of *Rhodopseudomonas spheroides* with altered carotenoids. - J. Bacteriol.
105: 701-709, 1971.

11499 - SEIBERT, M.: Spectral, kinetic and potentiometric studies of the laser-induced primary photochemical reactions in the photosynthetic bacterium, *Chromatium* D. - Diss. Abstr. int. B *32*: 2028-B - 2029-B, 1971.

11500 - SEIBERT, M., DeVAULT, D.: Photosynthetic reaction center transients, P_{435}
and P_{424}, in *Chromatium* D. - Biochim. biophys. Acta *253*: 396-411, 1971.

11501 - SEIBERT, M., CHANCE, B., DeVAULT, D.: The effect of glutaraldehyde fixation
on the primary photochemical processes in bacterial photosynthesis. - Arch.
Biochem. Biophys. *146*: 611-617, 1971.

11502 - SEIBERT, M., DUTTON, P.L., DEVAULT, D.: A low potential photosystem in *Chromatium* D. - Biochim. biophys. Acta *226*: 189-192, 1971.

11503 - SEĬFULINA, L.Ya.: Vliyanie impul'snogo kontsentrirovannogo solnechnogo sveta
na soderzhanie nekotorykh pigmentov v list'yakh khlopchatnika. [Effect of
pulsed concentrated sunlight on the level of some pigments in cotton leaves.]
- In: Svetoimpul'snaya Stimulyatsiya RasteniĬ. Pp. 228-246, 364. Nauka, Moskva 1971. [In R.]

11504 - SEITZ, K.: Die Ursache der Phototaxis der Chloroplasten: ein ATP-Gradient?
(Versuche zum Primärprozess der Starklichtbewegung bei *Vallisneria*.) - Z.
Pflanzenphysiol. *64*: 241-256, 1971.

11505 - SEKIGUCHI, S., YAMAMOTO, T.: Varietal difference of translocation of photosynthetic products between flue-cured and Burley tobacco. - Proc. Crop Sci.
Soc. Jap. *40*: 513-518, 1971.

*11506 - SEKIZAWA, Y., OHASHI, T.: Amperometric surveys on photosynthetic activity
of rice plant. - Sci. Rep. Meiji Seika Kaisha [Meiji Seika Kenkyu Nempo]
1970 (11): 1-14, 1970.

11507 - SELGA, M.P.: Elektronnomikroskopicheskie issledovaniya funktsional'nykh osobennosteĬ liposom v khloroplastakh list'ev. [Electron microscopic analysis
of functional peculiarities of liposomes in leaf chloroplasts.] - Izv. Akad.
Nauk latv. SSR *1971* (9): 80-85, 1971. [In R, ab: E.]

11508 - SELGA, M.P., RUD', M.S.: Adaptivnye izmeneniya assimiliruyushcheĬ tkani lista
pod vliyaniem vneshnikh faktorov. [Adaptive changes of the leaf assimilating
tissue under the influence of external factors.] - Izv. Akad. Nauk latv. SSR
1971 (9): 29-36, 1971. [In R, ab: E.]

11509 - SELGA, M.P., RUD', M.S.: ElektronnomikroskopicheskiĬ analiz nakopleniya assimilyatov v khloroplastakh pri vyrashchivanii rasteniĬ tabaka v razlichnykh
usloviyakh osveshcheniya i mineral'nogo pitaniya. [Electron microscopic analysis of photosynthate accumulation in chloroplasts of tobacco plants grown
under various illuminance and mineral nutrition.] - Izv. Akad. Nauk latv.
SSR *1971* (9): 46-53, 1971. [In R, ab: E.]

*11510 - SELGA, M.P., TAGEEVA, S.V.: ElektronnomikroskopicheskiĬ analiz deĬstviya
razlichnykh uchastkov ul'trafioletovogo spektra na fotosinteticheskiĬ apparat rasteniĬ tomatov i ogurtsov. [Electron microscopic analysis of the action
of various sections of the ultra-violet spectrum on the photosynthetic apparatus of tomato and cucumber plants.] - Izv. Akad. Nauk latv. SSR *1967* (8):
92-99, 1967. [In R.]

11511 - SELIRIO, I.S., BROWN, D.M., KING, K.M.: Estimation of net and solar radiation.
- Can. J. Plant Sci. *51*: 35-39, 1971.

11512 - SELMAN, B.R., BANNISTER, T.T.: Trypsin inhibition of Photosystem.II. - Biochim. biophys. Acta *253*: 428-436, 1971.

11513 - SEMENCHENKO, B.A., MELESHKO, K.E., KIRMALOV, R.L., VLADIMIROVA, I.A.: Pod-
vodnyĬ registriruyushchiĬ spektrometr s distantsionnym upravleniem. [A sub-
marine recording spectrometer with a remote control.] - Meteorol. Gidrol.
1971 (4): 94-99, 1971. [ChÌ; in R.]

*11514 - SEMENOVA, G.A., LADYGIN, V.G., TAGEEVA, S.V.: Vzaimosvyaz' ul'trastruktury,
fotosinteticheskoĬ aktivnosti i soderzhanie pigmentov u mutantov. II. Ul'-

trastrukturnaya organizatsiya khloroplasta pigmentnykh mutantov *Chlamydomonas reinhardti*. [Relation between ultrastructure and photosynthetic activity and pigment content in mutants. II. Ultrastructure organization of a chloroplast of pigment mutants of *Chlamydomonas reinhardti*.] - In: FRANK, G.M. (ed.): Biofizika Zhivoĭ Kletki. Vol. 1. Pp. 90-94. Pushchino 1970. [In R.]

11515 - SEMICHAEVSKIĬ, V.D., LOS', S.I., LOZOVAYA, G.I.: O vzaimodeĭstvii assotsiatov khlorofilla *a* s belkami. [Interaction of chlorophyll *a* associates with proteins.] - Biofizika *16*: 1117-1120, 1971. [In R, ab: E.]

11516 - SEMICHAEVSKIĬ, V.D., LOZOVAYA, G.I.: Kolichestvennye sootnosheniya pri vzaimodeĭstvii assotsiatov khlorofilla *a* s belkami. [Quantitative relations in the interaction between associates of chlorophyll *a* and proteins.] - Dokl. Akad. Nauk SSSR *199*: 965-967, 1971. [In R.]

11517 - SEMIKHATOVA, O.A., CHULANOVSKAYA, M.V., METZNER, H.: Manometric method of plant photosynthesis determination. - In: ŠESTÁK, Z., ČATSKÝ, J., JARVIS, P.G. (ed.): Plant Photosynthetic Production. Manual of Methods. Pp. 238-256. Dr. W. Junk N.V. Publ., The Hague 1971.

*11518 - SEMIN, A.S.: Vliyanie rannikh i preduborochnykh vnekornevykh podkormok azotom na velichinu i kachestvo urozhaya kukuruzy. [Effect of early and preharvest extra-root nitrogen application on the quantity and quality of maize yield.] - Tr. kharkov. sel'skokhoz. Inst. *90* (Issled. Fiziol. Biokhim. Rast.): 29-36, 1970. [Ps, Chl; in R.]

11519 - SENGER, H.: Quantum yield of photosynthesis in synchronous cultures of algae. - In: BRODA, E., LOCKER, A., SPRINGER-LEDERER, H. (ed.): Proceedings of the First European Biophysics Congress. Vol. 4. Pp. 33-36. Verlag wiener med. Adad., Wien 1971.

11520 - SEREBRENIKOV, V.S., ANISIMOV, B.V., PARFENOV, V.T., LIPSITS, D.V.: O deĭstvii γ- i elektronnogo izlucheniĭ na kartofel'. [Action of γ-and electron-radiation on potatoes.] - Radiobiologiya *11*: 426-430, 1971. [Ps; in R.]

11521 - SEREBRENIKOV, V.S., KRUPNOVA, L.V.: Issledovanie radiochuvstvitel'nosti kartofelya v razlichnye periody ontogeneza. [Radiosensitivity of potatoes during different periods of ontogenesis.] - Radiobiologiya *11*: 469-471, 1971. [Ps, Chl; in R.]

11522 - ŠESTÁK, Z.: Determination of chlorophylls *a* and *b*. - In: ŠESTÁK, Z., ČATSKÝ, J., JARVIS, P.G. (ed.): Plant Photosynthetic Production. Manual of Methods. Pp. 672-701. Dr. W. Junk N.V. Publ., The Hague 1971.

11523 - ŠESTÁK, Z., ČATSKÝ, J. (ed.): Bibliography of reviews and methodological papers on photosynthesis. - Photosynthetica *5*: 80-87, 175-182, 322-328, 433-443, 1971.

B11524 - ŠESTÁK, Z., ČATSKÝ, J., JARVIS, P.G. (ed.): Plant Photosynthetic Production. Manual of Methods. - Dr. W. Junk N.V. Publ., The Hague 1971.

11525 - ŠESTÁK, Z., JARVIS, P.G., ČATSKÝ, J.: Criteria for the selection of suitable methods. - In: ŠESTÁK, Z., ČATSKÝ, J., JARVIS, P.G. (ed.): Plant Photosynthetic Production. Manual of Methods. Pp. 1-48. Dr. W. Junk N.V. Publ., The Hague 1971. [Ps.]

11526 - SETCHENSKA, M., TOMOVA, N., DECHEV, G.: Comparative study of the Hill reaction activity and the light-induced uptake of molecular oxygen by isolated chloroplasts. - Dokl. bolg. Akad. Nauk *24*: 377-380, 1971.

11527 - ŠETLÍK, I., ŠESTÁK, Z.: Use of leaf tissue samples in ventilated chambers for long term measurements of photosynthesis. - In: ŠESTÁK, Z., ČATSKÝ, J., JARVIS, P.G. (ed.): Plant Photosynthetic Production. Manual of Methods. Pp. 316-324. Dr. W. Junk N.V. Publ., The Hague 1971.

11528 - SEUBERLING, H.-B.: Hohe Strahlenresistenz der Aktivität und der lichtinduzierten Aktivitätssteigerung der NADP-abhängigen Glycerinaldehyd-3-Phosphat-Dehydrogenase in *Ankistrodesmus braunii*. - Planta *98*: 270-273, 1971.

11529 - SHABEL'SKAYA, E.F., GVARDIYAN, V.N.: Sostoyanie fotosinteticheskogo apparata vysshikh zelenykh rasteniĭ v usloviyakh polnogo zatemneniya. [State of the photosynthetic apparatus in higher green plants under complete darkening.]

- In: Problemy Biosinteza Khlorofillov. Pp. 199-226. Nauka i Tekhnika, Minsk 1971.

11530 - SHAIN, Y., GIBBS, M.: Formation of glycolate by a reconstituted spinach chloroplast preparation. - Plant Physiol. *48*: 325-330, 1971.

11531 - SHAKHOV, A.A.: Teoreticheskie aspekty preobrazovaniya svetovoĭ energii v impul'snom rezhime. [Theoretical aspects of transformation of light energy in an impulse regime.] - In: Svetoimpul'snaya Stimulyatsiya Rasteniĭ. Pp. 9-44. Nauka, Moskva 1971. [Ps; in R.]

11532 - SHAKHOV, A.A., BALAUR, N.S.: Izmenenie membrannoĭ sistemy khloroplastov i mitokhondriĭ svetoimpul'snym oblucheniem i mutagenez. [Changes in membrane system of chloroplasts and mitochondria by light-impulse irradiation and mutagenesis.] - In: Svetoimpul'snaya Stimulyatsiya Rasteniĭ. Pp. 96-113. Nauka, Moskva 1971. [In R.]

11533 - SHAKHOV, A.A., KOTSUR, N.V., LIKHOLAT, Yu.A.: Biosintez pigmentov pri svetoimpul'snom obluchenii etiolirovannykh prorostkov kukuruzy. [Biosynthesis of pigments during the light-induced irradiation of etiolated maize shoots.] - In: Svetoimpul'snaya Stimulyatsiya Rasteniĭ. Pp. 210-221, 363. Nauka, Moskva 1971. [In R.]

*11534 - SHAKHOVA, M.F., SHNAĬDMAN, L.O.: Soderzhanie biologicheski aktivnykh veshchestv v proizvodstvennykh otkhodakh myaty perechnoĭ. [Content of biologically active substances in the industrial waste materials of *Mentha piperita* L.] - Rast. Resursy *4*: 53-62, 1968. [Car; in R.]

11535 - SHAKIROVA, S.V.: Kinetika vklyucheniya C^{14} v sostav produktov fotosinteza na raznykh uchastkakh svetovoĭ krivoĭ u gorokha Torsdag. [Kinetics of ^{14}C incorporation into photosynthates in different sections of the light curve in Torsdag peas.] - Dokl. Akad. Nauk-tadzh. SSR *14* (8): 65-69, 1971. [In R, ab: Tadzh.]

*11536 - SHAKIROVA, S.V., LEBEDEV, V.N.: Svetovye krivye fotosinteza nekotorykh sortov gorokha. [Light curves of photosynthesis of some pea cultivars.] - Izv. Akad. Nauk tadzh. SSR, Otd. biol. Nauk *1970* [3 (40)]: 58-61, 1970. [In R, ab: Tadzh.]

11537 - SHAMSIEV, A., SHAPTSEV, E.V.: Produktivnost' poseva sorgo v zavisimosti ot rezhima orosheniya v usloviyakh Gissarskoĭ doliny (Tadzhikistan). [Effect of irrigation on the sorghum crop productivity in Ghissar valley (Tadjikistan).] - In: ZALENSKIĬ, O.V. (ed.): Fotosintez i Ispol'zovanie Solnechnoĭ Energii. Pp. 66-69. Nauka, Leningrad 1971. [In R, ab: E.]

11538 - SHANIYAZOV, B.S.: Vliyanie kobal'ta na soderzhanie khlorofilla i gematina u bobovykh rasteniĭ. [Effect of cobalt on the content of chlorophylls and hematin in bean plants.] - Vestn. karkalp. fil. Akad. Nauk uzb. SSR *3* (45): 22-26, 1971. [In R.]

11539 - SHAPOSHNIKOVA, M.G., DROZDOVA, N.N., KRASNOVSKIĬ, A.A.: Izuchenie fotookisleniya khlorofilla v vodnom rastvore detergenta Triton X-100. [Photooxidation of chlorophyll in an aqueous solution of the detergent Triton X-100.] - Biokhimiya *36*: 704-711, 1971. [In R, ab: E.]

11540 - SHARMA, K.D., SEN, D.N.: Reversal of the effect of 2,4-dichlorophenoxy acetic acid by humic acid. - Z. Pflanzenphysiol. *65*: 81-84, 1971. [Chl.]

11541 - SHARMA, M.P., VANDEN BORN, W.H.: Effect of picloram on $^{14}CO_2$-fixation and translocation of ^{14}C-assimilates in Canada thistle, soybean and corn. - Can. J. Bot. *49*: 69-74, 1971.

11542 - SHATILOV, I.S., MALOFEEV, V.M., VAULIN, A.V., ABISALOV, R.S.: Intensivnost' fotosinteza i dykhaniya klevera krasnogo. [Rates of photosynthesis and respiration of red clover.] - Dokl. TSKhA *162*: 219-225, 1971. [In R.]

11543 - SHATILOV, I.S., MALOFEEV, V.M., VAULIN, A.V. ABISALOV, R.S.: Dinamika fotosinteza i dykhaniya otdel'nykh organov yachmenya. [Dynamics of photosynthesis and respiration of various organs of barley.] - Dokl. TSKhA *162*: 237-242, 1971. [In R.]

B11544 - SHATILOV, I.S., ROZOV, N.F., SHURYGINA, T.D.: Izmerenie intensivnosti Fotosinteza Rasteniĭ Kolorimetricheskim Metodom Slavika-Chatskogo. [Measuring

Photosynthetic Rate in Plants by Means of the Colorimetric Method of Slavík and Čatský.] - Mosk. sel'sko-khoz. Akad. K.A. Timiryazeva, Moskva 1971. [In R.]

B11545 - SHATILOV, I.S., ROZOV, N.F., SHURYGINA, T.D.: Tablitsy dlya izmereniya inten-sivnosti Fotosinteza Rastenii Kolorimetricheskim Metodom Slavika-Chatskogo. [Tables for Measuring Photosynthetic Rate by Means of the Colorimetric Method according to Slavík and Čatský.] - Mosk. sel'.-khoz. Akad. K.A. Timiryazeva, Moskva 1971. [In R.]

11546 - SHAVIT, N., DEGANI, H.: Ion translocation and the contribution of a membrane potential to ATP formation in chloroplasts. - In: QUAGLIARIELLO, E., PAPA, S., ROSSI, C.S. (ed.): Energy Transduction in Respiration and Photosynthesis. Pp. 1009-1010. Adriatica Editrice, Bari 1971.

11547 - SHAVIT, N., SHOSHAN, V.: Phosphorylation coupled to non-cyclic electron flow in photosystem I. - FEBS Lett. *14*: 265-267, 1971.

11548 - SHAW, M.A., RICHARDS, W.R.: Evidence for the formation of membranous chroma-tophore precursor fractions in *Rhodopseudomonas spheroides*. - Biochem. bio-phys. Res. Commun. *45*: 863-870, 1971.

11549 - SHEREVERYA, N.I., STOLYARENKO, V.S., ASEEVA, I.B., EKKERMAN, N.I.: Intensiv-nost' fotosinteza u gibridov kukuruzy v svyazi s yavleniem geterozisa. [Pho-tosynthetic rate in maize hybrids in relation to heterosis.] - Fiziol. Bio-khim. kul't. Rast. *3*: 7-13, 1971. [In R, ab: E.]

11550 - SHERMA, J.: Chromatography of leaf pigments on silica gel and aluminium hy-droxide loaded papers. - J. Chromatogr. *61*: 202-204, 1971.

11551 - SHEWRY, P.R., PINFIELD, N.J., STOBART, A.K.: The effect of 2,4-dichlorophen-oxyacetic acid and (2-chloroethyl)-trimethylammonium chloride on chlorophyll synthesis in barley leaves. - Planta *101*: 352-359, 1971.

11552 - SHIBATA, H., KONO, Y., OCHIAI, H.: [Effect of 4-thiouridine on chloroplast development in radish cotyledons (2). Photo-reductive activity and fine-struc-ture of chloroplasts.] - Bull. Fac. Agr. Shimane Univ. *1971* (5): 1-9, 1971. [In Jap., ab: E.]

11553 - SHIBATA, K.: Subchloroplast fragments: sodium dodecyl sulfate method. - In: COLOWICK, S.P., KAPLAN, N.O. (ed.): Methods in Enzymology. Vol. 23. Pp. 296-302. Academic Press, New York - London 1971.

11554 - SHIBLES, R.M.: Limitations to yield in soybeans. - Proc. 12th Pac. Sci. Congr. *1*: 83, 1971.

*11555 - SHIKHALIEV, S.S.: Issledovanie pigmentnoi sistemy ogurtsov i prevrashchenii ee pri tekhnicheskoi pererabotke. [Pigment system of cucumbers and its con-version during technical processing.] - Sb. nauch. Soobshch. tekhnol. Fak. dagestan. gos. Univ. (Makhachkala) *1*: 82-84, 1969. [In R.]

11556 - SHIMIZU, S.: Separation of chloroplast pigments on Sephadex LH-20. - J. Chromatogr. *59*: 440-443, 1971.

11557 - SHIMIZU, S.: [The breakdown of chlorophyll.] - Biol. Sci. *23*: 31-42, 1971. [In Jap.]

11558 - SHIN, M.: Ferredoxin-NADP reductase from spinach. - In: COLOWICK, S.P., KAPLAN, N.O. (ed.): Methods in Enzymology. Vol. 23. Pp. 440-447. Academic Press, New York - London 1971.

11559 - SHINOHARA, N.: [Real state of experiment of biological sciences and approach to future. How to study photosynthesis (1). Introduction to the real state of photosynthesis through experiments in biological sciences education.] - Science through Experiments [Kagaku No Jikken] 22 (9): 50-56, 1971. [In Jap.]

11560 - SHINOHARA, N.: [Real state of experiment of biological sciences and approach to future. How to study photosynthesis (2). How to study from qualitative to quantitative analysis.] - Science through Experiments [Kagaku No Jikken] 22 (10): 36-41, 1971. [Methods; in Jap.]

11561 - SHINOHARA, N.: [Real state of experiment of biological sciences and approach to future. How to study photosynthesis (3). Quantitative analysis of photo-synthesis.] - Science through Experiments [Kagaku No Jikken] 22 (11): 42-47, 1971. [Methods; in Jap.]

11562 - SHKOL'NIK, M.Ya., ALEKSEEVA, Kh.A.: O narusheniyakh v soderzhanii lipidov v
razlichnykh organakh i khloroplastakh l'na i kukuruzy pri bornom defitsite.
[Effect of boron deficiency on the lipid content in various organs and chloroplasts of flax and maize.] - Fiziol. Rast. 18: 582-587, 1971. [In R, ab: E.]

11563 - SHLYK, A.A.: Biosynthesis of chlorophyll b. - Annu. Rev. Plant Physiol. 22:
169-184, 1971.

11564 - SHLYK, A.A.: Sovremennoe sostoyanie voprosa o biosinteze khlorofilla b. [Pre-
sent views of biosynthesis of chlorophyll b.] - In: Problemy Biosinteza Khlorofillov. Pp. 53-77. Nauka i Tekhnika, Minsk 1971. [In R.]

11565 - SHLYK, A.A.: Opredelenie khlorofillov i karotinoidov v ekstraktakh zelenykh
list'ev. [Determination of chlorophylls and carotenoids in extracts of green
leaves.] - In: Biokhimicheskie Metody v Fiziologii Rastenii. Pp. 154-170.
Nauka, Moskva 1971. [In R.]

11566 - SHLYK, A.A., AKHRAMOVICH, N.I.: Sravnitel'noe issledovanie pigmentnykh fondov
pri fraktsionirovanii gomogenata i khloroplastov list'ev yachmenya s pomoshch'-
yu dezoksikholata. [Comparative investigation of pigments following fractionation of a homogenate and chloroplasts of barley leaves by means of deoxycho-
late.] - Dokl. Akad. Nauk SSSR 200: 473-476, 1971. [In R.]

11567 - SHLYK, A.A., CHKANIKAVA, R.A., ULASËNAK, L.I., MEL'NIKAŬ, S.S., AKULOVICH,
E.M.: Dasledavanne tsentraŭ biyasintezu khlarafilu pry fraktsyyaniravanni
pigmentnaga aparata khlarely. [Chlorophyll biosynthesis centres during fractionation of pigment apparatus of Chlorella.] - Vestsi Akad. Navuk BSSR,
Ser. biyal. Navuk 1971 (5): 30-37, 1971. [In Beloruss., ab: E.]

11568 - SHLYK, A.A., FRADKIN, L.I., RUDOŬ, A.B., VEZITSKIŬ, A.Yu., KALININA, L.M.,
SAVCHENKO, G.E., AVERINA, N.G., MALASHEVICH, A.V., AKHRAMOVICH, N.I.: Tsen-
try formirovaniya fotosinteticheskogo apparata. [Centres of formation of
the photosynthetic apparatus.] - In: Biokhimiya i Biofizika Fotosinteza. Pp.
8-14. Irkutsk 1971. [In R.]

11569 - SHLYK, A.A., PRUDNIKOVA, I.V., MALASHEVICH, A.V.: Sposobnost' gomogenata
etiolirovannykh prorostkov kukuruzy k temnovomu prevrashcheniyu vvedennogo
izvne khlorofilla a v khlorofill b. [Capacity of a homogenate of etiolated
maize sprouts for the dark transformation of externally introduced chlorophyll a into chlorophyll b.]- Dokl. Akad. Nauk SSSR 201: 1481-1484, 1971.
[In R.]

11570 - SHLYK, A.A., PRUDNIKOVA, I.V., SAVCHENKO, G.E., GROZOVSKAYA, M.S.: Nakoplenie
khlorofilla b i protokhlorofillida gomogenatom listovoi tkani v temnote.
[Accumulation of chlorophyll b and protochlorophyllide by a leaf tissue homogenate in the dark.] - Dokl. Akad. Nauk SSSR 200: 222-225, 1971. [In R.]

11571 - SHMUELI, E., ZIEGLER, H., KRAPF, G.: The ultrastructure of corn stomata on
leaves at different heights on the plant. - Israel J. Bot. 20: 330-331, 1971.
[Chloroplast.]

11572 - SHNEYOUR, A., AVRON, M.: Disproportionation of 1,5-diphenylcarbazone. A new
reaction catalysed by Photosystem I. - Biochim. biophys. Acta 253: 412-420,
1971.

*11573 - SHPOTA, L.A., PECHENOV, V.A.: Polevye metody i pribory dlya opredeleniya fo-
tosinteza. [Field methods and apparatuses for photosynthesis determination.]
- In: Tezisy Dokladov Vsesoyuznogo Soveshchaniya po Unifikatsii Metodov i
Priborov dlya Massovykh Izmerenii Intensivnosti Fotosinteza. Pp. 117-122.
Nauch.-issled. Inst. Rastenievod. N.I. Vavilova, Leningrad - Pushkin 1970.
[In R.]

11574 - SHUL'GA, A.M., SUBOCH, V.P.: Spektral'noe izuchenie produktov fotovosstanov-
leniya khlorinov. [Spectral study of products of chlorins photoreduction.]
- Biofizika 16: 214-220, 1971. [In R, ab: E.]

11575 - SHUVALOV, V.A., KRASNOVSKII, A.A.: Lyuminestsentsiya tsink porfirinov v mi-
kroorganizmakh i rasteniyakh: fosforestsentsiya i zamedlennaya fluorestsens-
iya. [Luminescence of zinc-porphyrins in microorganisms and plants: Phosphorescence and delayed fluorescence.] - Mol. Biol. (Moskva) 5: 698-710,
1971. [In R, ab: E.]

11576 - SHVETSOVA, V.M.: Zavisimost' fotosinteza nekotorykh rasteniĭ Zapadnogo Taĭ-myra ot intensivnosti osveshcheniya. [Dependence of photosynthesis on irradiance in some plants of Western Taĭmyr.] - Bot. Zh. *56*: 701-705, 1971. [In R.]

*11577 - SHVETSOVA, V.M., VOZNESENSKIĬ, V.L.: Sutochnye i sezonnye izmeneniya intensivnosti fotosinteza u nekotorykh rasteniĭ Zapadnogo Taĭmyra. [Diurnal and seasonal changes in photosynthetic rate of some plants of Western Taĭmyr.] - Bot. Zh. *55*: 66-76, 1970. [In R, ab: E.]

*11578 - SHVETSOVA, V.M., VOZNESENSKIĬ, V.L.: Intensivnost' fotosinteza nekotorykh rasteniĭ dominiruyushchikh v tundrakh tsentral'nogo Taĭmyra. [Photosynthetic rate of some dominant species of central Taĭmyr tundra.] - In: Produktivnost' Biotsenozov Subarktiki. Tezisy Dokladov. Pp. 94-96. Sverdlovsk 1970. [In R.]

11579 - SIDDIQUI, M.Q., MANNERS, J.G.: Some effects of general yellow rust (*Puccinia striiformis*) infection on ^{14}carbon assimilation, translocation and growth in a spring wheat. - J. exp. Bot. *22*: 792-799, 1971.

11580 - SID'KO, F.Ya., TERSKOV, I.A., BERESNEV, G.F., EROSHIN, N.S., ZAKHAROVA, V.A.: O vozmozhnosti ispol'zovaniya perekhodnykh protsessov dlya povysheniya effektivnosti fotosinteza odnokletochnoĭ vodorosli khlorelly. [Possibility of utilizing transient processes to incraese the photosynthetic efficiency of the unicellular alga *Chlorella*.] - Dokl. Akad. Nauk SSSR *199*: 1206-1208, 1971. [In R.]

11581 - SIEGELMAN, H.W., SCHOPFER, P.: Protochlorophyllide holochrome. - In: COLOWICK, S.P., KAPLAN, N.O. (ed.): Methods in Enzymology. Vol. 23. Pp. 578-582. Academic Press, New York - London 1971.

11582 - SIEGENTHALER, P.-A.: Conversion de l'energie chimique dans l'appareil photosynthétique. - Ann. Univ. Neuchatel *1970-1971*: 1-27, 1970-71.

11583 - SIEGENTHALER, P.-A., VAUCHER-BONJOUR, P.: Vieillissement de l'appareil photosynthétique. III. Variations et caractéristiques de l'activité *o*-diphénoloxydase (polyphénoloxydase) au cours du vieillissement *in vitro* de chloroplastes isolés d'épinard. - Planta *100*: 106-123, 1971.

11584 - SIKKA, H.C., CARROLL, J., ZWEIG, G.: Effect of certain quinone pesticides on acetate photometabolism and dark CO_2 fixation in *Chlorella*. - Pesticide Biochem. Physiol. *1*: 381-388, 1971.

*11585 - SILAYEVA, A.M., SHIRYAEV, A.I.: Three-dimensional plastid structure reconstructed on the basis of electron microscopic observations. - In: Sixth International Congress Electr. Microscopy, Kyoto. Pp. 375-376. Maruzen Co. Ltd. Nihonbashi, Tokyo 1966.

11586 - SILSBURY, J.H.: The effects of temperature and light energy on dry weight and leaf area changes in seedling plants of *Lolium perenne* L. - Aust. J. agr. Res. *22*: 177-187, 1971.

11587 - SINCLAIR, T.R., HOFFER, R.M., SCHREIBER, M.M.: Reflectance and internal structure of leaves from several crops during a growing season. - Agron. J. *63*: 864-868, 1971.

11588 - SINEGUB, O.A., EROKHIN, Yu.E.: Narushenie sostoyaniya bakteriokhlorofilla v khromatoforakh *Chromatium minutissimum* pri izmenenii pH, ionnoĭ sily i dobavlenii okisliteleĭ. [Alteration of bacteriochlorophyll state in *Chromatium minutissimum* chromatophores under the action of pH, ionic strength, and oxidants.] - Mol. Biol. (Moskva) *5*: 472-479, 1971. [In R, ab: E.]

11589 - SINGH, B.B.: Effect of gamma-irradiation on chlorophyll content of maize leaves. - Rad. Bot. *11*: 243-244, 1971.

11590 - SINGH, D.V., KRISHNAN, P.S.: Nuclease activity in chloroplasts from dodder filaments. - Phytochemistry *10*: 739-747, 1971.

11591 - SINGH, J., WASSERMAN, A.R.: The use of disc gel electrophoresis with nonionic detergent in the purification of cytochrome *f* from spinach grana membranes. - J. biol. Chem. *246*: 3532-3541, 1971.

11592 - SINGH, P., KRISHNAN, P.S.: Effect of root parasitism by *Orobanche* on the respiration and chlorophyll content of *Petunia*. - Phytochemistry *10*: 315-318, 1971.

11593 - SINGHAL, G.S., HEVESI, J.: The correlation between the absorption and the
fluorescence energy spectra, and the quantum yield of chlorophyll a in different solvents. - Photochem. Photobiol. *14*: 509-514, 1971.

11594 - SINYAKOV, G.N., SHUL'GA, A.M., GURINOVICH, G.P.: O svyazi protonirovannykh
form anionov porfirinovykh pigmentov s produktami reaktsii foto- i temnovogo
vosstanovleniya. [Relationship between protonated forms of anions of por-
phyrin pigments and reaction products of photo- and dark reduction.] - Biofizika *16*: 1110-1111, 1971. [In R, ab: E.]

*11595 - SIRENKO, L.A.: Uchastie dopolnitel'nykh pigmentov v fotosinteticheskikh re-
aktsiyakh sinezelenykh vodoroslei. [Participation of accessory pigments in
photosynthetic reactions of blue-green algae.] - In: II Vsesoyuznyi Biokhimicheskii S"ezd, Tezisy Sektsionnykh Soobshchenii. 19 Sektsiya: Problemy Fotosinteza. Pp. 108-109. Tashkent 1969. [In R.]

11596 - SIRONVAL, C.: The evolution of chlorophyll containing photoactive structures.
- In: SCHOFFENIELS, E. (ed.): Biochemical Evolution and the Origin of Life.
Molecular Evolution. Vol. 2. Pp. 236-258. North-Holland Publ. Comp., Amster-
dam 1971.

11597 - SIVTSEV, M.V., KOLPINA, L.S.: Sootnoshenie pokazatelei fotosinteticheskogo
apparata i produktivnosti kukuruzy pod vliyaniem mikroelementov. [Relation-
ship between indices of photosynthetic apparatus and productivity in maize
as affected by microelements.] - Fiziol. Biokhim. kul't. Rast. *3*: 434-440,
1971. [In R, ab: E.]

11598 - SJOLUND, R.D., WEIER, T.E.: An ultrastructural study of chloroplast structure and dedifferentiation in tissue cultures of *Streptanthus tortuosus*
(*Cruciferae*). - Amer. J. Bot. *58*: 172-181, 1971.

11599 - SKULACHEV, V.P.: Transformation of respiration and photosynthesis energy into
electric form of membrane potential: experimental proof. - In: QUAGLIARIELLO,
E., PAPA, S., ROSSI, C.S. (ed.): Energy Transduction in Respiration and Photosynthesis. Pp. 153-171. Adriatica Editrice, Bari 1971.

11600 - SLACK, C.R.: The C_4 pathway: assessment. - In: HATCH, M.D., OSMOND, C.B.,
SLATYER, R.O. (ed.): Photosynthesis and Photorespiration. Pp. 297-301. Wiley-
Interscience, New York - London - Sydney - Toronto 1971.

*11601 - SLATYER, R.O.: Measurement of the primary production of arid zone plant
communities. - In: MILNER, C., HUGHES, R.E. (ed.): Methods for the Measurement
of the Primary Production of Grassland. IBP Handbook No. 6. Pp. 53-61. Blackwell sci. Publ., Oxford - Edinburgh 1968.

11602 - SLATYER, R.O.: Effect of errors in measuring leaf temperature and ambient
gas concentration on calculated resistances to CO_2 and water vapor exchanges
in plant leaves. - Plant Physiol. *47*: 269-274, 1971.

11603 - SLATYER, R.O.: Relationship between plant growth and leaf photosynthesis in
C_3 and C_4 species of *Atriplex*. - In: HATCH, M.D., OSMOND, C.B., SLATYER, R.O.
(ed.): Photosynthesis and Photorespiration. Pp. 76-81. Wiley-Interscience,
New York - London - Sydney - Toronto 1971.

11604 - SLATYER, R.O., TOLBERT, N.E.: Photosynthesis and photorespiration. - Science
173: 1162-1167, 1971.

11605 - SLAVÍK, B.: Determination of stomatal aperture. - In: ŠESTÁK, Z., ČATSKÝ, J.,
JARVIS, P.G. (ed.): Plant Photosynthetic Production. Manual of Methods. Pp.
556-565. Dr. W. Junk N.V. Publ., The Hague 1971.

11606 - SLAVOV, N.: V"rkhu metoda za opredelyane na rastitelnata masa pri tsarevit-
sata. [Method of determining the vegetative mass of maize.] - Rastenievod.
Nauki (Sofia) *8* (3): 3-8, 1971. [In Bulg, ab: E, R.]

11607 - SMALL, L.F., RAMBERG, D.A.: Chlorophyll a, carbon and nitrogen in particles
from a unique coastal environment. - In: COSTLOW, J.D. Jr. (ed.): Fertility
of the Sea. Vol. 2. Pp. 475-492. Gordon and Breach Sci. Publ., New York 1971.

11608 - SMILLIE, R.M., ANDERSEN, K.S., BISHOP, D.G.: Plastocyanin-dependent photoreduction of NADP by agranal chloroplasts from maize. - FEBS Lett. *13*: 318-
320, 1971.

11609 - SMILLIE, R.M., BISHOP, D.G., GIBBONS, G.C., GRAHAM, D., GRIEVE, A.M., RAISON, J.K., REGER, B.J.: Determination of the sites of synthesis of proteins and lipids of the chloroplast using chloramphenicol and cycloheximide. - In: BOARDMAN, N.K., LINNANE, A.W., SMILLIE, R.M. (ed.): Autonomy and Biogenesis of Mitochondria and Chloroplasts. Pp. 422-433. North-Holland Publ. Comp., Amsterdam - London 1971.

11610 - SMILLIE, R.M., ENTSCH, B.: Phytoflavin. - In: COLOWICK, S.P., KAPLAN, N.O. (ed.): Methods in Enzymology. Vol. 23. Pp. 504-514. Academic Press, New York - London 1971.

*11611 - SMILLIE, R.M., SCOTT, N.S.: Organelle biosynthesis: The chloroplast. - Progr. mol. subcell. Biol. 1: 136-202, 1969.

11612 - SMIRNOVA, A.D., KOCHETOV, V.P.: Vliyanie NRV v sochetanii s borom i razdel'-no bora na ogurechnoe rastenie v usloviyakh levoberezh'ya Volgi Saratovskoĭ oblasti. [Effect of petroleum growth substance alone or combined with boron on cucumber plants on the eastern side of the Volga River in the Saratov region.] - In: NRV v Sel'skom Khozyaĭstve. Pp. 548-552. Elm, Baku 1971. [Ps; in R.]

11613 - SMITH, A.E., RAAB, K., EKPAHA-MENSAH, J.A.: Origin of enzymic and photosynthetic activity in a prebiotic system. - Experientia 27: 648-650, 1971.

11614 - SMITH, B.N.: $^{13}C/^{12}C$ ratios in grasses: A preliminary survey. - Plant Physiol. 47 (Suppl.): 11, 1971.

11615 - SMITH, B.N., EPSTEIN, S.: Two categories of $^{13}C/^{12}C$ ratios for higher plants. - Plant Physiol. 47: 380-384, 1971.

11616 - SMITH, D.W.: The effects of soil texture and soil moisture on photosynthesis, growth and nitrogen uptake of Scotch pine seedlings. - Diss. Abstr. int. B 31: 7025-B, 1971.

*11617 - SMITH, F.A.: The mechanism of chloride transport in characean cells. - New Phytol. 69: 903-917, 1970. [Ps.]

11618 - SMITH, F.A.: Transport of solutes during C_4 photosynthesis: assessment. - In: HATCH, M.D., OSMOND, C.B., SLATYER, R.O. (ed.): Photosynthesis and Photorespiration. Pp. 302-306. Wiley-Interscience, New York - London - Sydney - Toronto 1971.

11619 - SMITH, R.V., NOY, R.J., EVANS, M.C.W.: Physiological electron donor systems to the nitrogenase of the blue-green alga Anabaena cylindrica. - Biochim. biophys. Acta 253: 104-109, 1971.

*11620 - SNYDER, F.W., TOLBERT, N.E.: Effect of CO_2 concentration on photosynthetic product in sugar beets. - Amer. Soc. Sugar Beet Technol. 1966: 41, 1966.

11621 - SOROKIN, E.M.: Issledovanie funktsional'noĭ struktury pigmentnogo kompleksa fotosinteticheskogo apparata. I. Analiz kinetiki kvantovogo vykhoda fluorestsentsii khlorofilla s tochki zreniya unitsentral'noĭ i mul'titsentral'noĭ modeleĭ fotosinteticheskoĭ edinitsy. [Functional structure of the pigment complex of the photosynthetic apparatus. I. Analysis of the kinetics of quantum yield of chlorophyll fluorescence from the viewpoint of unicentral and multicentral models of the photosynthetic unit.] - Fiziol. Rast. 18: 264-274, 1971. [In R, ab: E.]

11622 - SOROKIN, E.M.: Issledovanie funktsional'noĭ struktury pigmentnogo kompleksa fotosinteticheskogo apparata. II. O metodakh izmereniya velichin, kharakteriziruyushchikh kinetiku izmeneniya fluorestsentsii khlorofilla in vivo. [Functional structure of the pigment complex of the photosynthetic apparatus. II. Methods for measuring the values characterizing the kinetics of changes of chlorophyll fluorescence in vivo.] - Fiziol. Rast. 18: 473-482, 1971. [In R, ab: E.]

11623 - SOROKIN, E.M.: Opredelenie kolichestva molekul v pigmentnoĭ matritse fotosinteticheskoĭ edinitsy. [Determination of number of molecules in the pigment matrix of a photosynthetic unit.] - Fiziol. Rast. 18: 874-886, 1971. [Apparatus; in R, ab: E.]

11624 - SOROKIN, E.M.: Dinamicheskie svoĭstva pigmentnoĭ matritsy fotosistemy II

vysshikh rasteniĭ. [Dynamic properties of the pigment matrix of the photosystem II in higher plants.] - Fiziol. Rast. *18*: 1098-1106, 1971. [In R, ab: E.]

11625 - SOROKIN, E.M., TUMERMAN, L.A.: Kooperativnyĭ kharakter vzaimodeĭstviya dvukh
fotokhimicheskikh sistem fotosinteza. [Cooperative character of the interaction between the two photochemical systems of photosynthesis.] - Mol. Biol.
(Moskva) *5*: 753-765, 1971. [In R, ab: E.]

11626 - SOROKIN, Ju.I.: On the role of bacteria in the productivity of tropical oceanic waters. - Int. Rev. ges. Hydrobiol. *56*: 1-48, 1971. [Ps.]

*11627 - SOROKIN, Yu.I.: Pervichnaya produktsiya i mikrobiologicheskie protsessy v
oz. Gek - Gel'. [Primary production and microbiological processes in the Geck-
Gel Lake.] - Mikrobiologiya *37*: 345-354, 1968. [In R, ab: E.]

11628 - SOROKIN, Yu.I.: Sravnitel'naya otsenka produktivnosti planktona melkovodiĭ
Volzhskogo plesa Rybinskogo vodokhranilishcha. [Comparative estimation of
plankton productivity of the Volga reach shallow waters of the Rybinsk reservoir.] - Tr. Inst. Biol. vnutr. Vod Akad. Nauk SSSR *21* (24) - Biologiya i
Produktivnost' Presnovodnykh Organizmov: 5-16, 295, 1971. [Ps; In R.]

11629 - SOSEBEE, R.E., WIEBE, H.H.: Effect of water stress and clipping on photosynthate translocation in two grasses. - Agron. J. *63*: 14-17, 1971.

11630 - SOUCHON, C.: Réouverture des stomates au cours de la fanaison de feuilles
coupées chez le *Cardamine pratensis* L.; relations avec la transpiration et
la photosynthèse. - Compt. rend. Acad. Sci. Paris, Sér. D *272*: 2892-2895,
1971.

11631 - SOURNIA, A.: Productivité primaire dans le Canal de Mozambique. - In: Symposium on Indian Ocean and Adjacent Seas. Pp. 24-25. Mar. biol. Ass. India,
Cochin 1971.

11632 - SOURNIA, A.: Mesure de la productivité primaire des océans par la méthode
du ^{14}C. - Terre malgache, N. spéc. *12* (L'énergie nucléaire et ses applica-
tions biologiques à Madagascar): 251-267, 1971.

11633 - SPANNER, D.C.: Transport in the phloem. - Nature *232*: 157-160, 1971. [Photo-
synthates.]

11634 - SPEKTOROV, K.S., KRYLOV, Yu.V., NIKOL'SKAYA, T.V., GROMAKOVSKIĬ, B.M.,
NICHIPOROVICH, A.A.: Izmenenie biologicheskikh i fiziologicheskikh osobennosteĭ kletok *Chlorella pyrenoidosa* PRINGSH. 82 T v ravnomernykh kul'turakh vy-
sokoĭ plotnosti. [Changes in the biological and physiological properties of
the cells of *Chlorella pyrenoidosa* PRINGSH. 82 T in steady-state cultures
of high density.] - Fiziol. Rast. *18*: 60-68, 1971. [Ps, Chl; In R, ab: E.]

11635 - SPENCE, J.A.: Cultivation of detached sweet potato (*Ipomoea batatas* (L.) LAM)
leaves with tuberous roots for photosynthetic studies. - Photosynthetica *5*:
424-425, 1971.

11636 - SPIERTZ, J.H.J., ten HAG, B.A., KUPERS, L.J.P.: Relation between green area
duration and grain yield in some varieties of spring wheat. - Neth. J. agr.
Sci. *19*: 211-222, 1971. [Growth analysis.]

11637 - SPIVAK, A.I.: Fotosinteticheskoe vklyuchenie ugleroda-14 u nekotorykh vidov
stepnykh rasteniĭ Zabaĭkal'ya. [Photosynthetic incorporation of carbon-14
in some species of Transbaikal steppe plants.] - In: Biokhimiya i Biofizika
Fotosinteza. Pp. 187-190. Irkutsk 1971. [In R.]

11638 - SPIVAK, A.I.: Raspredelenie C^{14} v produktakh fotosinteza u nekotorykh stepnykh rasteniĭ Zabaĭkal'ya. [^{14}C distribution in photosynthetic products of
certain steppe plants of Transbaikal.] - Bot. Zh. *56*: 722-727, 1971. [In R.]

11639 - SPLITTSTOESSER, W.E., MEYER, M.M. Jr.: Evergreen foliage contributions to
the spring growth of *Taxus*. - Physiol. Plant. *24*: 528-533, 1971. [Chl.]

11640 - SQUIRE, G.R., JONES, M.B.: Studies on the mechanisms of action of the antitranspirant phenylmercuric acetate and its penetration into the mesophyll.
- J. exp. Bot. *22*: 980-991, 1971. [Ps.]

11641 - SREERAMULU, K.: Chemical mutagenesis in sorghum. - Proc. Indian Acad. Sci.,
Sect. B *74* (4): 161-173, 1971. [Chl.]

11642 - SRIVASTAVA, B.S., KUMAR, H.D.: Pigment variations in ultraviolet-treated
strains of the blue-green alga *Anabaena doliolum*. - Arch. Mikrobiol. *77*:
247-251, 1971.

11643 - STABENAU, H.: Die Regulation des Photosyntheseapparates bei *Chlorogonium elon-
gatum* DANGEARD unter dem Einfluss von Licht und Acetat. - Biochem. Physiol.
Pflanzen *162*: 371-385, 1971.

11644 - STACY, W.T., MAR, T., SWENBERG, C.E., GOVINDJEE: An analysis of a triplet
exciton model for the delayed light in *Chlorella*. - Photochem. Photobiol.
14: 197-219, 1971.

11645 - ŠTAMBERA, J., PETŘÍKOVÁ, K.: Vliv giberelinu na produkci fotosyntézy u rajs-
kého jablíčka (*Solanum lycopersicum* L.). [Effect of gibberellin on photosynthetic production in *Solanum lycopersicum* L.] - Acta Univ. agr. Brno, Fac.
agron. *19*: 95-105, 1971. [In Czech, ab: E, G, R.]

11646 - STANEV, V.P., LINGOVA, S.: Izmeneniya FAR i intensivnost' fotosinteza u podsolnechnika pri razlichnoī gustote poseva. [PhAR fluctuations and photosynthetic rate in sunflower at various sowing density.] - Dokl. Akad. sel'skokhoz. Nauk Bolg. *4*: 323-329, 1971. [In R.]

11647 - STANHILL, G., FUCHS, M., OGUNTOYINBO, J.: The accuracy of field measurements
of solar reflectivity. - Arch. Meteorol. Geophys. Bioklimatol., Ser. B *19*:
113-132, 1971.

11648 - STANKO, S.A.: Stimuliruyushchee deīstvie impul'snogo kontsentrirovannogo solnechnogo sveta na semena i rasteniya. [Stimulating action of impulse concentrated sun light on seeds and plants.] - In: Svetoimpul'snaya Stimulyatsiya
Rasteniī. Pp. 144-209. Nauka, Moskva 1971. [Ps, Chl; in R.]

11649 - STANLEY, R.A.: Studies on nutrition, photosynthesis, and respiration in *Myriophyllum spicatum* L. - Diss. Abstr. Int. B *31*: 5843-B, 1971.

11650 - STAPLETON, H.N., MEYERS, R.P.: Modeling subsystems for cotton - the cotton
plant simulation. - Trans. ASAE *14*: 950-953, 1971. [Canopy.]

11651 - STARCK, Z.: Effect of light conditions on the photosynthesis and ^{14}C-assi-
milates distribution in bean plants. - Bull. Acad. pol. Sci., Sér. Sci. biol.
19: 807-814, 1971.

11652 - STARCK, Z.: Niektóre aspekty fizjologii plonowania roślin. [Some aspects of
physiology of the yielding ability of plants.] - Postępy Nauk roln. *18* (3):
21-32, 1971. [Ps; in Pol.]

11653 - STARCK, Z.: Pattern of ^{14}C-assimilates distribution in relation to their
supply and demand in sunflower. Part II. - Acta Soc. Bot. Pol. *40*: 653-667,
1971.

11654 - STAUFF, J., FUHR, H.: Chemiluminescence from the action of singlet oxygen
($^1\Delta_g$) on chlorophyll a and some other luminescing substances. - Z. Naturforsch. *26 b*: 260-263, 1971.

11655 - STEARNS, M.E., WAGENAAR, E.B.: Ultrastructural changes in chloroplasts of
autumn leaves. - Can. J. Genet. Cytol. *13*: 550-560, 1971.

11656 - STEEMANN NIELSEN, E., WILLEMOËS, M.: How to measure the illumination rate
when investigating the rate of photosynthesis of unicellular algae under various light conditions. - Int. Rev. ges. Hydrobiol. *56*: 541-556, 1971.

11657 - STEEMANN NIELSEN, E., WIUM-ANDERSEN, S.: The influence of Cu on photosynthesis and growth in diatoms. - Physiol. Plant. *24*: 480-484, 1971.

11658 - STEER, B.T.: The dynamics of leaf growth and photosynthetic capacity in *Capsicum frutescens* L. - Ann. Bot. *35*: 1003-1015, 1971.

11659 - STEFFEN, H., CALVIN, M.: Spectroscopic investigation of the inhibitory effect
of fatty acids on photosynthetic systems. - Nature - new Biol. *234*: 165-168,
1971.

11660 - STEIGER, E., ZIEGLER, I., ZIEGLER, H.: Unterschiede in der Lichtaktivierung
der NADP-abhängigen Glycerinaldehyd-3-phosphat-Dehydrogenase und der Ribu-
lose-5-phosphat-Kinase bei Pflanzen des Calvin- und des C_4-Dicarbonsäure-
Fixierungstypus. - Planta *96*: 109-118, 1971.

11661 - STEINEMANN, A., ALAMUTI, N., BRODMANN, W., MARSCHALL, O., LÄUGER, P.: Optical properties of artificial chlorophyll membranes. - J. Membrane Biol. 4: 284-294, 1971.

*11662 - STEINHÜBEL, G.: Nenáročná metóda gravimetrického určovania čistého výkonu fotosyntézy borovice čiernej a borovice sosny. [Simple method of gravimetric determination of dry matter increment of needles of Pinus nigra ARN. and Pinus silvestris L.] - Lesn. Časop. 16: 259-271, 1970. [In Slovak, ab: E, G, R.]

11663 - STEINHÜBEL, G.: Non-destructive estimation of the photoassimilating crown-surface of Pinus nigra ARN. - Photosynthetica 5: 376-383, 1971.

11664 - STEPANOV, K.I.: Chistaya produktivnost' fotosinteza rasteniĭ soi v raznykh usloviyakh mineral'nogo pitaniya i osveshchennosti. [Net photosynthetic productivity of soybean under various mineral nutrition and irradiance.] - Tr. kishinev. sel'skokhoz. Inst. M.V. Frunze 85 (Mineral'noe Pitanie i Svet kak Faktory Fotosinteticheskoĭ Deyatel'nosti Sel'skokhozyaĭstvennykh Rastenĭ): 29-36, 76, 1971. [In R.]

11665 - STEPHANSEN, K., ZALIK, S.: Inheritance and qualitative analysis of pigments in a barley mutant. - Can. J. Bot. 49: 49-51, 1971.

11666 - STEVENSON, K.R., SHAW, R.H.: Diurnal changes in leaf resistance to water vapor diffusion at different heights in a soybean canopy. - Agron. J. 63: 17-19, 1971. [Stomatal resistance.]

11667 - STEVENSON, K.R., SHAW, R.H.: Effects of leaf orientation on leaf resistance to water vapor diffusion in soybean (Glycine max L. MERR) leaves. - Agron. J. 63: 327-329, 1971. [Stomatal resistance.]

11668 - STEWARD, F.C., CRAVEN, G.H., WEERASINGHE, S.P.R., BIDWELL, R.G.S.: Effects of prior environmental conditions on the subsequent uptake and release of carbon dioxide in the light. - Can. J. Bot. 49: 1999-2007, 1971.

11669 - STEWART, I., WHEATON, T.A.: Continuous flow separation of carotenoids by liquid chromatography. - J. Chromatogr. 55: 325-336, 1971.

*11670 - STEWART, J.I., HAGAN, R.M.: Development of evapotranspiration - crop yield functions for managing limited water supplies. - In: Seventh Congress, International Commission on Irrigation and Drainage. R. 32, question 23. Pp. 23.505-23.530. Mexico City, Mexico 1969. [Ps.]

11671 - ŞTIRBAN, M.: Acumularea pigmenţilor şi randamentul fotosintetic la plantele de orz în lumina artificială şi naturala. [Pigment accumulation and photosynthetic production of barley plants in artificial and natural light.] - Stud. Cercet. Biol., Ser. bot. 23: 271-279, 1971. [In Roum., ab: E.]

11672 - STOCKER, O.: Der Wasser- und Photosynthese-Haushalt von Wüstenpflanzen der mauretanischen Sahara. II. Wechselgrüne, Rutenzweig- und stammsukkulente Bäume. - Flora 160: 445-494, 1971.

11673 - STOCKING, C.R.: Chloroplasts: nonaqueous. - In: COLOWICK, S.P., KAPLAN, N.O. (ed.): Methods in Enzymology. Vol. 23. Pp. 221-228. Academic Press, New York - London 1971.

11674 - STOCKING, C.R., BRADBEER, J.W.: Chloroplast development and enzyme changes in maize leaves. - Plant Physiol. 47 (Suppl.): 45, 1971.

11675 - STOKES, D.M., WALKER, D.A.: Phosphoglycerate as a Hill oxidant in a reconstituted chloroplast system. - Plant Physiol. 48: 163-165, 1971.

11676 - STOKES, D.M., WALKER, D.A.: Relative impermeability of the intact chloroplast envelope to ATP. - In: HATCH, M.D., OSMOND, C.B., SLATYER, R.O. (ed.): Photosynthesis and Photorespiration. Pp. 226-231. Wiley-Interscience, New York - London - Sydney - Toronto 1971.

11677 - STOLBOVA, A.V.: Geneticheskiĭ analiz pigmentnykh mutatsiĭ Chlamydomonas reinhardi. Soobshchenie I. Identifikatsiya osnovnykh pigmentov i opisanie kollektsii pigmentnykh form. [Genetic analysis of pigment mutations of Chlamydomonas reinhardi. I. Identification of basic pigments and characteristics of mutant collection.] - Genetika 7 (9): 90-94, 1971. [In R, ab: E.]

11678 - STOLBOVA, A.V.: Geneticheskiĭ analiz pigmentnykh mutatsiĭ Chlamydomonas rein-

hardi. Soobshchenie II. Analiz nasledovaniya mutatsiĭ beskhlorofil'nosti i svetochuvstvitel'nosti v skreshchivaniyakh s dikim tipom. [Genetic analysis of *Chlamydomonas reinhardi* pigment mutations. II. Pattern of inheritance mutations of chlorophyll deficiency and light-sensitivity in crosses with wild types.] - Genetika 7 (11): 124-129, 1971. [In R, ab: E.]

11679 - STOLBOVA, A.V., KHROPOVA, V.I.: Pigmentnye mutatsii khlorelly kak model' dlya izucheniya nekotorykh voprosov mekhanizma fotosinteza. [Pigment mutations of *Chlorella* as a model for the study of photosynthesis mechanism.] - In: ZALENSKIĬ, O.V. (ed.): Fotosintez i ispol'zovanie Solnechnoĭ Energii. Pp. 261-264. Nauka, Leningrad 1971. [In R, ab: E.]

11680 - STOYANOV, Zh.V., FLOROV, R.I.: Vozmozhnosti ispol'zovaniya entropii dissipatsii dlya opredeleniya kvantovogo raskhoda brutto-fotosinteza. [Possibility to utilize dissipation entropy for determination of the quantum yield of gross photosynthesis.] - Fiziol. Rast. 18: 483-487, 1971. [In R, ab: E.]

11681 - ST-PIERRE, J.-C.W.: The effect of defoliation on the accumulation of dry matter and the translocation of ^{14}C-photosynthates in timothy (*Phleum pratense* L.). - Diss. Abstr. Int. B 31: 5120-B, 1971.

11682 - STRAIN, H,H., COPE, B.T. Jr., McDONALD, G.N., SVEC, W.A., KATZ, J.J.: Chlorophylls a_1 and a_2. - Phytochemistry 10: 1109-1114, 1971.

11683 - STRAIN, H.H., COPE, B.T., SVEC, W.A.: Analytical procedures for the isolation, identification, estimation, and investigation of the chlorophylls. - In: COLOWICK, S.P., KAPLAN, N.O. (ed.): Methods in Enzymology. Vol. 23. Pp. 452-476. Academic Press, New York - London 1971.

11684 - STRAIN, H.H., SVEC, W.A., AITZETMÜLLER, K., COPE, B.T.Jr., HARKNESS, A.L., KATZ, J.J.: Mass fragmentation and structure of siphonaxanthin, siphonein, and derivatives. - Org. Mass Spectrom. 5: 565-572, 1971.

11685 - STRAIN, H.H., SVEC, W.A., AITZETMÜLLER, K., GRANDOLFO, M.C., KATZ, J.J., KJØSEN, J., NORGÅRD, S., LIAAEN-JENSEN, S., HAXO, F.T., WEGFAHRT, P., RAPOPORT, H.: The structure of peridinin, the characteristic dinoflagellate carotenoid. - J. amer. chem. Soc. 93: 1823-1825, 1971.

11686 - STRAIN, H.H., SVEC, W.A., AITZETMÜLLER, K., KATZ, J.J.: A comparison of the formulas proposed for heteroxanthin. - Tetrahedron Lett. 10: 733-736, 1971.

*11687 - STRAŠKRABA, M., PIECZYŃSKA, E.: Field experiments on shading effect by emergents on littoral phytoplankton and periphyton production. - In: STRAŠKRABA, M., PIECZYŃSKA, E., BRANDL, J., POŠTOLKOVÁ, M., DVOŘÁK, J., LIŠKOVÁ, E.: Relations of Aquatic Macroflora to Phytoplankton, Periphyton and Macrofauna. Rozpravy ČSAV, Ser. MPV (Praha) 80 (6): 7-32, 1970. [Ps.]

11688 - STRASSER, R.J.: Eine einfache Anlage zur kontinuierlichen Kultivierung von Lemnaceen met automatischer Probeentnahme. - Photosynthetica 5: 76-78, 1971. [Ps.]

11689 - STRAUB, O.: Lists of natural carotenoids. - In: ISLER, O. (ed.): Carotenoids. Pp. 771-850. Birkhäuser Verlag, Basel-Stuttgart 1971.

11690 - STRICHARTZ, G.R.: Absorbance changes in spinach chloroplasts resulting from salt addition. - Biophys. Soc. annu. Meet. Abstr. 15: 39A, 1971.

11691 - STRICHARTZ, G.R.: β-carotene involvement and the possibility of a membrane potential associated with the 520 nm absorbance change. - Diss. Abstr. int. B 31: 6445-B, 1971.

11692 - STRICHARTZ, G.R.: The necessity for β-carotene in the 518 nanometer absorbance change. - Plant Physiol. 48: 553-558, 1971.

11693 - STRICKLAND, J.D.H.: Microbial activity in aquatic environments. - In: Microbes and Biological Productivity. Symp. Soc. gen. Microbiol. 21: 231-253, 1971. [Ps.]

11694 - STUART, T.S.: Hydrogen evolution by algae the simplest *in-vivo* partial reaction of photosynthesis. - Biophys. Soc. annu. Meet. Abstr. 15: 37A, 1971.

11695 - STUART, T.S.: Hydrogen production by photosystem I of *Scenedesmus*: Effect of heat and salicylaldoxime on electron transport and photophosphorylation. - Planta 96: 81-92, 1971.

11696 - STUART, T.S., GAFFRON, H.: The kinetics of H_2 photoproduction by *Scenedesmus*. - Plant Physiol. *47* (Suppl.): 32, 1971.

11697 - STUART, T.S., GAFFRON, H.: The kinetics of hydrogen photoproduction by adapted *Scenedesmus*. - Planta *100*: 228-243, 1971.

11698 - SUD'INA, O.G.: Rozvytok doslidzhen' khlorofilu v Instytuti botaniky AN URSR. [Development of chlorophyll studies in the Institute of Botany of the Academy of Sciences of the Ukrainian SSR.] - Ukr. bot. Zh. *28*: 681-693, 1971. [In Ukr., ab: E.]

11699 - SUGAHARA, K., MURATA, N., TAKAMIYA, A.: Fluorescence of chlorophyll in brown algae and diatoms. - Plant Cell Physiol. *12*: 377-385, 1971.

*11700 - SUGENO, K., MATSUBARA, H.: The amino acid sequence of *Scenedesmus* ferredoxin. - Biochem. biophys. Res. Commun. *32*: 951-955, 1968.

11701 - SUGIYAMA, T., ITO, T., AKAZAWA, T.: Subunit structure of ribulose 1,5-diphosphate carboxylase from *Chlorella ellipsoidea*. - Biochemistry *10*: 3406-3411, 1971.

*11702 - SUGIYAMA, T., NAKAYAMA, N., TANAKA, Y., AKAZAWA, T.: Structure and function of chloroplast proteins. IV. Protective effect of substrates on the breakdown of spinach ribulose-1,5-diphosphate carboxylase. - Arch. Biochem. Biophys. *126*: 181-187, 1968.

11703 - SUN, A.S.K., SAUER, K.: Pigment systems and electron transport in chloroplasts. I. Quantum requirements for the two light reactions in spinach chloroplasts. - Biochim. biophys. Acta *234*: 399-414, 1971.

11704 - SUN, C.-N., ADAMS, R.S. Jr.: Effects of the phosphorus-manganese-atrazine interaction in soybean plants. - J. agr. Food Chem. *19*: 325-340, 1971. [Ps.]

*11705 - SURMATIS, J.D.: A new synthesis of β-carotene by the C_{20} + C_{20} scheme. - Tidsskr. Kjemi, Bergv., Metallurgi *26* (6-7): 131-132, 1966.

11706 - SURZYCKI, S.: Synchronously grown cultures of *Chlamydomonas reinhardi*. - In: COLOWICK, S.P., KAPLAN, N.O. (ed.): Methods in Enzymology. Vol. 23. Pp. 67-73. Academic Press, New York - London 1971.

11707 - ŠUŠLÍK, V.: Meranie plochy listov pomocou šablón. [Leaf area measurement using a stencil.] - Biológia (Bratislava) *26*: 349-353, 1971. [In Slovak, ab: E.]

11708 - SUSLOVA, V.V., NIKOLAEVSKIĬ, V.S.: Vliyanie kislykh gazov na pigmentnyĭ sostav list'ev drevesnykh i gazonnykh rasteniĭ. [Effect of acid gases on the pigment composition of the leaves of trees and lawn plants.] - Uch. Zap. perm. gos. Univ. *256*: 93-132, 1971. [In R.]

11709 - SÜZER, S., SAUER, K.: The sites of photoconversion of protochlorophyllide to chlorophyllide in barley seedlings. - Plant Physiol. *48*: 60-63, 1971.

11710 - ŠVACHULOVÁ, J.: $^{14}CO_2$ fixation, ribulose 1,5-diphosphate carboxylase activity and free sugar content of two chlorophyll mutants of *Arabidopsis thaliana* L. (HEYNH.). - Photosynthetica *5*: 249-257, 1971.

11711 - ŠVIHRA, J.: Fyziologická heterogenita orgánov ozimnej pšenice. [Physiological heterogeneity of organs of winter wheat.] - In: Súbor Referátov zo Sympózia o Intenzifikácii Rastlinnej Výroby v Rôznych Stanovištných Podmienkách. Pp. 437-444. Nitra 1971. [Ps, Chl; in Slovak, ab: G, R.]

11712 - SZABAD, J.: Migration of electron excitation energy in mixed solutions of beta-carotene and chlorophyll-*a*. - Acta biochim. biophys. *6*: 468, 1971.

11713 - SZALAY, L.: The migration of electron excitation energy between chlorophylls and other molecules in solutions. - Acta biochim. biophys. *6*: 467, 1971.

11714 - SZÁSZ, K.: Should the intensity of photosynthetic carbon dioxide fixation be related to the chlorophyll content? - Acta agron. Acad. Sci. hung. *20*: 442-443, 1971.

11715 - SZÁSZ, K., BARSI, E.Sz.: Stimulatory effect of red light on the polysaccharide accumulation in the leaves. - Photosynthetica *5*: 71-73, 1971.

11716 - SZCZYPA, E., WOJCIESKA, U.: Wzrost i produktywność fotosyntezy pszenic ozimych. Część I. Dynamika wzrostu i przyrost masy roślin. [Growth and productivity of photosynthesis in winter wheats. Part I. Growth dynamics and increase of plant matter.] - Pamięt. puław. - Prace IUNG *44*: 31-47, 1971. [In Pol., ab: E, R.]

11717 - SZÉKELY, Á.: Klorofill-mérések *in vivo*. [Measurement of chlorophyll *in vivo*.] - Az Erdö *1971*: 62-66, 1971. [In Hung., ab: E, R.]

11718 - SZUJKÓ-LACZA, J., RAKOVÁN, J.N., HORVÁTH, G., FEKETE, G., FALUDI-DÁNIEL, Á.: Anatomical, ultrastructural and physiological studies on one-year old *Euonymus europaeus* bark displaying photosynthetic activity. - Acta agron. Acad. Sci. hung. *20*: 247-260, 1971.

11719 - TAGEEVA, S.V., GENEROZOVA, I.P., DEREVYANKO, V.G., LADYGIN, V.G., SEMENOVA, G.A.: Raznoobrazie ul'trastrukturnoĬ organizatsii khloroplastov v zavisimosti ot funktsional'nogo sostoyaniya tkaneĬ rasteniĬ, geneticheskogo faktora i svetovykh usloviĬ. [Ultrastructural organization of chloroplasts as affected by the functional state of plant tissues and organs, genetic factors and light conditions.] - In: ZALENSKIĬ, O.V. (ed.): Fotosintez i Ispol'zovanie SolnechnoĬ Energii. Pp. 126-144. Nauka, Leningrad 1971. [In R, ab: E.]

11720 - TAĬLAKOV, N.: Vliyanie stimulyatorov rosta na nekotorye fiziologo-biokhimicheskie protsessy kukuruzy. [Effect of growth regulators on some physiological and biochemical processes in maize plants.] - Izv. Akad. Nauk turkm. SSR, Ser. biol. Nauk *1971* (5): 58-62, 1971. [Ps, Chl; In R, ab: E, Turkm.]

11721 - TAKACS, B.J., HOLT, S.C.: *Thiocapsa floridana*; cytological, physical and chemical characterization. I. Cytology of whole cells and isolated chromatophore membranes. - Biochim. biophys. Acta *233*: 258-277, 1971.

11722 - TAKACS, B.J., HOLT, S.C.: *Thiocapsa floridana*; a cytological, physical and chemical characterization. II. Physical and chemical characteristics of isolated and reconstituted chromatophores. - Biochim. biophys. Acta *233*: 278-295, 1971.

11723 - TAKAHASHI, M., SHIMURA, S., YAMAGUCHI, Y., FUJITA, Y.: Photo-inhibition of phytoplankton photosynthesis as a function of exposure time. - J. oceanograph. Soc. Jap. [Nippon Kaiyo Gakkai-Shi] *27* (2): 43-50, 1971.

11724 - TAKAMIYA, A.: Chlorophyll-protein complexes. - In: COLOWICK, S.P., KAPLAN, N.O. (ed.): Methods in Enzymology. Vol. 23. Pp. 603-613. Academic Press, New York - London 1971.

11725 - TAKAMIYA, K.: The light-induced oxidation-reduction reactions of menaquinone in intact cells of a green photosynthetic bacterium, *Chloropseudomonas ethylica*. - Biochim. biophys. Acta *234*: 390-398, 1971.

11726 - TAKEDA, T., FUKUYAMA, M.: [Studies on the photosynthesis of the *Gramineae*. I. Differences in photosynthesis among subfamilies and their relations with the systematics of the *Gramineae*.] - Proc. Crop Sci. Soc. Jap. *40*: 12-20, 1971. [In Jap., ab: E.]

11727 - TAKEDA, T., HIROTA, O.: [Relationship between spacing and grain yield of rice plant.] - Proc. Crop Sci. Soc. Jap. *40*: 381-385, 1971. [In Jap., ab: E.]

11728 - TAMÁS, I.A., ATKINS, B.: Stimulation of ATP synthesis by indoleacetic acid in illuminated chloroplasts. - Plant Physiol. *47* (Suppl.): 9, 1971.

11729 - TAMÀS, I.A., BIDWELL, R.G.S.: Metabolism of glycolic acid-1-^{14}C in barley leaves with or without added CO_2. - Can. J. Bot. *49*: 299-302, 1971.

11730 - TAN, C.K., BADOUR, S.S., WAYGOOD, E.R.: Photosynthesis of the unicellular green alga *Gloeomonas* in synchronous culture. - Plant Physiol. *47* (Suppl.): 9, 1971.

11731 - TANAKA, A., HARA, T.: [Photosynthetic rate of leaves. 2. Effect of phosphorus status on the photosynthetic rate in maize plants.] - J. Sci. Soil Manure Jap. [Nippon Dojo-Hiryogaku Zasshi] *42*: 300-303, 1971. [In Jap.]

11732 - TANAKA, T., MATSUSHIMA, S.: [Analysis of yield-determining process and its application to yield-prediction and culture improvement of lowland rice. XCVIII. Effects of the nitrogen content and thickness of a leaf-blade on the light-curve of carbon assimilation of the obverse, the reverse and both sides of the leaf-blade.] - Proc. Crop Sci. Soc. Jap. *40*: 164-169, 1971. [In Jap., ab: E.]

11733 - **TANAKA, T., MATSUSHIMA, S.:** [Analysis of yield-determining process and its application to yield-prediction and culture improvement of lowland rice. CII. Relations between the leafiness of rice plant communities and their light-curves of carbon assimilation.] - Proc. Crop Sci. Soc. Jap. *40*: 356-365, 1971. [In Jap., ab: E.]

11734 - **TANAKA, T., MATSUSHIMA, S.:** [Analysis of yield-determining process and its application to yield-prediction and culture improvement of lowland rice. CIII. On factors affecting the light-curves of carbon assimilation in rice plant communities.] - Proc. Crop Sci. Soc. Jap. *40*: 366-375, 1971. [In Jap., ab: E.]

11735 - **TANAKA, T., MATSUSHIMA, S.:** [Analysis of yield-determining process and its application to yield-prediction and culture improvement of lowland rice. CIV. Effects of light intensity and different shading methods during the ripening period on the percentage of ripened grains.] - Proc. Crop Sci. Soc. Jap. *40*: 376-380, 1971. [In Jap., ab: E.]

11736 - **TĂNASE, V.:** The influence of various sources of nitrogen on some products of photosynthesis. - Rev. roum. Biol. - Sér. Bot. *16*: 371-375, 1971.

*11737 - **TANAS'EV, V.K.:** Soderzhanie khlorofilla v list'yakh novykh form vegetativno razmnozhaemykh podvoev yabloni. [Chlorophyll content in the leaves of new types of vegetatively propagating apple seedling stocks.] - Tr. kishinev. sel'skokhoz. Inst. *69*: 59-65, 1970. [In R.]

*11738 - **TANAS'EV, V.K.:** Intensivnost' nakopleniya sukhogo veshchestva v list'yakh novykh form vegetativno razmnozhaemykh podvoev yabloni. [Rate of dry-matter accumulation in the leaves of new forms of vegetatively propagated apple stocks.] - Tr. kishinev. sel'skokhoz. Inst. *69*: 66-71, 1970. [In R.]

11739 - **TARASENKO, N.D., SOKOLOV, V.A.:** Mutagennyī effekt 5-bromuratsila na yachmene. [Mutagenic effect of 5-bromuracil on barley.] - Genetika *7* (8): 46-49, 1971. [Chl; in R, ab: E.]

*11740 - **TARASOV, V.M., KOVALENKO, V.F.:** Fiziologicheskie osobennosti yabloni v uslo-viyakh mednoī nedostatochnosti. [Physiological features of an apple tree under copper deficiency.] - Izv. TSKhA *1970* (1): 141-152, 1970. [Ps; in R, ab: E.]

11741 - **TARUSOV, B.N., DZHANUMOV, D.A., VESELOVSKIĬ, V.A., SHCHERBAKOV, A.A.:** Vliyanie soleī na dlitel'noe poslesvechenie list'ev rasteniī. [Effect of salts on long-term delayed light emission of plant leaves.] - Fiziol. Rast. *18*: 999-1003, 1971. [In R, ab: E.]

11742 - **TAUYA, M.F., FRISHENFELDE, A.Ya.:** Vliyanie usloviī pitaniya rasteniī na sos-toyanie plastid v epidermise list'ev nekotorykh dvudol'nykh rasteniī. [Effect of plant nutrition on the state of plastids in leaf epidermis of some decoty-ledons.] - Izv. Akad. Nauk latv. SSR *1971* (9): 86-92, 1971. [In R, ab: E.]

11743 - **TAYLOR, A.O., CRAIG, A.S.:** Plants under climatic stress. II. Low temperature, high light effects on chloroplast ultrastructure. - Plant Physiol. *47*: 719-725, 1971.

11744 - **TAYLOR, A.O., ROWLEY, J.A.:** Plants under climatic stress. I. Low temperature, high light effects on photosynthesis. - Plant Physiol. *47*: 713-718, 1971.

11745 - **TAYLOR, D.L.:** Photosynthesis of symbiotic chloroplasts in *Tridachia crispata* (BERGH). - Comp. Biochem. Physiol. *38A*: 233-236, 1971.

*11746 - **TAYLOR, J.:** Chlorosis of pear leaves caused by *Alternaria tenuis*. - Plant Dis. Reporter *54*: 871-872, 1970. [Chl.]

11747 - **TAYLOR, R.F., IKAWA, M.:** Gas chromatography of carotenoids. - Anal. Biochem. *44*: 623-627, 1971.

11748 - **TEARE, I.D., PETERSON, S.J.:** Surface area of chlorophyll-containing tissue on the inflorescence of *Triticum aestivum* L. - Crop Sci. *11*: 627-628, 1971.

*B11749 - Techniques d'Étude des Facteurs Physiques de la Biosphère. - Inst. nat. Rech. agron., Paris 1970. [Ps.]

11750 - TELFER, A., EVANS, M.C.W.: Photosynthetic control in broken spinach chloroplasts. - FEBS Lett. *14*: 241-244, 1971.

11751 - TEMPER, E.E.: Mutatsionnaya izmenchivost' *Scenedesmus obliquus* (TURP) KUTZ,
indutsirovannaya N-nitrozometilmochevinoT. [Mutational variability in *Scenedesmus obliquus* (TURP) KUTZ, induced by N-nitrosomethylurea.] - Genetika *7*
(5): 42-49, 1971. [Ps; in R, ab: E.]

11752 - TEMPER, E.E., KVITKO, K.V.: Kharakteristika pigmentnykh mutantov *Scenedesmus
obliquus* (TURP. KÜTZ.). [Characteristic of pigment mutants of *Scenedesmus
obliquus* (TURP. KÜTZ.).] - Nauch. Dokl. vyssh. Shkoly, biol. Nauki *14* (4):
106-111, 1971. [Chl, Car; in R.]

11753 - TERRY, N.: Changes in the distribution of ^{14}C-labelled assimilates in sugarbeet with variation of temperature. - J. exp. Bot. *22*: 472-480, 1971.

11754 - TERRY, N., WALDRON, L.J., ULRICH, A.: An apparatus for the measurement of
carbon dioxide and water vapor exchange of attached sugarbeet leaves. - J.
amer. Soc. Sugar Beet Technol. *16*: 471-478, 1971.

11755 - TEVINI, M.: Der Einfluss von Phosphat- und Nitratmangel auf die Synthese der
Phospho- und Glykolipide bei *Impatiens balsamina*. - Ber. deut. bot. Ges. *84*:
595-606, 1971. [Chl.]

11756 - TEVINI, M.: Die Phospho- und Glycolipidänderungen während des Ergrünens etio-
lierter *Hordeum*-Keimlinge. - Z. Pflanzenphysiol. *65*: 266-272, 1971. [Chl.]

11757 - TEVINI, M.: Der Einfluss von Phosphat-Mangel-Ernährung auf die Synthese der
Phospho- und Glykolipide bei *Impatiens*. - Z. Pflanzenphysiol *66*: 64-72, 1971.
[Chl.]

11758 - THAYER, G.W.: Phytoplankton production and the distribution of nutrients in
a shallow unstratified estuarine system near Beaufort, N.C. - Chesapeake
Sci. *12*: 240-253, 1971. [Ps, Chl.]

11759 - THEKAEKARA, M.P., DRUMMOND, A.J.: Standard values for the solar constant and
its spectral components. - Nature - phys. Sci. *229*: 6-9, 1971.

11760 - THEKAEKARA, M.P., DRUMMOND, A.J., MUCRRAY, D.G., GAST, P.R., LAUE, E.G.,
WILLSON, R.C.: Solar electromagnetic radiation. - NASA Space Vehicle Design
Criteria (Environment) SP-8005: 1-33, 1971. [Solar constant.]

11761 - THIBAULT, P.: Etude cinétique des capacités en oxygène photosynthétique en
liaison avec les voies de fixation du gaz carbonique à la lumière. - Compt.
rend. Acad. Sci. Paris, Sér. D *272*: 1361-1364, 1971.

11762 - van THINH, L., GRIFFITHS, D.J.: The contrasting effects of chloramphenicol
and cycloheximide on the recovery of cell division and chlorophyll synthesis
in "Giant" cells of *Chlorella vulgaris* (Emerson strain). - Plant Cell Physiol. *12*: 171-179, 1971.

11763 - THOMAS, A.S., Jr., DUNN, S., ROUTLEY, D.G.: Fluorescent lamp spectral emission
and carbohydrate formation in leaves. III. Effects of various factors on rate
of starch formation and CO_2 uptake. - Advancing Frontiers Plant Sci. *28*: 379-
386, 1971.

11764 - THOMAS, J., DAVID, K.A.V.: Differentiation, dedifferentiation and function
in a nitrogen-fixing blue-green alga. - In: Proceedings of Symposium on Cellular Processes in Growth, Development and Differentiation Held at Bhabha Atomic
Research Centre, November 22-24, 1971. Pp. 401-411. [Chl.]

11765 - THOMAS, J.B.: The approximate red absorption band of chlorophyll *b* in *Ulva
lactuca* at 77 °K. - FEBS Lett. *14*: 61-64, 1971.

11766 - THOMAS, J.B.: On the *in vivo* absorption spectrum of chlorophyll *b*. - In:
BRODA, E., LOCKER, A., SPRINGER-LEDERER, H. (ed.): Proceedings of the First
European Biophysics Congress. Vol. 4. Pp. 37-41. Verlag wiener med. Akad.,
Wien 1971.

11767 - THOMAS, J.R., NAMKEN, L.N., OERTHER, G.F., BROWN, R.G.: Estimating leaf water content by reflectance measurements. - Agron. J. *63*: 845-847, 1971.
[Leaf optical properties.]

11768 - THOMAS, W.H., OWEN, R.W. Jr.: Estimating phytoplankton production from ammonium and chlorophyll concentrations in nutrient-poor water of the eastern tropical Pacific Ocean. - Fishery Bull. nat. oceanic atmos. Admin. *69*: 87-92, 1971.

11769 - THOMMEN, H.: Metabolism. - In: ISLER, O. (ed.): Carotenoids. Pp. 637-668. Birkhäuser Verlag, Basel - Stuttgart 1971. [Car.]

11770 - THOMPSON, F.B., LEYTON, L.: Method for measuring the leaf surface area of complex shoots. - Nature *229*: 572, 1971.

11771 - THOMPSON, R.G., NELSON, C.D.: Photosynthetic assimilation and translocation of ^{3}H- and ^{14}C-organic compounds after ^{3}HHO and ^{14}CO$_2$ were simultaneously offered to a primary leaf of soybean. - Can. J. Bot. *49*: 757-766, 1971.

11772 - THORNBER, J.P.: Chlorophyll *a*-protein complex of blue-green algae. - In: COLOWICK, S.P., KAPLAN, N.O. (ed.): Methods in Enzymology. Vol. 23. Pp. 682-687. Academic Press, New York - London 1971.

11773 - THORNBER, J.P.: The photochemical reaction center of *Rhodopseudomonas viridis*. - In: COLOWICK, S.P., KAPLAN, N.O. (ed.): Methods in Enzymology. Vol. 23. 688-691. Academic Press, New York - London 1971.

11774 - THORNBER, J.P., OLSON, J.M.: Chlorophyll-proteins and reaction center preparations from photosynthetic bacteria, algae and higher plants. - Photochem. Photobiol. *14*: 329-341, 1971.

11775 - THORNE, G.N.: Physiological factors limiting the yield of arable crops. - In: WAREING, P.F., COOPER, J.P. (ed.): Potential Crop Production. A Case Study. Pp. 143-158. Heinemann Educational Books, London 1971.

11776 - THORNE, S.E.: The greening of etiolated bean leaves I. The initial photoconversion process. - Biochim. biophys. Acta *226*: 113-127, 1971.

11777 - THORNE, S.W.: The greening of etiolated bean leaves II. Secondary and further photoconversion processes. - Biochim. biophys. Acta *226*: 128-134, 1971.

11778 - THORNE, S.W.: The greening of etiolated bean leaves III. Multiple light/dark step photoconversion processes. - Biochim. biophys. Acta *253*: 459-475, 1971.

11779 - THORNE, S.W., BOARDMAN, N.K.: Formation of chlorophyll *b*, and the fluorescence properties and photochemical activities of isolated plastids from greening pea seedlings. - Plant Physiol. *47*: 252-261, 1971.

11780 - THORNE, S.W., BOARDMAN, N.K.: The effect of temperature on the fluorescence kinetics of spinach chloroplasts. - Biochim. biophys. Acta *234*: 113-125, 1971.

11781 - THROM, G.: Einfluss von Hemmstoffen und des Redoxpotentials auf die lichtabhängige Änderung des Membranpotentials bei *Griffithsia setacea*. - Z. Pflanzenphysiol. *64*: 281-296, 1971.

11782 - THROM, G.: Aktionsspektrum der Photosysteme II and I für die lichtabhängige Änderung des Membranpotentials bei *Griffithsia setacea*. - Z. Pflanzenphysiol. *65*: 389-403, 1971.

11783 - TILLBERG, J.-E., KYLIN, A., SUNDBERG, I.: Microcalorimetric measurements of the heat evolution and its sensitivity to dinitrophenol during different stages of synchronous cultures of *Scenedesmus*. - Plant Physiol. *48*: 779-782, 1971.

11784 - TIMANOVSKIĬ, D.F.: Polevoĭ pribor dlya poluavtomaticheskogo izmereniya profileĭ temperatury. [A field instrument for semi-automatic measurements of temperature profiles.] - Meteorol. Gidrol. *1971* (4): 105-108, 1971. [In R.]

11785 - TIMOFEEV, K.N., RUBIN, A.B.: O kinetike fotoindutsirovannogo signala EPR fotosinteziruyushchikh bakteriĭ *Rhodospirillum rubrum*. [Kinetics of photoinduced ESR signal of photosynthesizing bacteria *Rhodospirillum rubrum*.] - Biofizika *16*: 348-350, 1971. [In R, ab: E.]

11786 - TING, I.P.: Nonautotrophic CO$_2$ fixation and crassulacean acid metabolism. - In: HATCH, M.D., OSMOND, C.B., SLATYER, R.O. (ed.): Photosynthesis and Photorespiration. Pp. 169-185. Wiley-Interscience, New York - London-Sydney - Toronto 1971.

11787 - TING, I.P., JOLLEY, K., BEASLEY, C.A., POHL, H.A.: Dielectrophoresis of chloroplasts. - Biochim. biophys. Acta *234*: 324-329, 1971.

11788 - **TING, I.P., ROCHA, V.**: NADP-specific malate dehydrogenase of green spinach leaf tissue. - Arch. Biochem. Biophys. *147*: 156-164, 1971.

11789 - **TING, I.P., ROCHA, V., MUKERJI, S.K., CURRY, R.**: On the localization of plant cell organelles. - In: HATCH, M.D., OSMOND, C.B., SLATYER, R.O. (ed.): Photosynthesis and Photorespiration. Pp. 534-540. Wiley-Interscience, New York - London - Sydney - Toronto 1971. [Chloroplast.]

11790 - **TIRIMANNA, A.S.L., WICKREMASINGHE, R.L.**: Thin layer chromatographic studies of the carotenoids of tea. - Qualitas Plant. Materiae veg. *20*: 341-346, 1971.

11791 - **TKACHUK, E.S.**: Aktivnost' fotosinteticheskogo apparata i produktivnost' ozimoĭ pshenitsy i yarovogo yachmenya v usloviaykh orosheniya. [Activity of photosynthetic apparatus and productivity of winter wheat and spring barley under irrigation.] - Fiziol. Biokhim. kul't. Rast. *3*: 171-175, 1971. [In R, ab: E.]

11792 - **TOLBERT, N.E.**: Isolation of leaf peroxisomes. - In: COLOWICK, S.P., KAPLAN, N.O. (ed.): Methods in Enzymology. Vol. 23. Pp. 665-682. Academic Press, New York - London 1971.

11793 - **TOLBERT, N.E.**: Leaf peroxisomes and photorespiration. - In: HATCH, M.D., OSMOND, C.B., SLATYER, R.O. (ed.): Photosynthesis and Photorespiration. Pp. 458-471. Wiley-Interscience, New York - London - Sydney - Toronto 1971.

11794 - **TOLBERT, N.E.**: Microbodies - peroxisomes and glyoxysomes. - Annu. Rev. Plant Physiol. *22*: 45-74, 1971.

11795 - **TOLBERT, N.E., NELSON, E.B., BRUIN, W.J.**: Glycolate pathway in algae. - In: HATCH, M.D., OSMOND, C.B., SLATYER, R.O. (ed.): Photosynthesis and Photorespiration. Pp. 506-513. Wiley-Interscience, New York - London - Sydney - Toronto 1971.

11796 - **TOLIBEKOV, D.**: Deĭstvie UF-radiatsii na reaktsiyu dezepoksidatsii i epoksidatsii violaksantina. [Effect of UV on the reaction of violaxanthin epoxidation and desepoxidation.] - In: ZALENSKIĬ, O.V. (ed.): Fotosintez i ispol'zovanie Solnechnoĭ Energii. Pp. 219-222. Nauka, Leningrad 1971. [In R, ab: E.]

*11797 - **TOLIBEKOV, D., KRASICHKOVA, G.V.**: Ob ispol'zovanii iskusstvennykh pigment-belkovolipoidnykh kompleksov dlya modelirovaniya nekotorykh svoĭstv fotosinteticheskogo apparata. [Utilization of artificial pigment-protein-lipoid complexes for modelling some properties of the photosynthetic apparatus.] - Izv. Akad. Nauk tadzh. SSR, Otd. biol. Nauk *1970* [2 (39)]: 3-11, 1970. [In R, ab: Tadzh.]

11798 - **TOMBESI, L.**: Su alcuni aspetti fondamentali della fertilità. Bilanci Energetici, Idrologici e Nutritivi delle Colture. [Some fundamental aspects of fertility. Energy, water and nutrient balances of a stand.] - Ann. Ist. sperim. Nutrizione Piante *2*: 205-248, 1971. [In Ital.]

*11799 - **TOMBESI, L., CALÉ, M.T.**: Diagnostica Fogliare, Attività Enzimatiche e Potere Assimilante delle Colture. [Leaf Analysis, Enzymatic Activity and Photosynthetic Potential of a Culture.] - Roma 1969. [Ps, Hill reaction; in Ital.]

11800 - **TOMBESI, L., LAUCIANI, E.**: Ambiente pedoclimatico dell' Agro Romano e zone adiacenti. Nota I: Ricerche sulla bioclimatologia e sulla caratteristiche fisiche e chimiche dei principali tipi di terreno. [The lands and the climate of Agro Romano and surrounding zones. Note 1: Bioclimatological researches and physical and chemical characteristics of the most important types of lands.] - Ann. Ist. sperim. Nutrizione Piante *2*: 37-187, 1971. [Ps; in Ital.]

11801 - **TOMINAGA, H.**: Chlorophyll a and phaeophytin contents in the surface water of the Antarctic Ocean through the Indian Ocean. - Antarct. Rec. Rep. Jap. antarct. Res. Exped. *42*: 124-134, 1971.

11802 - **TOMITA, G., OKU, T.**: Protection of chlorophyll a from photo-oxidation by β-carotene in binary mixture of organic solvents. - Experientia *27*: 1406-1408, 1971.

11803 - **TOMOVA, N., DIMITRIEVA, L., SETCHENSKA, M., DIMOVA, O., DETCHEV, G.**: Spectrophotometric determination of the intermediates of glycolysis and the pentose phosphate cycle in *Chlorella* cells. - Arch. Microbiol. *76*: 204-211, 1971.

11804 - TOVMASYAN, A.S.: Vliyanie gustoty posevov i urovnya pitaniya na fotosinte-ticheskuyu deyatel'nost' yarovogo yachmenya. [Effect of seed spacing and level of nutrition on photosynthetic activity of spring barley.] - Biol. Zh. Arm. *24* (12): 20-26, 1971. [In R, ab: Arm.]

11805 - TÔYAMA, S., FUNAZAKI, K.: Electron microscope studies on the morphogenesis of plastids V. Concerning one-dimensional metamorphosis of the plastids in *Cryptomeria* leaves. - Bot. Mag. (Tokyo) *84*: 123-136, 1971.

11806 - TRACZYK, H.: Relation between productivity and structure of the herb layer in associations on "the Wild Apple-Tree Island" (Masurian lake district). - Ekol. pol. *19*: 333-363, 1971.

11807 - TRANQUILLINI, W.: Photosynthese an extremen Standorten. - Biol. unserer Zeit *1*: 43-49, 1971.

11808 - TRANQUILLINI, W., MACHL-EBNER, I.: Über den Einfluss von Wärme auf das Pho-tosynthesevermögen der Zirbe (*Pinus cembra* L.) und der Alpenrose (*Rhododen-dron ferrugineum* L.) im Winter. - Rep. Kevo Subarctic Res. Sta. *8*: 158-166, 1971.

11809 - TRAVIS, D.M., STEWART, K.D., WILSON, K.G.: Genetics and ultrastructure of chlorophyll mutants induced by N-methyl-N'-nitro-N-nitroso guanidine (MNNG) in *Mimulus cardinalis* (*Scrophulariaceae*). - Genetics *68*: s 68-s 69, 1971.

11810 - TREHARNE, K.J., PRITCHARD, A.J., COOPER, J.P.: Variation in photosynthesis and enzyme activity in *Cenchrus ciliaris* L. - J. exp. Bot. *22*: 227-238, 1971.

11811 - TRELEASE, R.N., GRUBER, P.J., BECKER, W.M., NEWCOMB, E.H.: Microbodies in fat-storing cotyledons: ultrastructural and enzymatic changes during green-ing. - In: HATCH, M.D., OSMOND, C.B., SLATYER, R.O. (ed.): Photosynthesis and Photorespiration. Pp. 523-533. Wiley-Interscience, New York - London - Sydney - Toronto 1971.

11812 - TRÉMOLIÈRES, A., LEPAGE, M.: Changes in lipid composition during greening of etiolated pea seedlings. - Plant Physiol. *47*: 329-334, 1971.

11813 - TRIBUTSCH, H.: Application of electrochemical kinetics to photosynthesis and oxidative phosphorylation. The redox element hypothesis and the prin-ciple of parametric energy coupling. - J. Bioenerg. *2*: 249-273, 1971.

11814 - TRIBUTSCH, H., CALVIN, M.: Electrochemistry of excited molecules: photo-electrochemical reactions of chlorophylls. - Photochem. Photobiol. *14*: 95-112, 1971.

*11815 - TROBISCH, S., SCHILLING, G.: Untersuchungen über Zusammenhänge zwischen Mas-senentwicklung und N-Umsatz während der generativen Phase bei *Sinapis alba* L - Albrecht-Thaer-Archiv *13*: 867-878, 1969. [Growth analysis.]

11816 - THROUGHTON, J.H.: Aspects of the evolution of the photosynthetic carboxy-lation reaction in plants. - In: HATCH, M.D., OSMOND, C.B., SLATYER, R.O. (ed.): Photosynthesis and Photorespiration. Pp. 124-129. Wiley-Interscience, New York - London - Sydney - Toronto 1971.

11817 - THROUGHTON, J.H.: The lack of carbon dioxide evolution in maize leaves in the light. - Planta *100*: 87-92, 1971.

11818 - THROUGHTON, J.H., HENDY, C.H., CARD, K.A.: Carbon isotope fractionation in *Atriplex* spp. - Z. Pflanzenphysiol. *65*: 461-464, 1971.

11819 - TROXLER, R.F.: Mechanism of bile pigment formation in plants. - Plant Physiol. *47* (Suppl.): 46, 1971.

*11820 - TRUDEL, M.J., OZBUN, J.L.: Relationship between chlorophylls and carotenoids of ripening tomato fruit as influenced by potassium nutrition. - J. exp. Bot. *21*: 881-886, 1970.

11821 - TSCHETWERGOW, D.I., GEORGIEV, G.D., SWENTIZKIJ, I.I.: Empfänger mit optisch korrigierter Spektralempfindlichkeit zur Messung von photosynthetisch effek-tiven Strahlungen. - In: Proceedings of the 5th IMEKO-Symposium on "Photon-Detectors", May 1971, Varna, Bulgaria. Pp. 339-345. IMEKO Secretariat, Buda-pest 1971.

11822 - TSEL'NIKER, Yu.L., MALKINA, I.S., KNYAZEVA, I.F.: Reaktsiya fotosinteticheskogo apparata klena ostrolistnogo na izmenenie rezhima osveshcheniya. [Effect of illumination on the photosynthetic apparatus of Norway maple.] - Fiziol. Rast. *18*: 1127-1133, 1971. [In R, ab: E.]

11823 - TSENOVA, E.N., FEDINA, I., VAKLINOVA, S.: Issledovanie svyazi mezhdu fotookisleniem gidroksilamina i fotofosforilirovaniem v izolirovannykh khloroplastakh. [Investigation of the relation between photooxidation of hydroxylamine and photophosphorylation in isolated chloroplasts.] - Dokl. Akad. Nauk SSSR *201*: 741-743, 1971. [In R.]

11824 - TSIKOV, D.: Bezkhlorofilni rasteniya pri khibridi na *Triticum timopheevi* ZHUK. s *T. turgidum* L. [Albina-hybrids from crosses of *Triticum timopheevi* ZHUK. with *T. turgidum* L.] - Genet. Selek.(Sofia) *4*: 437-444, 1971. [In Bulg., ab: E, R.]

11825 - TSIPA, L.: Intensivnost' fotosinteza u *Oscillatoria agardhii* v zavisimosti ot raznykh kontsentratsiī azota i vliyanie P, K, S, Mg na nakoplenie biomassy. [Photosynthetic rate in *Oscillatoria agardhii* as related to various nitrogen concentrations and the effect of P, K, S, Mg on biomass accumulation.] - Rev. roum. Biol., Sér. Bot. *16*: 47-61, 1971. [In R.]

11826 - TSUJI, T., FUJITA, Y.: Ascorbate photo-oxidation by a photochemically active chromoprotein isolated from the blue-green alga *Anabaena cylindrica*: the effect of monochromatic illumination. - Plant Cell Physiol. *12*: 807-811, 1971.

11827 - TSUJIMOTO, H.Y., CHAIN, R.K.: Photoreduction of ferredoxin-NADP in the presence and absence of ferredoxin-reducing substance (FRS). - Plant Physiol. *47* (Suppl.): 33, 1971.

*11828 - TSYBUL'KO, V.S.: Soderzhanie vitamina C i karotina v list'yakh rasteniī v svyazi s tempami ikh razvitiya. [Content of vitamin C and carotene in leaves as related to the rate of plant development.] - Tr. khar'kov. sel'skokhoz. Inst. *90* (Issledovaniya Fiziologii i Biokhimii Rasteniī): 101-106, 1970. [In R.]

11829 - TU, S.-I., TAN, Y.J., WANG, J.H.: Synthetic model complexes for studying light-driven electron transfer in photosynthesis. - Bioinorg. Chem. *1*: 79-95, 1971.

11830 - TURNER, N.C., INCOLL, L.D.: The vertical distribution of photosynthesis in crops of tobacco and sorghum. - J. appl. Ecol. *8*: 581-591, 1971.

11831 - TYANKOVA, L.: Die Wirkung von Aminosäuren auf Thylakoide beim Gefrieren in Abhängigkeit von der Stellung der Aminogruppe. - Ber. deut. bot. Ges. *84*: 437-444, 1971.

11832 - TZAPIN, A.I., MOLOTKOVSKY, Y.G., GOLDFIELD, M.G., DZJUBENKO, V.S.: Light-induced structural transitions of chloroplasts studied by the spin-probe method. - Europe. J. Biochem. *20*: 218-224, 1971.

11833 - UCHIJIMA, Z.: [The climate in growth chamber (5) - Simulated CO_2 environment and photosynthesis in a glasshouse.] - J. agr. Meteorol. *27*: 45-57, 1971. [In Jap., ab: E.]

11834 - UDOVENKO, G.V., SEMUSHINA, L.A., PETROCHENKO, N.G.: Kharakter i veroyatnye prichiny izmeneniya fotosinteziruyushcheī deyatel'nosti rasteniī pri zasolenii. [Character and possible explanation of the changed photosynthesizing activity of plants during salinization.] - Fiziol. Rast. *18*: 708-715, 1971. [In R, ab: E.]

11835 - UFFEN, R.L., SYBESMA, C., WOLFE, R.S.: Mutants of *Rhodospirillum rubrum* obtained after long-term anaerobic, dark growth. - J. Bacteriol. *108*: 1348-1356, 1971. [Chl.]

*11836 - UGULAVA, N.A.: Nekotorye osobennosti gerbitsidov ingibitorov protsessa fotosinteza i rosta na narushenie obmena veshchestv rasteniī. [Some features of the action of herbicides, inhibitors of photosynthesis and growth, on plant metabolism.] - In: Materialy II Vsesoyuznoī Konferentsii po Razrabotke i Primeneniyu Gerbitsidov v Sel'skom Khozyaīstve. Sektsiya II. Priroda Deīstviya Gerbitsidov. Pp. 38-39. Moskva 1969. [In R.]

11837 - UHM, K.B.: [Studies on the productive structure in some lakes in Korea.] -
Korean J. Bot. *14*: 15-23, 1971. [Chl, in Korean, ab: E.]

11838 - ULLRICH, W.R.: Nitratabhängige nichtcyclische Photophosphorylierung bei
Ankistrodesmus braunii in Abwesenheit von CO_2 und O_2. - Planta *100*: 18-30,
1971.

11839 - UMEDA, K., KAWASHIMA, K.: [Studies on citrus carotenoids. Part I. Systematic
separation of carotenoid groups by thin layer chromatography.] - J. Food
Sci. Technol. (Tokyo) [Nippon Shokuhin Kogyo Gakkai-Shi] *18* (4): 147-154,
1971. [in Jap., ab: E.]

11840 - UMEDA, K., KAWASHIMA, K.: [Studies on citrus carotenoids. Part II. Carotenoid pattern of *Citrus unshiu* peel at different mature stages.] - J. Food
Sci. Technol. (Tokyo) [Nippon Shokuhin Kogyo Gakkai-Shi] *18* (4): 155-160,
1971. [TLC; in Jap., ab: E.]

11841 - UMEDA, K., KAWASHIMA, K.: [Studies on citrus carotenoids. Part 3. Carotenoid patterns of *Citrus unshiu* flesh at different mature stages.] - J.
Food Sci. Technol. (Tokyo) [Nippon Shokuhin Kogyo Gakkai-Shi] *18* (8): 359-
365, 1971. [TLC; in Jap., ab: E.]

11842 - UMIEL, N., GABELMAN, W.H.: Analytical procedures for detecting carotenoids
of carrot (*Daucus carota* L.) roots and tomato (*Lycopersicum esculentum*)
fruits. - J. amer. Soc. hort. Sci. *96*: 702-704, 1971.

11843 - UMNOV, A.A.: Primenenie metoda matematicheskogo modelirovaniya pri issledovanii roli fotosinteticheskoi aeratsii v ozere. [Application of a mathematical
modelling method to the study of the role of photosynthetic aeration in a
lake.] - Ekologiya *2* (6): 5-12, 1971. [In R.]

11844 - USACHEVA, M.N.: O prirode vliyaniya dobavok kislot na spektry i fotoniku molekul feofitina. [Nature of the effect of acid additions on the spectra and
photonics of pheophytin molecules.] - Biofizika *16*: 983-990, 1971. [In R,
ab: E.]

11845 - USACHEVA, M.N., DOLIDZE, I.A., ASHKINAZI, M.S.: O vliyanii vlagi na protsessy fotokhimicheskogo vosstanovleniya khlorofilla i feofitina. [Effect of
moisture on the process of photochemical reduction of chlorophyll and pheophytin.] - Biofizika *16*: 195-200, 1971. [In R, ab: E.]

11846 - USIK, G.E.: Effektivnost' predposadochnoi obrabotki klubnei kartofelya rast-
vorami solei medi i tsinka. [Effectiveness of the preplanting treatment of
potato tubers with copper and zinc salt solutions.] - Agrokhimiya *1971* (9):
118-121, 1971. [Chl; in R.]

11847 - USMANOV, P.D.: Morfologicheskie osobennosti khloroplastov v zavisimosti ot
vozdeistviya na geneticheskii apparat. [Morphological characteristics of
chloroplasts in relation to changes in the genetic apparatus.] - In: ZALENS-
KII, O.V. (ed.): Fotosintez i ispol'zovanie Solnechnoi Energii. Pp. 241-244.
Nauka, Leningrad 1971. [In R, ab: E.]

*11848 - USMANOV, P.D., ABDULLAEV, Kh.A., BOBODZHANOV, V.A.: Tsitologicheskii analiz
khloroplastov i obshchaya kharakteristika mutantnykh form gorokha sorta
Torsdag. [Cytologic analysis of chloroplasts and general characteristics of
mutant forms of pea cultivar Torsdag.] - Izv. Akad. Nauk tadzh. SSR, Otd.
biol. Nauk *1970* [3 (40)]: 49-57, 1970. [In R, ab: Tadzh.]

11849 - USMANOV, P.D., KAS'YANENKO, A.G., BATALOV, R.B.: Vliyanie nekotorykh fizicheskikh agentov i etilmetansul'fonata na chastotu khlorofil'nykh mutatsii
u *Arabidopsis thaliana* (L.) HEYNH. [Effect of some physical agents and ethyl-
methanesulphonate on the frequency of chlorophyll mutations in *Arabidopsis
thaliana* (L.) HEYNH.] - In: NASYROV, Yu.S. (ed.): Geneticheskie Aspekty Fotosinteza. Pp. 24-43. Donish, Dushanbe 1971. [In R, ab: E.]

*11850 - USMANOV, P.D., STARTSEV, G.A., SHABALOV, V.V., NASYROV, Yu.S.: Lazernoe iz-
luchenie - novyi mutagennyi faktor. [Laser radiation - a new mutagen.] -
Dokl. Akad. Nauk tadzh. SSR *12* (7): 55-58, 1969. [Chl; in R, ab: Tadzh.]

11851 - USPENSKAYA, V.E.: Khlorofillaznaya aktivnost' fotosinteziruyushchikh bakterii. [Chlorophyllase activity of photosynthetic bacteria.] - Mikrobiologiya
40: 790-795, 1971. [In R, ab: E.]

11852 - USUDA, H., KANAI, R., TAKEUCHI, M.: Comparison of carbon dioxide fixation
and the fine structure in various assimilatory tissues of *Amaranthus re-
troflexus* L. - Plant Cell Physiol. *12*: 917-930, 1971.

11853 - UZZO, A.: The characterization of temperature induced inhibition of chloroplast replication in *Euglena*. - Diss. Abstr. int. B *31*: 7126-B, 1971.

*11854 - VAGANOV, A.P., LYASKOVSKIĬ, G.M.: Vysokochastotnyĭ analizator uglekislogo
gaza "Vesna". [High-frequency CO_2 analyser "Vesna".] - In: Tezisy Dokladov
Vsesoyuznogo Soveshchaniya po Unifikatsii Metodov i Priborov dlya Massovykh
Izmereniĭ Intensivnosti Fotosinteza. Pp. 21-24. Nauch. - issled. Inst. Rastenievod. N.I. Vavilova, Leningrad - Pushkin 1970. [In R.]

*11855 - VAGANOV, A.P., SOROKINA, A.P.: Vliyanie mikroelementov na semennuyu produk-
tivnost' ogurtsov. [Effect of microelements on seed production of cucumber.]
- Tr. khar'kov. sel'skokhoz. Inst. *90* (Issledovaniya po Fiziologii i Biokhimii Rasteniĭ): 11-15, 1970. [Chl; in R.]

11856 - VAKLINOVA, S.: P"rvichni reaktsii i mekhanizmi na fotosintezata (po sluchaĭ
200 godini ot otkrivaneto na protsesa fotosinteza). [Primary reactions and
mechanisms of photosynthesis (on the occasion of the 200th anniversary of
discovery of the photosynthetic process).] - Priroda (Sofiya) *20* (4): 38-41,
1971. [In Bulg.]

11857 - VALANNE, N.: The effects of prolonged darkness and light on the fine structure of *Ceratodon purpureus*. - Can. J. Bot. *49*: 547-554, 1971. [Chl.]

11858 - VANDENHEUVEL, F.A.: Structure of membranes and role of lipids therein. -
Advances Lipid Res. *9*: 161-248, 1971. [Chloroplasts.]

11859 - VANSEVEREN, J.P., AMBROES, P.: Note sur l'index foliaire et les quantités
de chlorophylles à l'hectare. - Bull. Soc. roy. Bot. Belg. *104*: 291-300, 1971.

11860 - VARLET GRANCHER, C., ARTIS, P., BONHOMME, R.: Increase of leaf area index and
photosynthetic rest in various sweet potato crops. - Proc. 9th Annual Meeting
of the Caribbean Food Crops Society, Guyana, June 1971. 10 pp.

*11861 - VARLET GRANCHER, C., BONHOMME, R.: Albedo et reflexion monochromatique de la
lumière solaire par une culture. - Météorologie *1969* (12): 45-46, 1969.

11862 - VARLET GRANCHER, C., BONHOMME, R.: Assimilation nette, utilisation de l'eau
et microclimatologie d'un champ de maïs. III. - Composition spectrale de la
lumière dans la culture. - Ann. agron. *22*: 515-525, 1971.

11863 - VASCONCELOS, A., POLLACK, M., MENDIOLA, L.R., HOFFMANN, H.-P., BROWN, D.H.,
PRICE, C.A.: Isolation of intact chloroplasts from *Euglena gracilis* by zonal
centrifugation. - Plant Physiol. *47*: 217-221, 1971.

11864 - VASIL'EVA, V.E., PINEVICH, V.V.: K kharakteristike karotinoidnykh pigmentov
i veroyatnosti funktsionirovaniya ksantofillovogo tsikla u *Anacystis nidulans*.
[Characterization of carotenoid pigments and functioning of xanthophyll cycle
in *Anacystis nidulans*.] - Vestn. leningrad. Univ. *1971* (15): 105-110, 1971.
[In R, ab: E.]

11865 - VÉBER, K., ONDOK, J.P.: Velikost celkové asimilace a respirace u okurek pěs-
tovaných ve skleníkových a venkovních podmínkách. [Gross photosynthesis and
respiration rates in cucumbers cultivated in a glasshouse and outdoors.] -
Rostlinná Výroba (Praha) *17*: 1275-1281, 1971. [In Czech, ab: E, R]

11866 - VECCHI, L. de: Fine structure of detached oat leaves senescing under different experimental conditions. - Israel J. Bot. *20*: 169-183, 1971. [Chloro-
plast.]

11867 - VECHAR, A.S., CHAĬKA, M.Ts., MAS'KO, A.A., KAVALCHUK, R.A.: Biyakhimichnyya
dasledavanni nekatorykh vidaŭ plastyd. [Biochemical studies of some types
of plastids.] - Vestsy Akad. Navuk belaruss. SSR, Ser. biyal. Navuk *1971*
(1): 36-41, 1971. [In Beloruss.]

11868 - VECHER, A.S., KOVAL'CHUK, R.A.: Izoprenoidnye khinony v khloroplastakh prorostkov yachmenya. [Isoprenoid quinones in chloroplasts of barley seedlings.]
- Vestsy Akad. Navuk belaruss. SSR, Ser. biyal. Navuk *1971* (5): 92-95, 133,
1971. [Chl; in R.]

11869 - VECHER, A.S., KOVAL'CHUK, R.A.: Plastokhinony v plastidakh s razlichnoĭ fo-
tokhimicheskoĭ aktivnost'yu. [Plastoquinones in plastids of various photochemical activity.] - Dokl. Akad. Nauk beloruss. SSR *15*: 359-361, 1971. [In
R.]

11870 - VEERASEKARAN, P., RAO, J.S.: Studies on the physiology of heterosis. Physiological and biochemical advantages in hybrid Bajra-1 (pearl millet) as com-
pared to its non-heterotic parents. - Madras agr. J. *58*: 260-263, 1971. [Ps,
Chl.]

11871 - VELYCHKO, I.M., MIGAL', O.K.: Do metodyky vyznachennya serednikh ob"emiv kli-
tyn vodorosteĭ. [Method of determination of the average volume of algal cells.]
- Ukr. bot. Zh. *28*: 251-253, 1971. [In Ukr., ab: E, R.]

11872 - VENEDIKTOV, P.S., MATORIN, D.N., KRENDELEVA, T.E., NIZOVSKAYA, N.V., RUBIN,
A.B.: O roli dlinnovolnovoĭ fotosistemy fotosinteza v generatsii poslesveche-
niya zelenykh rasteniĭ. [Role of long-wave photosynthetic system in the generation of delayed fluorescence of green plants.] - Nauch. Dokl. vyssh. Shkoly,
biol. Nauki *14* (11): 50-54, 1971. [In R.]

11873 - VENNESLAND, B., JETSCHMANN, C.: The nitrate dependence of the inhibition of
photosynthesis by carbon monoxide in *Chlorella*. - Arch. Biochem. Biophys. *144*:
428-437, 1971.

*11874 - VENTER, F.: Beobachtungen über die Temperatur und den Chlorophyllgehalt in
Tomatenfrüchten und das Auftreten von "Grünkragen". - Angew. Bot. *44*: 263-
270, 1970.

11875 - VERDUIN, J.: Phytoplankton energetics in a sewage-treatment lagoon. - Ecology *52*: 626-631, 1971.

11876 - VERNON, L.P., HUBER, E.P.: Current literature survey of photosynthesis -
XXIII. - Photochem. Photobiol. *13*: 299-307, 1971.

11877 - VERNON, L.P., SHAW, E.R.: Subchloroplast fragments: Triton X-100 method. -
In: COLOWICK, S.P., KAPLAN, N.O. (ed.): Methods in Enzymology. Vol. 23. Pp.
277-289. Academic Press, New York - London 1971.

11878 - VERNON, L.P., SHAW, E.R., OGAWA, T., RAVEED, D.: Structure of Photosystem
I and Photosystem II of plant chloroplasts. - Photochem. Photobiol. *14*:
343-357, 1971.

11879 - VERNOTTE, C.: Separation et caracterisation spectroscopique du monomère et
des polymères de la *C*. Phycocyanine. - Photochem. Photobiol. *14*: 163-173,
1971.

11880 - VETTER, W., ENGLERT, G., RIGASSI, N., SCHWIETER, U.: Spectroscopic methods.
- In: ISLER, O. (ed.): Carotenoids. Pp. 189-266. Birkhäuser Verlag, Basel -
Stuttgart 1971. [Car.]

11881 - VICKERY, P.J., BRINK, V.C., ORMROD, D.P.: Net photosynthesis and leaf area
index relationships in swards of *Dactylis glomerata* under contrasting defoliation regimes. - J. brit. Grassland Soc. *26*: 85-90, 1971.

11882 - VIGNES, D., CALMÉS, J., CARLES, J.: Éclairement et production végétale chez
la Vigne vierge (*Parthenocissus tricuspidata* PLANCHON). - Compt. rend. Acad.
Sci. Paris, Sér. D *273*: 872-875, 1971.

11883 - VĪTOLA, Ā., GROSA, V.: Ogļhidrātu uzkrāšanās augu lapās atkarībā no minerā-
lās barošanās līmeņa un apgaismojuma. [Accumulation of carbohydrates in
plant leaves depending on the level of mineral nutrients and illuminance.]-
Izv. Akad. Nauk latv. SSR *1971* (9): 54-62, 1971. [In Latv., ab: E, R.]

11884 - VĪTOLA, Ā., KREICBERGS, O., KRISTKALNE, S., GUBARE, G., SELGA, M.: Apgaismo-
juma intensitātes izmaiņu ietekme uz augu produktivitāti sējumu modeļos.
[The influence of illuminance in model sowing on plant productivity.] - Izv.
Akad. Nauk latv. SSR *1971* (9): 19-28, 1971. [In Latv., ab: E, R.]

11885 - VLASOVA, M.P., DROZDOVA, I.S., VOSKRESENSKAYA, N.P.: Izmenenie tonkoĭ struktury khloroplastov u rasteniĭ gorokha, zeleneyushchikh na sinem i krasnom
svetu. [Changes in chloroplast ultrastructure in pea plants greening under
blue and red illuminance.] - Fiziol. Rast. *18*: 5-11, 1971. [In R, ab: E.]

11886 - **VLASYUK, P.A.**: Opyt ispol'zovaniya yadernykh izlucheniĭ i radioaktivnykh izotopov'v issledovaniyakh po fiziologii rasteniĭ. [Use of nuclear radiation and radioactive isotopes in plant physiological studies.] - In: Primenenie izotopov i Yadernykh izlucheniĭ v Sel'skom KhozyaĭTstve. Pp. 114-123. Atomizdat, Moskva 1971. [Chl, Car; in R.]

11887 - **VLASYUK, P.A., BALAGANSKAYA, V.E., GORYACHEVA, L.O.**: Vliyanie predposevnogo obogashcheniya semyan mikroelementami na zharostoĭkost' rasteniĭ i kachestvo plodov arbuzov v usloviyakh yuzhnykh chernozemov. [Effect of presowing enrichment of seeds with trace nutrients on heat tolerance of plants and quality of watermelons on southern chernozem.] - Fiziol. Biokhim. kul't. Rast. *3*: 575-580, 1971. [Car; in R, ab: E.]

11888 - **VLASYUK, P.A., GULYAEV, B.I., OKANENKO, A.S., MANUIL'SKIĬ, V.D.**: Ustanovka dlya avtomaticheskoĭ zapisi pogloshcheniya CO_2 rasteniyami. [Apparatus for automatic recording of CO_2 absorption by plants.] - Fiziol. Biokhim. kul't. Rast. *3*: 93-98, 1971. [In R, ab: E.]

11889 - **VLASYUK, P.A., LISNIK, S.S., KLIMOVITSKAYA, Z.M.**: Lipidy membran khloroplastov i intaktnykh korneĭ pri margantsevoĭ nedostatochnosti. [Membrane lipids of chloroplasts and intact roots under manganese deficiency.] - Fiziol. Biokhim. kul't. Rast. *3*: 398-403, 1971. [In R, ab: E.]

11890 - **VOGL, M., NEUWIR⁺H, G., LEONHARDT, U.**: Gaswechseluntersuchungen an abgeschnittenen Zweigen aus einem älteren Kiefernbestand. - Biol. Zentralbl. *90*: 769-777, 1971.

11891 - **VOLODARSKIĬ, N.I.**: Intensivnost' fotosinteza v ontogeneze. [Photosynthetic rate during ontogenesis.] - In: RUBIN, B.A. (ed.): Fiziologiya Sel'skokhozyaĭTstvennykr Rasteniĭ. Vol. XI. Fiziologiya Tabaka. Pp. 92-106. Izdat. moskov. Univ., Moskva 1971. [In R.]

11892 - **VOLODARSKIĬ, N.I.**: Fiziologo-biokhimicheskaya kharakteristika list'ev tabaka. [Physiological and biochemical characteristics of tobacco leaves.] - In: RUBIN, B.A. (ed.): Fiziologiya Sel'skokhozyaĭTstvennykh Rasteniĭ. Vol. XI. Fiziologiya Tabaka. Pp. 113-160. Izdat. moskov. Univ., Mosvka 1971. [Chl, Car; in R.]

11893 - **VORONKOV, N.A., NEZGOVOROV, V.M.**: O sezonnoĭ dinamike khlorofilla v khvoe kul'tur sosny. [Seasonal dynamics of chlorophyll in pine needles.] - Izv. vyssh. ucheb. Zaved., lesnoĭ Zh. *14* (1): 132-134, 1971. [In R.]

11894 - **VORONKOVA, N.M., NARBUT, N.A., SMOLEĬ, V.Ya., SEMKIN, B.I.**: Raspredelenie assimilyatov u mnogoyarusnogo luka *Allium proliferum* SCHRAD. [Distribution of photosynthates in the multilayer onion *Allium proliferum* SCHRAD.] - Fiziol. Biokhim. kul't. Rast. *3*: 628-631, 1971. [In R, ab: E.]

11895 - **VOSKRESENSKAYA, N.P., DROZDOVA, I.S., GOSTIMSKIĬ, S.A.**: Osobennosti organizatsii elektrontransportnoĭ tsepi fotosinteza u letal'nogo khlorofil'nogo mutanta gorokha (*Pisum sativum*). [Peculiarities of organization of the photosynthetic electron transport chain in a lethal chlorophyll mutant of pea (*Pisum sativum*).] - In: ZALENSKIĬ, O.V. (ed.): Fotosintez i Ispol'zovanie Solnechnoĭ Energii. Pp. 236-230. Nauka, Leningrad 1971. [In R, ab: E.]

11896 - **VOSKRESENSKAYA, N.P., VIĬL, Yu.A., GRISHINA, G.S., PYARNIK, T.P.**: Raspredelenie mechenogo ugleroda v produktakh fotosinteza list'ev fasoli na krasnom i sinem svetu pri razlichnom soderzhanii v srede kisloroda. [Distribution of labelled carbon in photosynthates of bean leaves under red and blue light with different oxygen contents in the medium.] - Fiziol. Rast. *18*: 488-493, 1971. [In R, ab: E.]

11897 - **VOTINTSEV, K.K.**: Pervichnaya produktsiya BaĭKala i ee znachenie v biolimnicheskikh protsessakh v ozere. [Primary production of lake BaĭKal and its significance for the biolimnological processes in the lake.] - Izv. Akad. Nauk SSSR, Ser. biol. *1971* (6): 892 - 900, 1971. [In R, ab: E.]

*11898 - **VOZNESENSKIĬ, V.L.**: PolevoĭT konduktometricheskiĭT pribor dlya izmereniya intensivnosti fotosinteza i dykhaniya rasteniĭ. [Field conductimetric apparatus for determining plant photosynthetic and respiration rates.] - In: Tezisy Dokladov Vsesoyuznogo Soveshchaniya po Unifikatsii Metodov i Priborov dlya Massovykh IzmereniÏ Intensivnosti Fotosinteza. Pp. 25-28. Nauch. -issled. Inst. Rastenievod. N.I. Vavilova, Leningrad - Pushkin 1970. [In R.]

11899 - VOZNESENSKIÏ, V.L.: UglekislotnyÏ kompensatsionnyÏ punkt gazoobmena u khlo-relly i kukuruzy. [CO_2 compensation point of gas exchange in *Chlorella* and maize.] - In: ZALENSKIÏ, O.V. (ed.): Fotosintez i Ispol'zovanie SolnechnoÏ Energii. Pp. 105-118. Nauka, Leningrad 1971. [In R, ab: E.]

*11900 - VOZNESENSKIÏ, V.L., REINUS, R.M., ZALENSKIÏ, O.V.: Vliyanie temperatury na fotosintez pustynnykh rasteniÏ Karakumov. [Influence of temperature on pho-tosynthesis of desert plants of Karakums.] - Problemy Osvoeniya Pustyn' *5*: 21-29, 1970. [In R.]

11901 - VOZNESENSKIÏ, V.L., ZALENSKIÏ, O.V., AUSTIN, R.B.: Methods of measuring rates of photosynthesis using carbon-14 dioxide. - In: ŠESTÁK, Z., ČATSKÝ, J., JARVIS, P.G. (ed.): Plant Photosynthetic Production. Manual of Methods. Pp. 276-293. Dr. W. Junk N.V. Publ., The Hague 1971.

11902 - VREDENBERG, W.J.: Changes in membrane potential associated with cyclic and non-cyclic electron transport in photochemical system 1 in *Nitella translu-cens*. - Biochem. biophys. Res. Commun. *42*: 111-118, 1971.

11903 - VREDENBERG, W.J.: The potential response of plasmalemma tonoplast and cell wall upon photosynthetic energy conversion in *Nitella*. - In: BRODA, E., LOCKER, A., SPRINGER-LEDERER, H. (ed.): First European Biophysics Congress. Vol. 3. Pp. 435-439. Verlag wiener med. Akad., Wien 1971.

*11904 - VRHOVEC, B., WRISCHER, M.: The effect of amitrole on the fine structure of developing chloroplasts. - Acta bot. croat. *29*: 43-49, 1970.

11905 - VSEVOLZHSKAYA, G.T., PETROVA, N.L.: Vliyanie NRV pri razlichnykh usloviyakh pitaniya na fiziologicheskie protsessy, urozhaÏ kukuruzy i podsolnechnika. [Effect of petroleum growth substance on the physiological processes and yield of maize and sunflower under various mineral nutrition.] - In: NRV v Sel'skom KhozyaÏstve. Pp. 386-392. ELM, Baku 1971. [Chl, Car; in R.]

11906 - VULCHEV, P., GETOV, G.: Varietal differences in the optical properties of po-tato leaves and possibilities for their use in breeding. - Dokl. Akad. sel'-skokhoz. Nauk Bolg. *4*: 215-221, 1971. [Chl, Car.]

11907 - VULCHEV, P., GETOV, G.: Absorption maxima of carotene and lutein *in vivo*. - Dokl. Akad. sel'skokhoz. Nauk Bolg. *4*: 257-260, 1971.

11908 - VYARK, E., KEERBERG, O., KEERBERG, Kh., PYARNIK, T.: Vliyanie intensivnosti sveta na metabolizm serina v list'yakh fasoli i tabaka. [Effect of illuminan-ce on serine metabolism in leaves of kidney bean and tobacco.] - Izv. Akad. Nauk est. SSR, Biol. *20*: 179-182, 1971. [Photosynthates; in R.]

11909 - VYAS, L.N., AGARWAL, S.K., GARG, R.K.: Biomass production by *Erythrina subero-sa* ROXB. - In: International Symposium on Tropical Ecology Emphasizing Organic Production. Pp. 195-200. New Delhi 1971.

11910 - VYAS, L.N., AGARWAL, S.K., RANAWAT, M.P.S.: Relation between above ground bio-mass, girth and number of growth rings in three species of the deciduous for-est of Udaipur, Rajasthan. - Jap. J. Ecol. *21*: 52-54, 1971.

11911 - VYAS, L.N., GARG, R.K., AGARWAL, S.K.: Net aboveground production in the mon-soon vegetation at Udaipur. - In: International Symposium on Tropical Ecolo-gy Emphasizing Organic Production. Pp. 95-99. New Delhi 1971.

11912 - VYAS, L.N., GARG, R.K., RANAWAT, M.P.S.: Observations on the dry matter pro-duction in *Mitragyna parvifolia* KORTH. - Jap. J. Ecol. *21*: 227-230, 1971.

11913 - VYAS, L.N., GARG, R.K., RANAWAT, M.P.S., DAS, R.R.: Method for estimation of biomass and rate of production in a tropical tree. - Jap. J. Ecol. *21*: 244-246, 1971.

11914 - WADA, K., ARNON, D.I.: Three forms of cytochrome b_{559} in chloroplasts. - Plant Physiol. *47* (Suppl.): 33, 1971.

11915 - WADA, Y.: Changes in activities of ribulosediphosphate carboxylase and phospho-pyruvate carboxylase during leaf growth of tobacco. - Bot. Mag. (Tokyo) *84*: 159-168, 1971.

11916 - WADE, N.L., BRADY, C.J.: Effects of kinetin on respiration, ethylene production, and ripening of banana fruit slices. - Aust. J. biol. Sci. *24*: 165-167,
1971. [Chl.]

11917 - WAGGONER, P.E., TURNER, N.C.: Transpiration and its control by stomata in a
pine forest. - Bull. Connecticut agr. exp. Sta., New Haven *726*: 1-87, 1971.
[Diffusion parameter.]

11918 - WAISEL, Y., SHAPIRA, Z.: Functions performed by roots of some submerged hy-
drophytes. - Israel J. Bot. *20*: 69-77, 1971. [Chl.]

11919 - WAKAMATSU, K., KOBAYASHI, Y., SASA, T.: Effect of chlorophyllase on photochemical activities of isolated chloroplasts. - Bot. Mag. (Tokyo) *84*: 101-105,
1971.

11920 - WALEWSKI, R.: Wielkość powierzchni próbnej do oceny plonu pastwiska. [Sample
area size for pasture yield estimation.] - Rocz. Nauk rolnicz. F *78*: 159-180,
1971. [In Pol., ab: E, R.]

11921 - WALKER, D.A.: Chloroplasts (and grana): aqueous (including high carbon fixation ability). - In: COLOWICK, S.P., KAPALN, N.O. (ed.): Methods in Enzymology. Vol. 23. Pp. 211-220. Academic Press, New York - London 1971.

11922 - WALKER, D.A.: CO_2 fixation: assessment. - In: HATCH, M.D., OSMOND, C.B.,
SLATYER, R.O. (ed.): Photosynthesis and Photorespiration. Pp. 294-296. Wiley-
Interscience, New York - London - Sydney - Toronto 1971.

11923 - WALKER, D.A.: The site of sucrose synthesis in green plants. - In: YUDKIN, J.,
EDELMAN, J., HOUGH, J. (ed.): Sugar. Chemical, Biological and Nutritional Aspects of Sucrose. Pp. 103-109. D. Davey & Co., Hartford, Conn. 1971. [Photo-
synthates.]

11924 - WALKER, D.A., McCORMICK, A.V., STOKES, D.M.: CO_2-dependent oxygen evolution
by envelope-free chloroplasts. - Nature *233*: 346-347, 1971.

11925 - WALKER, R.B., SALO, D.J.: Light intensity and temperature effects on net photosynthesis in Douglas fir. - Plant Physiol. *47* (Suppl.): 30, 1971.

*11926 - WALLACE, A.: Some effects of 5-fluorouracil on corn seedlings. - In: WALLACE,
A. (ed.): Current Topics in Plant Nutrition. Pp. 29-31. Los Angeles, Calif.
1966. [Ps, Chl.]

*11927 - WALLACE, A.: Magnesium and manganese differences in inhibitor effects on
$C^{14}O_2$ fixation with phosphoenolpyruvate as a substrate. - In: WALLACE, A.
(ed.): Current Topics in Plant Nutrition. Pp. 183-184. Los Angeles, Calif.
1966.

B11928 - WALLACE, A., ABOU-ZAMZAM, A.M., ALEXANDER, G., BRINKERHOFF, F., CARMACK, C.R.,
ELGAZZAR, A., FROLICH, E.F., HALE, V.Q., JOVEN, C., HSIEH (YOUNG), H.,
MOTOYAMA, E., MUELLER, R.T., ROMNEY, E.M., SUFI, S.M.: Regulation of the Mi-
cronutrient Status of Plants by Chelating Agents and Other Factors. - Arthur
Wallace, Los Angeles, Calif. 1971. [Chl.]

11929 - WALLACE, A., HALE, V.Q., KLEINKOPF, G.E., HUFFAKER, R.C.: Carboxydismutase
and phosphoenolpyruvate carboxylase activities from leaves of some plant
species from the Northern Mojave and Southern Great Basin Deserts. - Ecology *52*: 1093-1095, 1971.

*11930 - WALLACE, A., ROMNEY, E.M.: Effect of beryllium on *in vitro* carboxylation reactions. - In: WALLACE, A. (ed.): Current Topics in Plant Nutrition. Pp. 185-
188. Los Angeles, Calif. 1966.

11931 - WALLEN, D.G.: The effect of light quality on growth rates, photosynthetic
rates and metabolism in plankton algae. - Diss. Abstr. Int. B *32*: 749-B, 1971.

11932 - WALLEN, D.G., GEEN, G.H.: Light quality in relation to growth, photosynthetic
rates and carbon metabolism in two species of marine plankton algae. - Mar.
Biol. 10: 34-43, 1971.

11933 - WALLEN, D.G., GEEN, G.H.: Light quality and concentration of proteins, RNA,
DNA and photosynthetic pigments in two species of marine plankton algae. -
Mar. Biol. *10*: 44-51, 1971.

11934 - WALLEN, D.G., GEEN, G.H.: The nature of the photosynthate in natural phytoplankton populations in relation to light quality. - Mar. Biol. *10*: 157-168, 1971.

11935 - WALLES, B.: Chromoplast development in a carotenoid mutant of maize. - Protoplasma *73*: 159-175, 1971.

11936 - WALLES, B.: Plastid inheritance and mutations. - In: GIBBS, M. (ed.): Structure and Function of Chloroplasts. Pp. 51-88. Springer-Verlag, Berlin - Heidelberg - New York 1971.

11937 - WALZ, D., SCHULDINER, S., AVRON, M.: Photoreactions of chloroplasts in a glycine medium. - Europe. J. Biochem. *22*: 439-444, 1971.

11938 - WANG, C.-T., NOBEL, P.S.: Permeability of pea chloroplasts to alcohols and aldoses as measured by reflection coefficients. - Biochim. biophys. Acta *241*: 200-212, 1971.

11939 - WANG, J.H., YANG, C.-S., TU, S.-I.: Effect of buffer concentration on the efficiency of photosynthetic energy conversion. - Biochemistry *10*: 4922-4930, 1971.

11940 - WANG, R., HEALEY, F.P., MYERS, J.: Amperometric measurement of hydrogen evolution in *Chlamydomonas*. - Plant Physiol. *48*: 108-110, 1971.

11941 - WANG, R.T., CLAYTON, R.K.: The absolute yield of bacteriochlorophyll fluorescence *in vivo*. - Photochem. Photobiol. *13*: 215-224, 1971.

11942 - WARD, F.J., NAKANISHI, M.: A comparison of Geiger-Mueller and liquid scintillation counting methods in estimating primary productivity. - Limnol. Oceanogr. *16*: 560-563, 1971.

11934 - WAREING, P.F.: Potential crop production in Britain - some conclusions. - In: WAREING, P.F., COOPER, J.P. (ed.): Potential Crop Production. A Case Study. Pp. 362-378. Heinemann Educational Books, London 1971.

B11944 - WAREING, P.F., COOPER, J.P. (ed.): Potential Crop Production. A Case Study. - Heinemann Educational Books, London 1971.

11945 - WATSON, D.J.: Size, structure, and activity of the productive system of crops. - In: WAREING, P.F., COOPER, J.P. (ed.): Potential Crop Production. A Case Study. Pp. 76-88. Heinemann Educational Books, London 1971.

11946 - WATSON, D.J., FRENCH, S.A.W.: Interference between rows and between plants within rows of a wheat crop, and its effects on growth and yield of differently-spaced rows. - J. appl. Ecol. *8*: 421-445, 1971. [Growth analysis.]

11947 - WAYGOOD, E.R., ARYA, S.K., MACHE, R.: Carbon dioxide fixation by maize chloroplasts isolated by the "laceration technique". - In: HATCH, M.D., OSMOND, C. B., SLATYER, R.O. (ed.): Photosynthesis and Photorespiration. Pp. 246-254. Wiley-Interscience, New York - London - Sydney - Toronto 1971.

11948 - WAYGOOD, E.R., ARYA, S., PAN, D.: Intracellular localization of P-enolpyruvate acid carboxylase in C_3 and C_4 plants. - Plant Physiol. *47* (Suppl.): 10, 1971.

11949 - WEAVER, P.: Temperature-sensitive mutations of the photosynthetic apparatus of *Rhodospirillum rubrum*. - Proc. nat. Acad. Sci. USA *68*: 136-138, 1971.

11950 - WEBB, W.L.: Photosynthetic response models for a terrestrial plant community. - Diss. Abstr. Int. B *32*: 2116-B, 1971.

*11951 - WEEDON,, B.C.L.: Some studies on carotenoid synthesis. - Tidsskr. Kjemi, Bergv., Metallurgi *26*: 130-131, 1966.

11952 - WEEDON, B.C.L.: Occurrence. - In: ISLER, O. (ed.): Carotenoids. Pp. 29-59. Birkhäuser Verlag, Basel - Stuttgart 1971. [Car.]

11953 - WEEDON, B.C.L.: Stereochemistry. - In: ISLER, O. (ed.): Carotenoids. Pp. 267-323. Birkhäuser Verlag, Basel - Stuttgart 1971. [Car.]

11954 - WEEDON, B.C.L.: Stereokemija karotenoida. [Stereochemistry of carotenoids.] - Kemija Industriji *20*: 627-630, 1971. [In Croat.]

11955 - WEGMANN, K.: Osmotic regulation of photosynthetic glycerol production in *Dunaliella*. - Biochim. biophys. Acta *234*: 317-323, 1971.

11956 - **WEGMANN, K., METZNER, H.:** Synchronization of *Dunaliella* cultures. - Arch. Mikrobiol. *78*: 360-367, 1971. [Ps, Chl.]

11957 - **WEICHMANN, J.:** Automatische Anlage für Gasstoffwechselmessungen von Gemüse. - Gartenbauwissenschaft *36*: 251-258, 1971.

11958 - **WEISS, C. Jr., SOLNIT, K.T., VON GUTFELD, R.J.:** Flash activation kinetics and photosynthetic unit size for oxygen evolution using 3-nsec light flashes. - Biochim. biophys. Acta *253*: 298-301, 1971.

11959 - **WEITZMAN, P.D.J., KENNEDY, I.R., CALDWELL, R.A.:** Polarographic activity and electrolytic reduction of ferredoxin. - FEBS Lett. *17*: 241-244, 1971.

11960 - **WELLBURN, F.A.M., WELLBURN, A.R.:** Chlorophyll synthesis by isolated intact etioplasts. - Biochem. biophys. Res. Commun. *45*: 747-750, 1971.

11961 - **WENZEL, M., HOFFMANN, K.:** Automatic and nondestructive scanning of ^{14}C and ^{3}H on two-dimensional chromatograms. - Anal. Biochem. *44*: 97-105, 1971.

11962 - **WESSELS, J.S.C., VAN LEEUWEN, M.J.F.:** Photosystem-II activity in subchloroplast fragments prepared by the action of digitonin. - In: QUAGLIARIELLO, E., PAPA, S., ROSSI, C.S. (ed.): Energy Transduction in Respiration and Photosynthesis. Pp. 537-550. Adriatice Editrice, Bari 1971.

*11963 - **WEST, M.L.:** Physiological ecology of three species of *Artemisia* in the White Mountains of California. - Diss. Abstr. int. B *30*: 4940-B - 4941-B, 1970. [Ps.]

11964 - **von WETTSTEIN, D., HENNINGSEN, K.W., BOYNTON, J.E., KANNANGARA, G.C., NIELSEN, O.F.:** The genetic control of chloroplast development in barley. - In: BOARDMAN, N.K., LINNANE, A.W., SMILLIE, R.M. (ed.): Autonomy and Biogenesis of Mitochondria and Chloroplasts. Pp. 205-223. North-Holland Publ. Comp., Amsterdam - London 1971.

11965 - **WHATLEY, F.R.:** Recycling through higher plants. - Proc. roy. Soc. London, Ser. B *179*: 193-200, 1971. [Ps.]

11966 - **WHATLEY, J.M.:** Ultrastructural changes in chloroplasts of *Phaseolus vulgaris* during development under conditions of nutrient deficiency. - New Phytol. *70*: 725-742, 1971.

11967 - **WHATLEY, J.M.:** The chloroplasts of *Equisetum telmateia* ERHR.: a possible developmental sequence. - New Phytol. 70: 1095-1102, 1971.

11968 - **WHITE, R.A.:** Photo-redox reactions of chlorophyll and chlorophyll analogs in the presence of quinones and hydroquinones. - Diss. Abstr. int. B *32*: 136-B, 1971.

11969 - **WHITE, R.A., TOLLIN, G.:** Chlorophyll one-electron photochemistry: light-induced absorbance changes and ESR signals for various porphyrin-quinone and hydroquinone systems in alcohol solvents. - Photochem. Photobiol. *14*: 15-42, 1971.

11970 - **WHITE, R.A., TOLLIN, G.:** Chlorophyll-one electron photochemistry: flash photolysis studies of the chlorophyll-quinone reaction in alcohol solvents. - Photochem. Photobiol. *14*: 43-63, 1971.

11971 - **WHITEHOUSE, D.G., LUDWIG, L.J., WALKER, D.A.:** Participation of the Mehler reaction and catalase in the oxygen exchange of chloroplast preparations. - J. exp. Bot. *22*: 772-791, 1971.

11972 - **WHITTAKER, R.H., WOODWELL, G.M.:** Measurement of net primary production of forests. - In: DUVIGNEAUD, P. (ed.): Productivity of Forest Ecosystems. Pp. 159-175. Unesco, Paris 1971.

11973 - **WHITTENBURY, R.:** Enrichment and isolation of photosynthetic bacteria. - In: SHAPTON, D.A., BOARD, R.G. (ed.): Isolation of Anaerobes. Pp. 241-249. Academic Press, New York - London 1971.

11974 - **WHITTINGHAM, C.P.:** The potentialities of chloroplast fragments. - Proc. roy. Soc. London, Ser. B *179*: 237-246, 1971.

11975 - **WHITTLE, C.M.:** The behaviour of ^{14}C profiles in *Helianthus* seedlings. - Planta *98*: 136-149, 1971.

11976 - **WHITTON, B.A., CARR, N.G., CRAIG, I.W.:** A comparison of the fine structure and nucleic acid biochemistry of chloroplasts and blue-green algae. - Protoplasma 72: 325-327, 1971.

11977 - WIĘCKOWSKI, S.: Photosynthetic activity of bean leaves and the relative content of chlorophyll in particles isolated from chloroplasts after irradiation of etiolated seedlings. - Photosynthetica 5: 44-49, 1971.

11978 - WIEDENROTH, E.-M., WEEGE, K.-H.: Die Bestimmung von Blattflächen bei Phaseolus vulgaris L. über lineare Messungen. - Photosynthetica 5: 406-413, 1971.

11979 - WIESSNER, W., FORK, D.C.: The development of system 2 activity in Chlamydobotrys during transition from photoheterotrophic to autotrophic nutrition. - Carnegie Inst. Year Book 69: 695-699, 1971.

11980 - WILD, A., ZICKLER, H.-O., GRAHL, H.: Weitere Untersuchungen zur Variabilität der photosynthetischen Einheit. - Planta 97: 208-223, 1971.

11981 - WILDMAN, S.G.: An approach towards defining the role of chloroplast DNA in the reproduction and differentiation of chloroplasts in higher plants. - In: HARRIS, P.J. (ed.): Biological Ultrastructure: The Origin of Cell Organelles. Pp. 91-105. Oregon State Univ. Press, Corvallis 1971.

11982 - WILKINSON, T.G.: Studies on the effects of ozone on photosynthetic processes in Pinus strobus. - Diss. Abstr. Int. B 32: 1307-B, 1971.

11983 - WILLIAMS, G.J. III: Populational differentiation in the Hill reaction of United States, Mexico, and Central America Liquidambar styraciflua L. - Photosynthetica 5: 139-145, 1971.

11984 - WILLIAMS, G.J. III: Levels of soluble sugars and ATP in ecotypes of Liquidambar styraciflua L. - Photosynthetica 5: 399-405, 1971.

11985 - WILLIAMSON, A.L., ELLSWORTH, R.K.: Identification by gas chromatography-mass spectrometry of methylvinylmaleimide and fromylethylmaleimide, obtained by oxidation of heme and chlorophylls a and b. - Anal. Biochem. 43: 633-639, 1971.

*11986 - WINGET, G.D.: The inhibition of photophosphorylation by phlorizin and closely related compunds. - Diss. Abstr. Int. B 29: 3203-B, 1969.

11987 - WINKLER, E.: Kartoffelbau in Tirol II: Photosynthesevermögen und Respiration von verschiedenen Kartoffelsorten. - Potato Res. 14: 1-18, 1971.

11988 - WINTERMANS, J.F.G.M.: On the galactolipid composition of subchloroplast fragments. - Biochim. biophys. Acta 248: 530-535, 1971. [Chl.]

11989 - WISHNICK, M., LANE, M.D.: Ribulosa diphosphate carboxylase from spinach leaves. - In: COLOWICK, S.P., KAPLAN, N.O. (ed.): Methods in Enzymology. Vol. 23. Pp. 570-577. Academic Press, New York - London 1971.

11990 - WISHNICK, M.M.: Studies on the mechanism of action of ribulosa diphosphate carboxylase from spinach. - Diss. Abstr. Int. B 31: 4502-B-4503-B, 1971.

11991 - DE WIT, C.T., BROUWER, R., PENNING DE VRIES, F.W.T.: A dynamic model of plant and crop growth. - In: WAREING, P.F., COOPER, J.P. (ed.): Potential Crop Production. A Case Study. Pp. 117-142. Heinemann Educational Books, London 1971.

11992 - WITT, H.T.: Coupling of quanta, electrons, fields, ions and phosphorylation in the functional membrane of photosynthesis. Results by pulse spectroscopic methods. - Quart. Rev. Biophys. 4: 365-477, 1971.

11993 - WITTENBERG, T., SISTROM, W.R.: Mutant of Rhodopseudomonas spheroides unable to grow aerobically. - J. Bacteriol. 106: 732-738, 1971. [Chl.]

11994 - WIUM-ANDERSEN, S.: Photosynthetic uptake of free CO_2 by the roots of Lobelia dortmanna. - Physiol. Plant. 25: 245-248, 1971.

11995 - WOJCIESKA, U., SZCZYPA, E.: Wzrost i produktywność fotosyntezy pszenic ozimych. Część II. Zmiany w aktywności fotosyntetycznej i jej produktywności w miare starenia się roślin. [Growth and productivity of photosynthesis in winter wheats. Part II. Changes in photosynthetic activity and its productivity as plants age.] - Pamięt. puławski - Prace IUNG 44: 49-70, 1971. [In Pol., ab: E, R.]

11996 - WOLEDGE, J.: The effect of light intensity during growth on the subsequent rate of photosynthesis of leaves of tall fescue (Festuca arundinacea SCHREB.). - Ann. Bot. 35: 311-322, 1971.

11997 - WOLF, D.D., BLASER, R.E.: Photosynthesis of plant parts of alfalfa canopies. - Crop Sci. *11*: 55-58, 1971.

11998 - WOLF, D.D., BLASER, R.E.: Leaf development of alfalfa at several temperatures. - Crop Sci. *11*: 479-482, 1971.

11999 - WOLF, E.G., CUSHING, C.E., RABE, F.W.: An automated system for multiple recording of diurnal pH. - Limnol. Oceanogr. *16*: 577-580, 1971. [Conversion to CO_2 for primary productivity and respiration data.]

12000 - WOLF, F.A., NEASE, F.R.: Assimilatory pigments in leaves of pleistocene age from Eastern North Carolina. - J. Elisha Mitchell sci. Soc. *87*: 18-19, 1971.

12001 - WOLF, F.T.: Effects of light on the enzymatic activities of green plants. - Advancing Frontiers Plant Sci. *29*: 19-96, 1971. [Ps.]

12002 - WOLF, F.T.: The development of photosynthetic capacity by etiolated seedlings upon illumination. - Plant Physiol. *47* (Suppl.): 44, 1971.

12003 - WOLF, F.T.: The development of photosynthetic capacity by etiolated seedlings upon illumination. - Z. Pflanzenphysiol. *64*: 124-129, 1971.

12004 - WOLTZ, S.S., ENGELHARD, A.W.: Physiological disorders of leaves of *Chrysanthemum* cultivars relative to accumulation of excess carbohydrate. - Proc. Fla. State hort. Soc. *84*: 370-374, 1971. [Chl.]

12005 - WOO, K.C., PYLIOTIS, N.A., DOWNTON, W.J.S.: Thylakoid aggregation and chlorophyll *a*/chlorophyll *b* ratio in C_4-plants. - Z. Pflanzenphysiol. *64*: 400-413, 1971.

12006 - WOO, S.-C., NG, C.-M., TUNG, I.-J.: Induction of mutations in cultivated rice by chemical and physical agents: II. The effect of M_1 damage and M_2 mutations. - Bot. Bull. Acad. sin. (Taipei) *12* (2): 66-73, 1971. [Chl.]

12007 - WOOD, K.G.: Self-absorption corrections for the ^{14}C method with $BaCO_3$ for measurement of primary productivity. - Ecology *52*: 491-498, 1971.

*12008 - WOOD, P.M.: A study of the redox state of plastoquinone *A* and the importance of the redox state in photosynthetic reactions. - Diss. Abstr. int. B *29*: 3619-B-3620-B, 1969.

12009 - WOODCOCK, C.L.F., BOGORAD, L.: Nucleic acids and information processing in chloroplasts. - In: GIBBS, M. (ed.): Structure and Function of Chloroplasts. Pp. 89-128. Springer-Verlag, Berlin - Heidelberg - New York 1971.

12010 - WOODMAN, J.N.: Variation of net photosynthesis within the crown of a large forest-grown conifer. - Photosynthetica *5*: 50-54, 1971.

12011 - WOODS, D.B., TURNER, N.C.: Stomatal response to changing light by four tree species of varying shade tolerance. - New Phytol. *70*: 77-84, 1971.

12012 - WOOLEY, J.T.: Reflectance and transmittance of light by leaves. - Plant Physiol. *47*: 656-662, 1971.

12013 - WRAIGHT, C.A., CROFTS, A.R.: Delayed fluorescence and the high-energy state of chloroplasts. - Europe. J. Biochem. *19*: 386-397, 1971.

12014 - WREN, J.D.: The effect of radiant energy transfer and chlorophyll concentration on the growth kinetics of *Chlorella pyrenoidosa* TX71105 in a laminar flow photosynthetic gas exchanger during low intensity illumination. - Diss. Abstr. int. B *32*: 2142-B-2143-B, 1971.

12015 - WRIGHT, R.D.: Local photosynthetic ecotypes in *Pinus attenuata* as related to altitude. - Madroño *21*: 254-257, 1971.

12016 - WU, B.-F.: Partitioning of photosynthates in maize during the ear-filling stage . - Diss.Abstr. int. B *32*: 39-B-40-B, 1971.

12017 - WU, J.H.: Retardation of ultraviolet light accelerated leaf senescence by a cytokinin: N^6 benzyladenine. - Photochem. Photobiol. *13*: 179-181, 1971. [Chl.]

12018 - WUENSCHER, J.E., KOZLOWSKI, T.T.: Relationship of gas-exchange resistance to tree-seedling ecology. - Ecology *52*: 1016-1023, 1971.

12019 - WUENSCHER, J.E., KOZLOWSKI, T.T.: The response of transpiration resistance

to leaf temperature as a desiccation resistance mechanism in tree seedlings.
- Physiol. Plant. *24*: 254-259, 1971. [Stomatal resistance.]

*12020 - WYATT, J.T.: Selected physiological and biochemical studies on blue-green
algae. - Diss. Abstr. Int. B *30*: 5166-B, 1970. [Ps.]

12021 - YAGSHIEV, A.: Vliyanie mikroelementov na nekotorye fiziologicheskie protsessy
i urozhaTnost' lyutserny. [Effect of trace elements on some physiological
processes and yield of alfalfa.] - Izv. Akad. Nauk turkm. SSR, Ser. biol.
Nauk *1971* (2): 41-46, 1971. [Ps; in R, ab: E, Turkm.]

12022 - YAKOVENKO, G.M., MIKHNO, A.I.: Metod vydeleniya i razdeleniya po klassam lipidov list'ev i khloroplastov rasteniT. [Method of isolation and separation
into classes of lipids of plant leaves and chloroplasts on the silica gel
columns.] - Fiziol. Biokhim. kul't. Rast. *3*: 651-656, 1971. [In R, ab: E.]

12023 - YAKUBOVA, M.M., LEBEDEVA, G.P., NASYROV, Yu.S.: DeTstvie UF-radiatsii na
fotosinteticheskoe fosforilirovanie. [Effect of UV-radiation on photophosphorylation.] - In: ZALENSKII, O.V. (ed.): Fotosintez i Ispol'zovanie SolnechnoT Energii. Pp. 217-219. Nauka, Leningrad 1971. [In R, ab: E.]

12024 - YAKUBOVA, M.M., LEBEDEVA, G.P., NASYROV, Yu.S.: Issledovanie fotokhimiches-
koT aktivnosti khloroplastov mutantov gorokha. [Photochemical activity of
chloroplasts from pea mutants.] - In: NASYROV, Yu.S. (ed.): Geneticheskie
Aspekty Fotosinteza. Pp. 178-185. Donish, Dushanbe 1971. [In R, ab: E.]

12025 - YAKUSHIJI, E.: Cytochromes: algal. - In: COLOWICK, S.P., KAPLAN, N.O. (ed.):
Methods in Enzymology. Vol. 23. Pp. 364-368. Academic Press, New York -
London 1971.

12026 - YAKUSHKINA, N.I., PUSHKINA, G.P.: Nekotorye osobennosti vliyaniya gibberellina i kinetina na soderzhanie khlorofilla i na protsess fotofosforilirovaniya
v prorostkakh kukuruzy. [Effect of gibberellin and kinetin on chlorophyll
content and photophosphorylation in maize seedlings.] - Fiziol. Rast. *18*: 898-
903, 1971. [In R, ab: E.]

12027 - YAMAGUCHI, J.: [Method of measuring photosynthetic rate of corn leaf.] - J.
Sci. Soil Manure, Jap.] Nippon Dojo-Hiryogaku Zasshi *42*: 26-28, 1971. [In
Jap.]

*12028 - YAMAGUCHI, J., TANAKA, A.: The effect of light on respiratory rate of the
rice plant. - Plant Cell Physiol. *8*: 343-346, 1967. [Ps.]

12029 - YAMAMOTO, H.Y., WANG, Y., KAMITE, L.: A chloroplast absorbance change from
violaxanthin de-epoxidation. A possible component of 515 nm changes. -
Biochem. biophys. Res. Commun. *42*: 37-42, 1971.

12030 - YAMASHITA, T., TSUJI, J., TOMITA, G.: Reactivation of the Hill reaction of
Tris-washed chloroplasts. - Plant Cell Physiol. *12*: 117-126, 1971.

12031 - YANAGI, K., KOYAMA, T.: Thin layer chromatographic method for determining
plant pigments in marine particulate matter, and ecological significance
of the results. - Geochem. J. *5*: 23-37, 1971.

12032 - YARTSEVA, I.A., SOLOV'EVA, O.L.: Nekotorye dannye o vliyanii fosfora na fotosintez ChernomorskoT fillofory. [Effect of phosphorus on photosynthesis of
Black Sea *Phyllophora*.] - Gidrobiol. Zh. *7* (5): 75-78, 1971. [In R.]

12033 - YASNIKOV, A.A., BERSHTEIN, B.I., VOKOVA, N.V., VOLOVIK, O.I., ZAITSEVA, N.A.,
OKANENKO, A.S., POLISHCHUK, A.I., PSHENICHNAYA, A.I., REINGARD, T.A.,
SEMENYUK, I.I.: DeTstvie fosfoenolpiruvata i piruvatkinazy na kinetiku fotofosforilirovaniya v khloroplastakh gorokha. [Effect of phosphoenolpyruvate
and pyruvate kinase on the kinetics of photophosphorylation in pea chloroplasts.] - Fiziol. Biokhim. kul't. Rast. *3*: 468-473, 1971. [In R, ab: E.]

12034 - YASNIKOV, A.A., VOLKOVA, N.V., MEL'NICHENKO, I.I., BOIKO, T.S., BERSHTEIN,
B.I., OKANENKO, A.S., REINGARD, T.A.: Primenenie metoda modeleT dlya izucheniya mekhanizma otdel'nykh stadiT fotosinteza. [Models for investigation of
the mechanism of specific stages of photosynthesis.] - In: ZALENSKII, O.V.
(ed.): Fotosintez i Ispol'zovanie SolnechnoT Energii. Pp. 232-235. Nauka,
Leningrad 1971. [In R, ab: E.]

12035 - YASNIKOV, O.O., OKANENKO, A.S., OSTROVS'KA, L.K.. VOLKOVA, N.V., KOZLOVA,

I.Ya., MEL'NICHENKO, I.V., BERSHTEĬN, B.Ĭ., PETRENKO, S.G., REINGARD, T.A., SEMENYUK, I.I.: Pro mekhanizm zaklyuchnykh etapiv peretvorennya sonyachnoī energiī v khimichnu v roslynakh. [Mechanism of final stages of transformation of solar energy into chemical energy in plants.] - Visnyk Akad. Navuk URSR *1971* (5): 10-16, 1971. [In Ukr.]

12036 - YATSENKO, G.K., KOSTYLEV, E.F.: Soderzhanie karotinoidnykh pigmentov v giponeīstone severo-zapadnoī chasti Chernogo morya. [Level of carotenoid pigments in the hyponeuston of the northwestern part of the Black See.] - Biol. Morya *22*: 129-138, 1971. [In R.]

12037 - YAU, H.F.: Action spectra for the absorbance change at 880 nm and for P870 fluorescence from a photosynthetic reaction center. - Photochem. Photobiol. *14*: 475-482, 1971.

12038 - YORDANOV, I.T.: Photophosphorylation activity *in vivo* and intensity of the photosynthesis of bean leaves of different physiological state. - Dokl. bolg. Akad. Nauk *24*: 1551-1554, 1971.

12039 - YOSHIDA, F., WATANABE, H.: Effects of mineral nutrients on chlorophyll contents, friability and yield of cultured tobacco-callus. - Bull. Fac. Agr., Tamagawa Univ. *11*: 3-14, 1971.

12040 - YOSHIZAWA, T., HORIUCHI, S.: [Low temperature spectrophotometry of biological materials.] - Bunko Kenkyu *20*: 206-219, 1971. [Car; in Jap.]

12041 - YUR'EVA, K.V.: Vliyanie NRV na urozhaī i kachestvo sel'skokhozyaīstvennykh kul'tur. [Effect of petroleum growth substance on the yield and quality of farm crops.] - In: NRV v Sel'skom Khozyaīstve. Pp. 322-325. Elm, Baku 1971. [Chl; in R.]

*12042 - YURKOVA, G.N.: Spektrofotometricheskoe izuchenie pigmentnykh mutantov khlorelly. [Spectrophotometric study of pigment mutants of *Chlorella*.] - Tsitol. Genet. *4*: 356-358, 1970. [In R,ab: E.]

12043 - YUSUPOVA, G.A., VAKHIDOVA, L.R., GILLER, Yu.E.: O fotokhimicheskoī aktivnosti khlorofilla v gomogenatakh list'ev mutantnykh form *Arabidopsis thaliana*. [Photochemical activity of chlorophyll in leaf homogenates of mutant forms of *Arabidopsis thaliana*.] - Dokl. Akad. Nauk tadzh. SSR *14* (2): 66-69, 1971. [In R, ab: Tadzh.]

*12044 - ZADONTSEV, A.I., PIKUSH, G.R., PYKHTIN, N.I.: Osobennosti morfo-fiziologicheskikh izmeneniī u rasteniī ozimoī pshenitsy pod vliyaniem CCC. [Morpho-physiological changes in winter wheat plants under the action of CCC.] - Fiziol. Biokhim. kul't. Rast. *2*: 574-580, 1970. [Chl; in R, ab: E.]

*12045 - ZADOROZHNIĬ, V.Z.: Vliyanie khlorkholinkhlorida na soderzhanie khlorofilla v list'yakh kartofelya. [Effect of chlorcholinechloride on the chlorophyll content in potato leaves.] - Tr. khar'kov. sel'skokhoz. Inst. *90* (Issledovaniya po Fiziologii i Biokhimii Rasteniī: 25-28, 1970. [In R.]

12046 - ZANKEL, K.L.: Rapid delayed luminescence from chloroplasts: Kinetic analysis of components; the relationship to the O_2 evolving system. - Biochim. biophys. Acta *245*: 373-385, 1971.

*12047 - ZASLONKIN, V.P.: Vliyanie molibdena i nitragina na intensivnost' fotosinteza list'ev gorokha. [Effect of molybdenum and nitragin on the rate of pea leaf photosynthesis.] - Nauch. Tr., orlovskaya obl. sel'skokhoz. opyt. Sta. *5*: 97-104, 1970. [In R.]

12048 - ZATYKÓ, J.M.: Effect of benzyladenine on the amount of leaf pigments in bean. - Acta agron. Acad. Sci. hung. *20*: 427-437, 1971.

12049 - ZATYKÓ, J.M.: Parallelism in the photobleaching of chlorophylls and carotenoids in leaves of *Phaseolus vulgaris* L. - Phyton *28*: 11-13, 1971.

12050 - ŻELAWSKI, W., KUCHARSKA, J., KINELSKA, J.: Relationship between dry matter production and carbon dioxide absorption in seedlings of Scots pine (*Pinus silvestris* L.) in their second vegetation season. - Acta Soc. Bot. Pol. *40*: 243-256, 1971.

12051 - ŻELAWSKI, W., NALBORCZYK, E.: Productivity of photosynthesis in Scots pine

(*Pinus silvestris* L.) seedlings grown from seed irradiated by X-rays. - Acta Soc. Bot. Pol. *40*: 413-421, 1971.

12052 - ZELDIN, M.H., RADKE, K.: Low temperatures and chloroplast development in *Euglena*. - Plant Physiol. *47* (Suppl.): 45, 1971.

12053 - ZELENSKIĬ, M.I., CHERNYAEVA, I.I.: Sopostavlenie spektrofotometricheskogo i potentsiometricheskogo metodov izmereniya skorosti reaktsii Khilla. [Comparison of spectrophotometric and potentiometric methods for the determination of the rate of Hill reaction.] - Fiziol. Rast. *18*: 1276-1282, 1971. [In R, ab: E.]

B12054 - ZELITCH, I.: Photosynthesis, Photorespiration, and Plant Productivity. - Academic Press, New York - London 1971.

*12055 - ZEN'KEVICH, E.I.: Izuchenie migratsii energii v rastvorakh nekotorykh fotosinteticheskikh pigmentov. [Energy migration in solutions of some photosynthetic pigments.] - In: Materialy I Respublikanskoĭ Konferentsii Molodykh Uchenykh. Pp. 29-31. Akad. Nauk belorus. SSR, Minsk 1970. [In R.]

12056 - ZHUKOV, O.S., TUROVSKIĬ, I.I.: Osmiofil'nye globuly v khloroplastakh plodovykh rastenĭĭ. [Osmiophilic globules in chloroplasts of fruit plants.] - Tr. tsentr. genet. Lab. I.V. Michurina *12*: 182-187, 1971. [In R.]

12057 - ZHUKOVA, P.S.: Effektivnost' primeneniya NRV pri vyrashchivanii ovoshchnykh kul'tur. [Effectivity of using petroleum growth substance for growing vegetables.] - In: NRV v Sel'skom Khozyaĭstve. Pp. 30-40. ELM, Baku 1971. [Chl; in R.]

12058 - ZIEMER, R.R.: Translocation of ^{14}C in ponderosa pine seedlings. - Can. J. Bot. *49*: 167-171, 1971.

12059 - ZILINSKY, J.W.: Studies on the physiology and biochemistry of the photosynthetic bacterium, *Rhodopseudomonas capsulata*. - Diss. Abstr. int. B *31*: 5521-B-5522-B, 1971.

12060 - ZILINSKY, J.W., SOJKA, G.A., GEST, H.: Energy charge regulation in photosynthetic bacteria. - Biochem. biophys. Res. Commun. *42*: 955-961, 1971.

12061 - ZIMINA, T.A., FEDOROV, I.S., KRYSHNYAYA, S.V.: Vliyanie nekotorykh klimaticheskikh faktorov na rost, razvitie i produktivnost' ovoshchnykh kul'tur v usloviyakh yuga Sakhalina. [Effect of some climatic factors on the growth, development and productivity of vegetables in the south of Sacchaline.] - Tr. sakhalin. kompleks. nauch.-issled. Inst. *23* (Rasteniya i Faktory Vneshneĭ Sredy): 3-18, 1971. [In R.]

12062 - ZIMINA, T.A., KRYSHNYAYA, S.V.: Izmenenie pigmentnogo kompleksa v list'yakh introdutsirovannykh zernobobovykh i drugikh kul'tur reproduktsii semyan v usloviyakh yuga Sakhalina. [Pigment changes in leaves of introduced pulses and other crops during seed reproduction in the south of Sacchaline.] - Tr. sakhalin. kompleks. nauch.-issled. Inst. *23* (Rasteniya i Faktory Vneshneĭ Sredy): 65-79, 1971. [In R.]

*12063 - ZINOV'EVA, S.D.: Produktivnost' khlorelly v zavisimosti ot form azotnogo pitaniya i pH sredy. [*Chlorella* productivity in dependence on the forms of nitrogen nutrition and pH of the medium.] - Uch. Zap. ural'sk. gos. Univ. 113, Ser. biol. *8* (Voprosy Regulyatsii Fotozinteza, Sb.2): 131-138, 1970. [Ps, Chl; in R.]

12064 - ZINSOU, C.: Dégradation enzymatique et autooxydative du β-carotène. - Physiol. vég. *9*: 149-167, 1971.

12065 - ZLOBIN, V.S.: Modelirovanie gidrokhimicheskikh protsessov i tsiklicheskogo vosproizvodstva pervichnogo organicheskogo veshchestva v Evropeĭskom basseine severnogo Ledovitogo okeana. [Simulation of hydrochemical processes and cyclic reproduction of primary organic matter in the European basin of the Arctic Ocean.] - Okeanologiya *11*: 1033-1040, 1971. [Ps; in R, ab: E.]

12066 - ZLOBIN, V.S., PERLYUK, M.F.: Fotosintez i mekhanizm deĭstviya tsianida na kletochnoe dykhanie i nakoplenie plutoniya-239 morskimi vodoroslyami. [Photosynthesis and mechanism of cyanide action on cellular respiration and accumulation of plutonium-239 by marine algae.] - Tr. polyar. nauch.-issled. Proekt.

Inst. morsk. ryb. Khoz. Okeanogr. *29* (Vliyanie Ioniziruyushcheĭ Radiatsii na Organizm): 159-168, 1971. [In R, ab: E.]

12067 - ZUMFT, W.G., SPILLER, H.: Characterization of a flavodoxin from the green alga *Chlorella*. - Biochem. biophys. Res. Commun. *45*: 112-118, 1971.

*12068 - ZUNUNOV, Ch.Z., LIPKIND, I.M.: Vliyanie udobreniĭ na soderzhanie v list'yakh khlopchatnika N, P, K i pigmentov. [Effect of fertilizers on the content of N, P, K and pigments in cotton leaves.] - Dokl. Akad. Nauk tadzh. SSR *12* (9): 56-58, 1969. [Chl, Car; in R, ab: Tadzh.]

12069 - ZURZYCKI, J., STARZECKI, W.: Volumetric methods. - In: ŠESTÁK, Z., ČATSKÝ, J., JARVIS, P.G. (ed.): Plant Photosynthetic Production. Manual of Methods. Pp. 257-275. Dr. W. Junk N.V. Publ., The Hague 1971.

Authors' names are presented in the form in which they appear in the respective publication. The names from papers published in Cyrillic characters are transcribed as shown
on p. III of this volume. Alternative spellings and forms of the name of the same
author are usually cross-indexed. The numbers in italics refer to publications in which
the respective author acts as an editor.

A

AACH, H.G.   9088
AASE, J.K.   9089
ABDUL-BAKI, A.A.   9090-1
ABDULLAEV, Kh. A.   9092, 10982, *11848
ABDURAKHMANOVA, Z.N.   9093-4, 9791-2,
   10981, 10983
ABE, H.   *10817
ABIDOV, A.   9201
ABISALOV, R.S.   10767-8, 11542-3
ABOLINA, G.I.   9095-6
ABOU-HAÏDAR, M.   *11307, *11313
ABOU-ZAMZAM, A.   B11928
ABRAMYAN, A.G.   *9097
ABRAROV, A.A.   9098
ACOCK, B.   9099
ADACHI, M.   9100
ADAMOVA, N.P.   11369
ADAMS, M.S.   9101-3
ADAMS, R.S. Jr.   11704
ADEDIPE, N.O.   9104
ADEÏSHVILI, N.I.   *9990
ADLER, K.   10742
AEROV, I.L.   9105
AFANAS'EVA, T.A.   9106
AGARWAL, S.K.   11909-11
AGATA, W.   9107-8, 10560-1
AGEEVA, O.G.   9109
AITZETMÜLLER, K.   11684-6
AKANOV, E.N.   10635
AKAZAWA, T.   11701, *11702
AKHMEDOV, A.Ya.   *10686
AKHMEDOV, B.   10275
AKHMEDOV, N.   11283
AKHRAMOVICH, N.I.   *9110, 11566, 11568
AKHUNDOVA, E.M.   10086
AKOSU, F.I.   10435
AKOYUNOGLOU, G.   9111, 9165-6
AKSENOVA, E.I.   9112
AKULOVA, E.A.   9113
AKULOVICH, E.M.   11567
AKULOVICH, N.K.   9114-5, *9992, 9993
AL-ABBAS, H.   9255
ALAMUTI, N.   11661
AL-ANI, T.A.   9116
ALBERDA, T.   9117
ALBERTSSON, P.-Å.   10620
ALDERFER, R.G.   9118
ALEEM, M.I.H.   10484-5
ALEKHINA, S.K.   10961
ALEKPEROVA, M.S.   9119
ALEKSEEV, S.I.   10893
ALEKSEEVA, Kh. A.   11562
ALEKSEEVA, N.R.   11391

ALEXANDER, D.J.   9686
ALEXANDER, G.   B11928
ALIEV, E.A.   9120
ALIEV, K.   10981-2
ALIEV, K.A.   9845
ALIEV, N.A.   *10571
ALIEV, S.A.   9121
ALIEV, Z.Sh.   9457, 10533
ALIEVA, S.A.   9122-3
ALLEN, C.F.   9124
ALLEN, H.L.   9125
ALLEN, L.H. Jr.   9126-7
ALLEN, M.B.   9128
ALLEN, W.A.   *9933-4, 9935-7
ALLISON, J.C.S.   9129
ALSCHER, R.G.   9130
AL'SHEVSKII, N.G.   *9131
AMBLER, R.P.   9132
AMBROES, P.   9732, 11859
AMESZ, J.   9133-5, 10525
AMIR, S.   9136
AMIRDZHANOV, A.G.   *9137, 9138
ANALYTIS, S.   9139
ANDERSEN, K.S.   9140-1, 9354-7, 11608
ANDERSEN, W.R.   9142
ANDERSON, B.T.   10107
ANDERSON, D.E.   *9143
ANDERSON, G.C.   *9144
ANDERSON, J.M.   9145-7, 9384-6, 10184
ANDERSON, L.E.   9148-9
ANDERSON, M.C.   9150-1
ANDERSON, R.   9897
ANDERSSON, F.   9152
ANDO̅, M.   *10267
ANDREEVA, R.A.   9153
ANDREEVA, T.F.   *B9154-5
ANDREÏTSEV, A.P.   *10506
ANDREW, R.H.   *10449
ANDREWS, T.J.   9156-8, 10324
ANDRIANOV, V.K.   *9159, 9491-2
ÅNGSTRÖM, A.K.   9718
ANIKUSHKIN, N.F.   9160
ANISIMOV, B.V.   11520
ANTIPOVA, A.I.   *11236
ANTOSZEWSKI, R.   9161, *9747, 11238
ANTSUPOVA, L.V.   9162
d'AOUST, A.   9163
APEL, P.   11160
APPELQVIST, L.-Å.   10373
APPLEBY, A.P.   9454
APPLEMAN, D.   9286
AQUINO, O.   9164
ARBOLEDA, O.P.   10445

DIMITRIEVA, L. 11803
DIMOVA, O. 11803
DINER, B.A. 9681
DIRKS, M.P. 9135
DJAVANCHIR, A. 9682
DJENDOV, C. *9683
DMITRIEVA, L.V. 10635
DOBRENZ, A.K. 9779
DOBROVOL'SKAYA, T.G. 9194
DOCHINGER, L.S. 11340
DODD, W.A. 9684-5
DODGE, A.D. 9686, 9698
DOEMEL, W.N. 9687
DÖHLER, G. 9688
DOLIDZE, I.A. 11845
DOMAN, N.G. 9214, 9689, 10588
DOMASH, V.I. *9992, 9993
DOMES, W. 9690
DONAHER, D.J. 9691
DÖRFLING, P. 9692-4
DÖRING, G. 10092
DORNHOFF, G.M. 9695
DOROBANȚU, N. *9949
DORODNEVA, V.I. *9696
DOROKHOV, B.L. 9697
DOROZHKINA, L.N. *10288, 10291
DOWDELL, R.J. 9698
DOWNES, R.W. 9699-700
DOWNEY, L.A. 9701
DOWNTON, J. 9702
DOWNTON, W.J.S. 9233, 9703-7, 11260,
    12005
DOZIER, W.A. Jr. 9708-9
DRABKOVA, V.G. *9710
DRĂGHICI, L. *11189
DRAPER, S.R. 9711
DREW, E.A. 10211
DREWS, G. 9712, *11477-8
DRIESSCHE, R. van den 9594, 9713-4
DRING, M.J. 9715
DROKOVA, I.G. 9716-7
DROZDOVA, I.S. 11241, 11885, 11895
DROZDOVA, N.N. 11539
DRUMMOND, A.J. 9718, 11759-60
DUBOVA, K.P. *9719
DUBROVIN. V.N. 11368-9
DUBROVSKAYA, A.A. 9916
DUDA, M. 9771
DUDOK, E.P. *10047
DUDZIAK, B. 9271
DUGGER, W.M. 10303
DUMBADZE, V.Z. 11146
DUMFORD, S.W. 10903
DUMMLER, W. 9692-4
DUNCAN, W.G. 9720-1, 10173, 10483
DUNHAM, W.R. 9722-3, 11129
DUNIWAY, J.M. 9724
DUNN, E.L. 9725, 10911
DUNN, S. 11763
DUPAIGNE, F. 9726
DURANTON, H. 11442
DURANTON, J. 10598
DUS, K. 9727, 10543
DUSEK, D. 9282

DUTHIE, H.C. *9728
DUTTON, P.L. 9729-30, 11502
DUVIGNEAUD, P. *9152, 9362, 9659, 9679,*
    9731-2, 9742, 9899, 9914, 10429,
    10457, 10601, 10645, 10664-5, 10699,
    10869, 11087, 11429, 11483, 11972
DUYSENS, L.N.M. 9733-4
DVOŘÁK, J. *9447, 9735, *11687*
DWYER, M.R. 9736
DYBING, C.D. *9737
DYER, T.A. 9738
DYKYJOVÁ, D. 9739-41
DYLIS, N. 9742
DYSON, P.W. 9743
DZHAGAROV, B.M. 9744
DZHANUMOV, D.A. 9745, 11741
DZHASHI, R.G. *9990
DZHOKHADZE, G.K. 9746
DZIĘCIOŁ, U. *9747
DZJUBENKO, V.S. 11832

E

EAGLES, C.F. 9748-9, 11113
EBATA, T. 9750, 9905
EBREY, T.G. 9751
EBRINGER, L. *9752
ECKARDT, F.E. 9753, 10309, 10312
EDELMAN, J. 9754-7, 10112, *11923*
EDGE, H. 10099
EDMONDSON, D.E. 9627
EDREVA, A. *9758
EDWARDS, D.I. 9759
EDWARDS, G.E. 9760-2, 10826-7
EDWARDS, M.R. 9763, 9920
EFIMOV, A.K. *11383, 11384, *11385
EFIMOV, M.V. *9764-7
EFIMOVA, N.A. 9768
EFREMOV, V.V. *9769-70
EGAMBERDIEV, A.M. 11248-9
EGED, Š. 9771
EGLINTON, G. *9607
EGUNJOBI, J.K. 9772
EHEART, M.S. *9773
EHL, K. 9464
EICHENBERGER, W. 9631
EIDEL'MAN, Z.M. *B9774, 9775
EILAM, Y. 9776
EILMANN, I. 9975
EINOR, L.O. 9777-8, *10876
EISA, H.M. 9779
EKKERMAN, N.I. 11549
EKPAHA-MENSAH, J.A. 11613
ELEY, D.D. 9780
ELEY, J.H. 10484-5, 11435
EL-FOULY, M.M. 9781, *10341
EIGAZZAR, A. B11928
EL-HAMAWI, H.A. 9781
ELIZAROVA, V.A. 11261
ELLENBERG, H. *10157, 10616, 11373*
ELLIOTT, W.B. 9411
ELLIS, R.J. 9782

J

JACINI, G. 9820
JACKSON, J.B. 9619, 10301
JACKSON, W.A. 10937
JACOBI, G. 9872, 10302, 11240
JACOBSON, B.S. 10303
JAGELS, R.H. *9283
JAGENDORF, A.T. 9587, 11036, 11214-5,
  11386-7
JAIN, T.C. 10304-5, *10306
JAKABFI FRIGYESNÉ *11327
JAKŠINA, A.M. *10307
JAMES, B. *10308
JAMES, G.B. 10613
JANÁČ, J. 10309
JARVIS, P.G. *9151, *9534, *9664*, 10309,
  *10310, 10311-2, *10313, 10316, *10559*,
  10590, 10593, 10709-10, *11091*, *11183*,
  11517, *11522*, B11524, 11525, *11527*,
  11605, *11901*, *12069*
JAŠA, B. 10314
JAUNEAU, E. 11320
JAWORSKI, E.G. *10777
JAYARAMI REDDY, A. *10315
JEFFREE, C.E. 10316
JEFFREY, S.W. 9257
JENSEN, A. 10656
JENSEN, R.G. 9614, 10317-8
JENSEN, S.E. 9127
JERUZEL, G. 11454
JEŠKO, T. 10319
JETSCHMANN, C. 11873
JI, T.-H. 9316
JOHANNES, B. 10320
JOHN, J.B.St. 10188
JOHNSON, C.E. 9521, *9799, *10321, 10322,
  11292
JOHNSON, H.S. 9157, 10323-4
JOHNSON, R.P.C. 10316
JOHNSTON, T.D. 10325
JOLCHINE, G. 11320
JOLIOT, A. 10326-7
JOLIOT, P. 10326-7
JOLIFFE, P.A. 10330
JOLIVET, E. 10328, *10329
JOLLEY, K. 11787
JONES, L.H. 10331
JONES, M.B. 11640
JONES, R. *10332-4
JORDAN, C.F. 10335
JOSEPH, B. *10336
JOSHI, G.V. 9340, *10337, 10338, 11149
JOURDAN, F. 11405
JOVEN, C. B11928
JOYARD, J. 10339-40
JUNG, J. *10341
JUNGE, W. 10342
JUPIN, H. 10343
JURÁŠEK, A. *9752
JURČÍK, F. 11209-10

K

KABANOVA, Yu.G. 10344
KABYSH, V.A. *10345
KACHARAVA, N.F. 10346
KADANIKOVA, V. 10347
KADOTA, K. *11034
KADYSH, A.G. 9120
KAFALIEVA, D.N. *9381, 10348
KAGAWA, T. 10349
KAHN, A. 9386, 10163
KAHN, J.S. 10350-1
KAKHNOVICH, L.V. 10352, 10474
KAKIE, T. *10353-5
KAKUMOTO, S. 10934
KAKUNO, T. 9264, 10356
KAL"CHEVA, I. 10359
KALBE, L. 10357-8
KALBERER, P.P. 9482
KALER, V.L. 10360-1
KALFF, J. 10362
KALICHAVA, G.S. 10368
KALINA, M. 10099
KALININA, L.M. *10363, 11568
KALISHEVICH, S.V. *11235-6
KALLIO, P. 10364
KALMANSON, A.E. *10365
KAMATA, E. 9107-8, 10560-1
KAMEN, M.D. 9264, 9727, 9816, 10423-4,
  10858, 11049, 11320, 11457-8
KAMENSKIĬ, V.V. 10566
KAMENTSEVA, I.E. 9835
KAMEYA, T. 10366
KAMITE, L. 12029
KAMMERMEYER, K. *10870
KAMP-NIELSEN, L. 10367
KANAI, R. 9762, 11852
KANCHAVELI, L.A. 10368
KANDAUROV, V.I. 10369
KANEMASU, E.T. 10370
KANIUGA, Z. 9884, 10371
KANNANGARA, C.G. *10372, 10373-4, *10375
KANNANGARA, G.C. 11964
KAO, O. 10376
KAPLAN, N.O. *9124, 9143, 9224, 9234*,
  9254, 9262-3, 9304, 9347, 9359-60,
  9374-5, 9382, 9395, 9461, 9474, 9481,
  9581, B9590, 9595, 9628, 9800, 9875,
  9893, 9906, 9925, 10042, 10044, 10223-
  4, 10302, 10350, 10401, 10408, 10650,
  10656, 10662, 10796, 10829-30, 10882,
  11086, 11133, 11243, 11334, 11455,
  11553, 11558, 11581, 11610, 11673,
  11683, 11706, 11724, 11772-3,
  11792, 11877, 11921, 11989, 12025
KAPLAN, Ş. 9886
KAPLANOVÁ, M. 10377-8
KAPLER, R. 10379-81
KARANOV, E.N. 10382
KARAPETYAN, N.V. 10383-6
KARIEV, A.U. 9098
KARLANDER, E.P. 9569
KARLISH, S.J.D. 10387
KARNAUKHOV, V.V. 10388

This index contains a selection of primary items chosen according to their interest for photosynthesis researchers and to their relative importance and occurrence. The word "Photosynthesis" is only rarely regarded as a main item, but the individual factors affecting photosynthesis are listed. Numbers of selected references containing a more comprehensive information (review articles *etc.*) are printed in italics.

Assimilation chamber see Chamber, assimilation

ATP see Photophosphorylation ...

Autotrophy see Heterotrophy, autotrophy, mixotrophy

B

Bacteriochlorophyll, bacterioviridin, *Chlorobium* chlorophylls *9249, 9256, 9415,
     *9418, 9514, 9712, 9729-30, 9734, 9751, 9816, *9821-2, 9839, 9849, 9854-5,
     9912, 9929, *9945-6, *9959, 9965, 9995, *10079, 10273-4, 10356, *10506, 10509,
     10621, 10915, 11053, 11061, *11086*, 11115, 11166, 11182, 11315, 11320, 11457-8,
     *11477-8, 11588, 11722, 11773-4, 11785, 11851, 11941, 11993

Bacterioviridin see Bacteriochlorophyll ...

Bibliographies of photosynthesis *B9185-7, *9535, 11523*, B11524, 11876

Biliproteins 9289, 9309-10, 9464, 9643, 9763, 9796, 9885, 9920, 9984-5, 10026, 10090,
     10254, 10376, 10738-9, 10765, 10909, 11046, *11058*, 11474, *11595, 11764, 11819,
     11879, 11976

Bioclimatological methods (*cf.* also Aerodynamic methods, theory) *9664*, 9768, 9972-3,
     10700, 10794, 10904-5, 11295, *B11749, 11784, 11800, 11862

Biopotentials of photosynthetic apparatus *9159, 9250, *9381, 9491-2, 10227, 10301,
     10342, *11383, 11384, *11385, 11691, 11781-2, 11902-3

Boundary layer of leaf and canopy (*cf.* also Bioclimatological methods) *10311*, 11139

Bowen ratio see Bioclimatological methods

Bundle sheaths 9145-7, 9233, 9284, 9354-7, 9601, 9704, 9707, 9760-1, 9890, 10324,
     10388, 10597, 10648, 10826-7

C

$^{13}C/^{12}C$ ratio 10234, 11335, 11614-5

^{14}C, ^{13}C, use in photosynthesis measurement 9183, 9306, 9511, *9513, 9540, 9689, 9792,
     *9832-3, 9853, 9882, 10009, *10236, *10329, 10432, 10435, *10437, 10602, 10714,
     10973, 10978, 11027, 11060, 11074, *11255, *11257, 11352, 11431, 11453, 11505,
     11632, 11771, *11901*, 11942, 11961, 11975, 12007

C_3 and C_4 plants, differences (*cf.* also Bundle sheaths; Carbon fixation pathways; Photo-
     synthates formation patterns) 9156, *9383*, 11108-10, 11260, 11469, 11603, 11818,
     11948

Calibration of infra-red gas analyser see Infra-red gas analyser, calibration

Calorimetry, caloric values 9296, 9480, 10149, 10326, 10512, *11118, 11119, 11127,
     11373, 11783

Calvin cycle see C_3 and C_4 plants ...; Carbon fixation pathways; Photosynthates for-
     mation patterns

CAM 9208, 9300, *9602-3, 10208, 10476-9, 11604, 11786

Canopy architecture see Canopy structure and density

Canopy boundary layer see Boundary layer of leaf and canopy

Canopy, CO_2 profile see CO_2 profile in canopy

Canopy photosynthesis 9099, 9118, 9553, 9720-1, 9731-2, 9739-42, 9847, 10175, 10222,
     10457, 10491, 10545, *10556, 10560-1, *10586, *10599, 10601, 10604, *10619, 10628,
     10646, 10665, *10692*, *10873, 10904, *10921*, 11013, 11019, *11030, 11087, 11103,
     11126, 11143, 11219, 11222, 11252-3, 11295, 11326, 11347, 11646, 11650, 11727,
     11733-4, 11775, 11804, 11806, 11830, *11943*, *B11944*, 11945-6, 11991, 11997, 12010

Canopy structure and density (*cf.* also Foliage formation and heterogeneity) 9099,
     9117, 9553, 9571, 9593, 9664, 9720-1, 9739-41, 9768, 9815, 9869, 10174-5, 10345,
     10452, *10586, *10599, 10646, 10794, 11038, 11090, 11219, 11222, 11253, 11295,
     11326, 11333, 11357, 11378, 11646, 11667, 11727, 11804, 11806, 11943, B11944,
     11945-6, 11991

Canopy temperature see Temperature of leaf and canopy, measurement

Carbon-14 see ^{14}C

Carbon dioxide see CO_2

Carbon fixation pathways (cf. also C_3 and C_4 plants ...; Enzymes of carbon fixation;
    Photosynthates formation patterns; Photosynthetic bacteria, carbon fixation
    pathways) 9146-7, 9156-7, 9204, 9208, *9215, 9216-9, 9266-8, 9306-7, 9325-7,
    9363-7, 9383, 9489, *9499, 9556, 9584-6, 9596-7, 9599, 9703, 9704, 9705, 9706-7,
    9760-2, 9798, 9802-3, 9817, 9908, 9962, 9964, 9966, 9975, *9981, 10039, 10041,
    B 10052, 10126, 10135-6, B10137, 10186, 10266, *10267, 10268, 10317, 10323-4,
    10564, 10597, 10622-3, 10708, 10722, 10725, 10734, 10826, 10845-7, 10920, 11000,
    11093, 11158, 11435, 11469, 11489, 11600, 11603, 11614-5, 11660, 11726, 11761,
    11816, 11922, 11929, *11930, 11947-8, 11974-5

Carbonic anhydrase 9206, *9246, 9538, 9802-3, 10038-9, 10041, 10311, 11152

Carboxydismutase see Ribulose-1,5-diphosphate carboxylase

Carboxylation see Carbon fixation pathways

Carboxylation resistance (cf. also Resistance to CO_2 and water vapour transfer) 9099,
    10311, 10688, 10708, 11246

Carotene see Carotenoids

Carotenoids biosynthesis *9392, 9460-1, 9497, 10025-7, 10467, 10720, *10824, 10825,
    11305, *11951

Carotenoids, chemical structure (cf. also Carotenoids biosynthesis) 9483, 9716, 9866,
    10171, 10654-5, *10824, 10825, 11175, 11337-8, 11689, *11705, 11952-4

Carotenoids degradation see Pigment degradation

Carotenoids, effect of air humidity on see Humidity ...

Carotenoids, effect of antibiotics on see Antibiotics, effect on pigments

Carotenoids, effect of CO_2 on see CO_2 and bicarbonates, effect on pigments

Carotenoids, effect of growth regulators on see Growth regulators, effect on carote-
    noids

Carotenoids, effect of ionizing radiation on see Ionizing radiation (gamma, X), effect
    on pigments

Carotenoids, effect of irradiance on see Irradiance, effect on carotenoids

Carotenoids, effect of leaf and plant age on see Age of tissue and plant, effect on
    carotenoids

Carotenoids, effect of mineral elements on see Mineral elements, effect on carotenoids

Carotenoids, effect of oxygen on see Oxygen, effect on pigments

Carotenoids, effect of pesticides on see Pesticides, effect on pigments

Carotenoids, effect of temperature on see Temperature, effect on carotenoids

Carotenoids, effect of water supply on see Water supply and pigments

Carotenoids genetics see Genetics of pigment content

Carotenoids in algae see Algae carotenoids

Carotenoids in lipoprotein complexes see Carotenoids in vivo

Carotenoids in model systems see Carotenoids in photosynthesis ...

Carotenoids in photosynthesis and model photosynthetic reactions 9241-3, 9302-3, 9331,
    9484, 9619, 9680, 9729, 9881, 9955, 9970, 10182, 10301, 10550, 10611, *10806,
    10812, 10889-91, 10994, *11032, 11061, 11225, *11227, 11228, 11388-91, 11418-20,
    11461, 11497, 11501, *11595, 11691-2, 11712, 11796, 11802, 11864, 12029-30

Carotenoids in photosynthetic bacteria see Photosynthetic bacteria, carotenoids in

Carotenoids in seeds see Seed, pigments in

Carotenoids in tissue culture see Tissue culture, pigments in

Carotenoids *in vivo*   10408, 11285-6, 11906-7

Carotenoids, methods of analysis *9236, 9247-8, 9278, *9644, 9667, *9696, 9716, 9866,
    10061, *10250, 10320, 10408, 10486-7, 10610, 10615, 10665-6, 10792, 10802, 10852,
    11175, 11406, 11414-5, 11443, 11490, 11496, *11534, *11550*, 11556, *11565*, 11669,
    11677, 11684, 11747, 11790, 11839-42, 11880, 12031, 12040

Carotenoids, pathological effects on see Pathological effects on pigments

Carotenoids, seasonal changes *9131, *9279, *9623, 9675, *9766, *9918, 9977, *9989,
    *10997, *11191-2, 11193, *11232, *11235, 11415, 11490, 11708

Carotenoids, transformations of see Carotenoids in photosynthesis ...

Chamber, assimilation 9203, 9413, 9552, 9942, 10062, 10190, 10271, 10280, *10312*,
    *10413, *10499, 10616, 10767, 10906, 10937, *11081, 11469, 11830

Chemiosmotic hypothesis, proton transport 9133, 9718-9, 9588, *9697*, 9870, 9883,
    **10056*, 10145-6, 10252, 10351, 10387, 10667, 10723, 10725, 11003, 11009, *11034-
    5, 11057, 11264, *11296-7, 11298-302, 11361, 11387, 11479, *11486*, 11546-7, 11599,
    *11617, 11618, 11676, 11782

Chlorophyll and plankton production see Plankton standing crop ...

Chlorophyll biosynthesis (*cf.* also Irradiance, effect on chlorophyll) 9111, 9114-5,
    9140, 9166, *9181*-2, 9256, 9285-6, 9399, 9479, 9539, *9604, 9692, 9784, 9921-2,
    *9923, 9929, 9994, 10044, 10050-1, 10067, 10120, 10148, *10162, 10163, 10339-
    40, 10360-1, *10363, 10400, 10403, 10458-60, *10462, 10534, *10580, 10614, 10672
    3, 10697, 10719, 10814, 10955, 11024, 11047-8, 11071, 11073, *11075-7, 11078,
    11100, 11206, *11307-8, 11309-11, *11312-3, 11323, 11359, *11360, *11385, 11457,
    11462-4, 11487, 11533, *11563-4*, 11566-70, 11581, 11709, 11756, 11776-9, 11960,
    11977

Chlorophyll(s) chemical structure 9463, *9487, 9488, 9930, 10105, *10164, 10211, 10697

Chlorophyll, *Chlorobium* see Bacteriochlorophyll ...

Chlorophyll degradation see Pigment degradation

Chlorophyll delayed light emission see Chlorophyll fluorescence ...

Chlorophyll derivatives see Chlorophyll(s) chemical structure; Chlorophyll(s),
    methods of analysis; Pigment degradation; Chlorophyllase

Chlorophyll dimers see Chlorophyll-lipoprotein complexes ...

Chlorophyll, diurnal variation 12062

Chlorophyll, effect of air humidity on see Humidity ...

Chlorophyll, effect of antibiotics on see Antibiotics, effect on pigments

Chlorophyll, effect of CO_2 on see CO_2 and bicarbonates, effect on pigments

Chlorophyll, effect of growth regulators on see Growth regulators, effect on chloro-
    phyll

Chlorophyll, effect of ionizing radiation on see Ionizing radiation (gamma, X), effect
    on pigments

Chlorophyll, effect of irradiance on see Irradiance, effect on chlorophyll

Chlorophyll, effect of leaf and plant age on see Age of tissue and plant, effect on
    chlorophyll

Chlorophyll, effect of mineral elements on see Mineral elements, effect on chlorophyll

Chlorophyll, effect of oxygen on see Oxygen, effect on pigments

Chlorophyll, effect of pesticides on see Pesticides, effect on pigments

Chlorophyll, effect of photoperiod on see Photoperiod and chlorophyll

Chlorophyll, effect of temperature on see Temperature, effect on chlorophyll

Chlorophyll, effect of water supply on see Water supply and pigments

Chlorophyll energetics in model systems (*cf.* also Chlorophyll fluorescence ...; Chloro-
    phyll monolayers...)   9184, [9251-2, 9301, 9390, 9744, 9805-8, 9995, *10002,

Chlorophyll unit see Photosynthetic unit

Chlorophyllase 9257, *9377-8, 9570, 9784, 10835-6, *11028, 11851, 11919

Chloroplast (and chromatophore) chemical composition 9124, 9148-9, 9169, 9225, 9232-
    3, 9267, 9269, 9271, *9315-6*, 9337, 9356, 9361, 9445-6, 9479, 9503, 9510, *9568,
    9601, 9608, 9711, 9778, 9788, 9845, *9978*, 10026-7, 10059, 10165, 10179, 10188-9,
    10339-40, 10373-4, 10431, *10462, 10463, 10497, 10537, 10598, 10641, 10719,
    17740-1, 10749, 10789, *B10790*, 10848-9, 10871, 10901, *10932-3, 10941, 10944,
    10981-2, 11024, 11066, 11213-5, 11233, 11270-1, 11287, 11363, 11417, 11442,
    *11477-8, 11479, 11491, 11498, 11507, 11509, 11529-30, 11562, 11583, 11590,
    11609, *11611, 11674, 11704, 11756, 11858, 11867-9, 11889, 11914, 11981, 11988,
    12022

Chloroplast development see Chloroplast ontogenesis ...

Chloroplast dimensions see Chloroplast number ...

Chloroplast (and chromatophore) fragments 9179, 9346, 9382, 9476, 9517, 9581, 9613,
    9626, 9730, 9789, 9871, 9873-4, 9883, 9891, 9907, 9925, 10018, 10091, 10209,
    10260, 10263, 10302, 10348, 10409, 10661, 10761-2, 10830, *10932-3, 10934-5,
    10950-2, 11007-8, 11059, 11114, 11318, 11371, *11553*, 11566-70, 11675, 11750,
    11877, 11962, 11974, 11977, 11988, 12059

Chloroplast genetics see Genetics of photosynthetic apparatus

Chloroplast (and chromatophore) isolation 9192, 9231, 9293, 9320, 9359, 9374, 9451,
    9477, 9562, 9761, *9841, 10042, 10066, 10287, *10288, 10289, 10317-8, 10528,
    10620, 10648, 10734, 10829-30, 11022, 11033, 11068, 11070, *11101*, *11132-3*,
    11673, 11721, 11787, 11863, 11877, *11921*, 11947

Chloroplast membrane transport 9267, 9349, 9757

Chloroplast movement 9858, 9865, 9932, 10138-9, 10359, *10823*, 11010, 11472-3, 11504

Chloroplast mutation see Genetics of photosynthetic apparatus

Chloroplast number and dimensions 9517, 9634, 9661, 10366, 10671, *11232, 11234,
    *11235, *11319, *B11799, *11848

Chloroplast ontogenesis, replication, development (*cf.* also Chloroplast structure)
    9160, 9269, 9386, 9398-9, 9442, 9444, 9501, 9507, 9539, 9608-9, 9707, 9736,
    9801, 9895, 9967, *10019, 10040, 10067, 10089, 10131, *10162, 10188, 10215,
    10266, 10339-40, 10366, 10373, 10439, 10614, 10719, 10735, 10814, 10853, 10871,
    10887-8, 10901, 11024, 11047-8, 11057, *11313, 11356, *11385, 11405, 11481,
    11487, 11489, 11674, 11850, 11853, 11866, *11904, 11960, 11981, 12005, 12052

Chloroplast replication see Chloroplast ontogenesis ...

Chloroplast shrinkage see Chloroplast volume changes

Chloroplast structure (*cf.* also Chloroplast ontogenesis ...) 9122-3, 9141, 9145-7,
    9155, 9178-9, 9225, 9227, 9261, 9291-2, 9307, 9326, 9337, 9349, 9352-7, 9376,
    9386, 9398, 9425, *9532, 9547-8, 9565, *9613, 9642, 9653, 9668, 9763, 9801,
    9804, 9860, 9865, 9870, 9883, 9886, 9920, *9948, *B9961*, 10016-8, *10019, 10020-
    3, 10048, 10050-1, *10077-8, 10099, *10102, 10119, 10123, 10125-6, 10131-2,
    10147, 10154-5, *10162, 10187, 10209, *10231, 10251, 10342, 10352, 10430, 10438,
    10441, *10462, 10463, 10468-9, 10498, 10526-7, 10564, 10576, 10597, 10617, 10633,
    10675, 10719, 10723, 10734-5, *10784, *10807, 10827, 10853-4, 10879, 10887-8,
    10901, *10926*, 10944, 10946-7, 10983, 11011, 11021, 11024, 11095, 11100, *11134-5*,
    11204, 11216, 11220, 11231, 11260, 11304, 11325, 11332, 11336, *11339, 11356,
    11382, 11402-3, 11405, 11412, 11442, 11447-8, 11481, 11489, 11494, 11507,
    *11510, 11514, 11532, 11552-3, 11571, *11585, 11598, 11608, 11655, 11719, 11742-
    3, 11755, 11789, 11805, 11809, 11847, 11857-8, 11866, 11878, 11885, *11904, 11936,
    11966-7, 11976, 12005, 12056

Chloroplast volume changes 9932, 10287, *10288, 10289, 10359, *10823*, 11467

Chromatography of carotenoids and chlorophylls see Carotenoids, methods of analysis;
    Chlorophyll(s), methods of analysis

Chromatophore see also Chloroplast (and chromatophore) chemical composition; C
    (and c.) fragments; C. (and c.) isolation

Diffusion (diffusive) resistance see Resistance ...

Distribution of photosynthates see Photosynthates, translocation, exudation and distribution

Drought resistance see Water supply ...

E

Ear photosynthesis see Fruit photosynthesis

Ecotypes, differences in photosynthesis *9283, 9317-8, 9363-8, *9373*, 9748-9, 10478, 10566, *10687, 10710, 10863-4, 11482, *11963, 11983-4, 12015

Electric field, effect on chloroplast membrane 9173-4

Electron spin resonance, NMR, paramagnetic resonance 9806, 9828, 9897, 9986, 10322, *10365, 10368, 10371, 10381, 10386, 10443, 10543, 10763, 10976, 11039, 11211, 11328, 11785, 11969

Electron transport chain (*cf.* also Cytochromes, Ferredoxin ..., NADP ..., P700 ..., Plastocyanin, Quinones ...) *9133*, 9135, 9141, *9175-7*, 9192, 9200, 9210, 9308, 9313-4, 9316, 9354-5, 9357, *9358*, 9383-6, 9397-8, 9400-1, 9433, 9504, 9506, 9521, 9587, 9611, *9612, 9613-5, 9654, 9670, *9678*, 9733-4, 9794, 9859-60, 9871-5, 9883-4, 9888, 9909, 9974, 10005, 10021, 10035-6, *B10052*, 10089, 10092, 10098, *10100, 10101, 10140, 10143-5, 10179, 10184-5, 10193-5, 10197, 10216, *10217, 10252, 10260-3, 10284, 10303, 10326-7, 10351, 10387, 10402, 10410, 10450, *10494, 10527, 10537-8, 10544, 10563, 10564, 10578, 10632, 10650, 10705, 10722-3, 10725, 10737, 10764, 10781, 10826-7, 10831, 10855, 10885, 10894, 10919, 10931, 10946-52, 10964, 10976, 11007-8, 11054, 11120-1, 11133, 11179, 11181, 11203, 11256, *11262, B11263, 11264, 11267, 11269, 11272-3, 11284, 11292, 11342, 11366, 11370, 11388-93, 11434, 11480, *11506, 11572, 11582, 11608, 11625, 11659, 11679, 11694-7, 11703, 11750, 11827, 11829, 11856, *11878*, 11895, 11902-3, 11914, 11937, 11939-40, 11971, 11974, 11979, *12008, 12014, 12029-30, 12034-5, 12067

Electron transport chain components, methods of analysis 9254, 9262-4, 9304, 9359, 9627, 9926, 10401, 10882, 10885, *10960, *11558*, 11591, 11610, 12025

Electron transport chain in photosynthetic bacteria see Photosynthetic bacteria, electron transport chain

Emerson effect 10368, *10964*, 11256

Energy balance of canopies and cultures (*cf.* also Aerodynamic methods, theory; Bioclimatological methods) 9117-8, 9478, 9486, 9576, *9664*, 9914, 9931, 11064-51, 11283, 11798

Energy budget method see Bioclimatological methods

Energy trap pigment see P 700 ...

Environmental and internal factors affecting photosynthesis see Age of tissue and plant ...; Altitude and pressure ...; CO_2 and bicarbonates ...; Genetics of photosynthesis; Humidity of air ...; Inhibitors ...; Ionizing radiation ...; Irradiance ...; Leaf properties (anatomical *etc.*) ...; Mineral elements ...; Oxygen...; Pathological effects on ...; Photoperiod ...; Temperature ...; Water and osmotic potential of cell and tissue ...; Water supply ...; Wind ...

Enzymes of carbon fixation (*cf.* also Carbon fixation pathways; PEP carboxylase; Photosynthates formation patterns; Ribulose-1,5-diphosphate carboxylase) *9143, 9148-9, 9157, *9215, 9216-9, *9234-5, 9307, 9326-7, 9482, 9485, 9556, 9574, 9599, 9762, 10127, 10266, 10323, *10375, 10479, 10688, 10704, 10969, 10993, 11489, 11528, 11660, 11788, 11810

Enzymes of carbon fixation, methods 9214, *9234-5, 9237, 9311, 9482, 9485, 9800, 10011, 10405, 10910, 10927, 11130, *11341, *11380, 11701, *11989*

Enzymes of chlorophyll biosynthesis see Chlorophyll biosynthesis

Enzymes of electron transport chain and photophosphorylation (*cf.* also Electron transport chain; Photophosphorylation) *9403, 9875, 9884, 9888, 10219, *10414, 10632, 10661-2, 10681, 10827, 10829-31, 10931, 11154, 11287-90, 11361-3, 12033

Enzymes of electron transport chain and photophosphorylation, methods 9906, 10223-4,
      10350, 10662, *10829*, *11243*, 11288, *11334*

Enzymes of photophosphorylation 10464-5

Evolution of photosynthesis see Phylogenesis of photosynthesis

Excretion of photosynthates see Photosynthates, translocation, exudation and distribu-
      tion

Exposure chamber see Chamber, assimilation

Exudation of photosynthates see Photosynthates, translocation, exudation and distribu-
      tion

F

Ferredoxin isolation and determination 9263, *9481*, 9614, 9916, 10179, 10883-4

Ferredoxin-NADP (NAD)-reductase (*cf.* also Electron transport chain) 9169, 9519,
      10971-2, 11272-3, 11558, 11827

Ferredoxin, role in photosynthesis (*cf.* also Electron transport chain) 9169, 9396-7,
      9468, 9482, 9518-21, 9614, 9722-3, 9727, 9798, 9827-8, 9835, 9888, 9897, 9916,
      9932, 9941, 9986, 10098, *10100*, 10179, *10321, 10322, 10423-4, 10564, 10763,
      *10813, 10880, 10883-4, 10931, 10971-2, 11129, 11165, 11211, 11292, *11506,
      *11700, 11959

Field studies of photosynthesis see Photosynthesis and dry matter production

Flavodoxin 9627, 12067

Flooding see Rain ...

Flowers, photosynthesis and chlorophylls in 10165, 11748

Fluorescence, methods of measuring see Spectrophotometric ...

Foliage formation and heterogeneity (*cf.* also Canopy structure and density) 9108,
      9319, 9869, 10054-5, *10345, 10561, *11232, 11233, 11357, 11586, 11859-60

Fraction I protein see Ribulose-1,5-diphosphate carboxylase

Fruit photosynthesis 10304, 10557, 10881, *11174, 11543

Fucoxanthin see Algae carotenoids; Carotenoids ...

Fungi see Pathological effects ...

G

Gasometric methods, systems and circuits (*cf.* also Chamber, assimilation; CO_2 meas-
      urement ...; infra-red gas analyser ...; Water-plant photosynthesis, methods) 9107,
      9203, *9223, 9453, 9469, 9641, *9824, *10080, 10124, 10208, 10280, *10300, *10309*,
      10312, *10413, 10492, 10515, *10540-1, *10570, *10606-7, 10616, 10684, 10714, 10911,
      10937, 11029-30, *11045, 11139, *11207, 11352, 11439, 11483, 11525, *11573, 11602,
      11668, 11754, 11888, 11957, 12027

Genetics of photosynthesis (*cf.* also Ecotypes, differences in photosynthesis) 9123,
      9129, 9164, *9193, *9212, 9123, *9215, 9216-21, 9228, 9305, 9307, *9334, 9359-
      61, 9364-8, 9383, 9385, 9526, 9551, 9574, 9616-7, *9622, 9639, 9695, 9700, 9789,
      9837, 9898, *9900-1, 9902-3, 9956-8, 9987, *10028, *10034, 10062, 10113, *10114,
      10115, *10141, 10150, 10152, 10177, 10184, 10269-70, 10275, 10325, *10329, 10356,
      10369, *10372, 10396-8, *10399, 10404, 10469, 10498, *10510, 10516, 10566, *10586,
      *10595, 10630, 10650, 10663, 10703, 10767, 10771-3, 10774, *10820, 10833, 10922,
      10928-30, *10959, 10968-9, *B10980*, 10985, 10989, *10996, 11002, 11004, *11016,
      11017, 11057, *11116, 11117, 11157, 11247, 11303, 11351, 11378-9, 11424-5, 11440,
      11455-6, 11505, *11536, 11549, 11636, 11652, 11710, 11719, 11732, 11751, 11810,
      11835, 11895, 11949, 11964, 11980, 11987, 11995, 12024

Genetics of photosynthetic apparatus 9307, 9424-5, 9634, 9660, 10021-3, 10093, *10170*,
      10425, *10462, 10463, *B10980*, 10982-3, 11024, *11137, *11339, *11514, 11532,
      11719, 11809, 11847, *11848, *11936*, 11949, *11964*, 11981, *12009*

Mixotrophy see Heterotrophy, autotrophy, mixotrophy

Model of leaf and canopy photosynthesis (*cf*. also Resistance to CO_2 and water vapour
       transfer) 9127, 9537, 9549, 9606, 9624-5, 10096-7, 10452, 10502, 10603-4, 10646,
       10665, 10688, 10708, 10794, 10902, 11139, 11246, 11253, *11262, 11409, 11476,
       11650, 11950, *11991*

Mutagens, induction of mutants (*cf*. also Genetics of photosynthesis; G. of photosyn-
       thetic apparatus; G. of pigment content) 11291

Mutation see Genetics of photosynthesis; G. of photosynthetic apparatus; G. of
       pigment content

N

NADP (NAD) photoreduction (*cf*. also Electron transport chain) 9354, 9397, 9898, 10063,
       10754, 10761, 10763, *10913, 10931, 10947-8, 11240, 11375, 11413, 11608, 11919

Neoxanthin see Carotenoids ...

Net assimilation rate see Growth analysis

Nuclear magnetic resonance see Electron spin resonance ...

O

Optical properties, leaf (*cf*. also Chlorophyll *in vivo*; Irradiance ...) *9255, 9527-
       8, 9697, 9701, 9932, *9933-4, 9935-7, *9938, 10085, 10518, 10549, 10766, 10925,
       11140, 11587, 11647-8, 11767, 11906, 12012

Oscillations in photosynthesis 9554

Osmotic potential see Water and osmotic potential ...

Oxygen analysers and methods (*cf*. also Oxygen electrode) 9814, 10060, 10106, *10309*,
       10422, *10642, 10938, *10940, *10974, 11029

Oxygen, effect on photosynthesis (*cf*. also Warburg effect; Photorespiration) 9327,
       9363-4, 9366-8, 9373, *9406, 9490, 9494-5, 9684, 9861, 10097, 10107, 10151,
       10538, 10651, 10710, 10716, 10840, 11015, 11160, 11238, 11409, *11627, 11838,
       11843, 11875, 11896

Oxygen, effect on pigments 10385, 11418, 11475

Oxygen electrode 9452, 9530-1, 9544, 9658, 10130, 10667, 10670, 11072, *11245, 11277

Oxygen evolution mechanism and kinetics 9297, 9557-8, 9681, 9867, 10144, 10262, *10326-
       7*, 10577-8, 10855, 10967, 11269, 11317, 11322, 11324, 11393

Ozone see Pollution of air ...

P

P700, P750, P870, P890, *etc*. 9391, 9533, 9677, 9859, 9871-4, 9880, 10092, 10193-5,
       10216, 10410, 10761-2, 10796, 10827, 10952, 11375, 11458, 12037

P700, P750, P870, P890, *etc*., methods *10796*

Paramagnetic oxygen analyser see Oxygen analysers and methods

Paramagnetic resonance see Electron spin resonance ...

Parasitism (*cf*. also Heterotrophy ...; Pathological effects ...) 10158-9, 10942, 11592

Pathological effects on photosynthesis 9545, 9662, 9724, 9818, 9894, 9951, 10161,
       *10217-8, 10269-70, 10368, 10572, 10987, 11125, 11155, *11274, 11329, 11579

Pathological effects on pigments 9570, 9779, *9934, 10117, 10133-4, 10269-70, 11592,
       *11746

PEP carboxylase (*cf*. also Enzymes of carbon fixation) 9326, 9431, 9584, 9599, 9762,
       10426, 10704, 10927 11130, 11157, 11178, 11915, 11929, *11930, 11948

Peroxisome, glyoxysome (*cf.* also Glycolate metabolism) *9289*, 9887, 10186, 10349, 10803, *11014*, 11132, 11782-4, 11811

Pesticides, effect on photosynthesis see Inhibitors ...

Pesticides, effect on pigments *9377-8, 9497, *9637, 9694, 9708, *10555, *11028, 11099, 11150, 11186, 11254, 11323, 11359, 11399, *11518

pH, effect on photosynthesis 9205, 9685, 9808, 10038-9, 10528, 10934, 11361, 11526, 12063

pH, effect on pigments 11588

PhAR see Irradiance ...; Radiation regime in canopy

Pheophytin see Pigment degradation

Phosphodoxin 9375

Phosphoenolpyruvate carboxylase see PEP carboxylase

Phosphorescence, methods of measuring see Spectrophotometric ...

Photoperiod and chlorophyll *9279, *9992, 10040, 10579, 11128, 11332

Photoperiod and photosynthesis 9164, 9213, 9274, *9280, 9665, 10579, 11983

Photophosphorylation 9113, 9123, 9133, 9145-7, 9177-9, *9209*, 9127, 9224-5, *9260, *9313*-4, 9328, 9395, 9401, 9433, 9450, 9517, 9542, 9559, 9630, 9649, 9775, 9870, 9876-7, 9884, 9911, 9917, 9932, *9963*, 9975, *9981, 9982, *10004, 10005, *10006-7, 10015, 10036, *10056*, 10063-5, 10068, 10070, 10076, 10091, 10140, 10146-7, *10200, 10201, *10219, 10252, 10291-2, 10371, 10387, 10470, 10474, 10537, 10544, 10618, 10638, 10661-3, 10668, 10678, 10681, 10724, 10755, 10830-1, 10844-5, 10861-2, *10913, 10914, 10928-30, 11003, 11007-8, 11049-51, 11057, 11061, 11121, 11162, *11172, 11298-9, 11318, 11324, 11361-3, 11386-92, 11400-1, 11417, 11424-5, 11436, *11470, *11486*, 11504, 11546-7, 11676, 11694, 11695, 11728, *B11799, 11823, 11838, *11986, *11992*, 12023-4, 12033, 12038, 12066

Photophosphorylation, methods 9224, 9395, 10076, *B10461, 10596

Photoreduction 10578, 11694, 11696-7, 11940

Photorespiration (*cf.* also Glycolate metabolism; Oxygen, effect on photosynthesis; Warburg effect) 9158, 9203-4, 9289-90, *9341, 9439, 9490, 9495, 9505, 9526, 9549-50, 9574, 9600, 9684, 9724, 9887, 10096-7, *B10137*, 10143, 10151, 10187, 10234, 10256, 10330, 10349, 10464-5, 10524, 10564, 10602, 10715, 10936, 11015, 11055, 11107-10, 11160, 11238, *11257, 11287, 11404, 11408-9, 11604, 11792-5, 11817, *B12054*

Photorespiration, methods 9290, *10709*, 11408-9

Photosynthates, effect on photosynthesis see Sink, source of photosynthates

Photosynthates formation patterns (*cf.* also C₃ and C₄ plants....; Carbon fixation pathways; Glycolate metabolism) 9093-4, 9229, 9276, 9288, 9342, *9343, 9432, 9449, 9515, *9555, 9556, 9567, 9597, 9610, 9684-5, 9736, 9791-2, 9811, 9815, *9834, 9846, 9853, *10007, .10008, 10136, B10137, 10209, *10218, 10228, *10229-30, 10266, *10267, 10268, 10272, 10277, 10281-3, 10328, *10329, *10353-5, 10373-4, 10412, 10432, 10440, 10442, 10447, *10482, 10514, 10564, *10622*-3, 10651, 10689, 10744, 10752, *10771-3, 10774, *10822, 10841, *10900, 10906, 10916, *10958-9, *10977, 10983-5, 11027, *11044, 11074, 11149, 11162, 11169, 11200, 11259, 11265, *11274, 11276, 11364, 11399, 11411, 11435, 11468-9, 11509, 11530, 11535, 11541, 11579, *11620, 11637-8, 11651, 11681, 11710, 11715, 11736, 11763, 11883, 11896, 11908, 11923, *11927, 11932, 11934, 11955, 11975, 11982, 11984

Photosynthates, methods of determination 9220-3, 9652, 10210-1, 10743

Photosynthates, sink, source of see Sink, source of photosynthates

Photosynthates, translocation, exudation and distribution (*cf.* also Photosynthesis and dry matter production) 9136, 9161, 9228, *9351, 9380, *9421, 9422, 9524, 9259, 9645, *9719, 9747, 9812, 9818, *9842, 9843-4, *9950, 9976, 10103, 10112, 10146, 10154-5, 10196, 10228, *10229-30, 10233, 10435, 10493, 10573, 10652, 10676, 10703, 10715, 10752-3, 10757-8, 10798, 10898, *10958-9, 10978, 10984, 11104, *11156, 11248-9, 11303, 11355, 11428, 11482, 11505, 11541, 11579, 11629,

Photosynthetic unit 10036, 10721, 10965, 11114, 11456, 11621-5, 11958, 11980

Photosynthetically active radiation see Irradiance ...; Radiation regime in canopy

Photosystem I 9270, 9391, 9445-6, 9475, 10018, 10193-5, 11039, 11547, 11572, 11695, 11902

Photosystem II 9190, 9305, 9361, 9443, 9557-8, 9658, 9707, 9789, 9859, 10252, 10261-3, 10284, 10326-7, 10480, 10644, 10705, 10761-2, 10764, 10834, 11624, 11962, 11979

Photosystems (*cf.* also Electron transport chain) 9146-7, 9175-7, 9233, 9284, 9391, 9417, 9472, 9546, *9580*, 9613, 9643, *9932*, 10016, 10036, 10185, 10197, 10216, 10260, 10388, 10481, 10563, 10567, 10706, 10826-7, 10946-8, 10951-2, 11059, *11172, 11256, 11412, 11501-2, 11512, 11625, 11703, 11782, *11878*, 11988

Phycobilins see Biliproteins

Phycobiliproteins see Biliproteins

Phycobilisome 9763, 11976

Phycocyanin see Biliproteins

Phycoerythrin see Biliproteins

Phylogenesis of photosynthesis 9317-8, *9703*, 9798, *9908*, 10098, 11613, 11816

Phytoflavin 11610

Phytol see Chlorophyll biosynthesis

Pigment degradation (*cf.* also Chlorophyllase) 9090-1, 9093, 9105, 9207, 9298, 9320, *9377-9, 9570, *9666, *9752, 9759, *9773, 9784, 9787, 9796, 9805, 9924, 9930, 10116, 10346, 10382, 10435, *10449, 10453, 10489-90, 10612, *10626, 10720, 10756, *10760, 10769, *10783, 10835-6, 10943, 10970, *11028, 11043, *11056, *11067, 11100, *11164, *11184, 11286, *11327, *11376-7, 11414-6, 11529, *11555, 11557, 11574, *11746, 11801-2, 11844-5, 11851, 11891-3, 11985, 12031, 12049, 12064

Plankton standing crop and chlorophyll 9112, 9125, *9144, 9162, *9189, *9321, *9429, *9493, 9836, 9944, 10178, 10357-8, 10523, 10551, 10625-6, 10769, 10787, 11145, 11261, 11330, 11396, 11452-3, 11607, 11631, 11758, 11768, 11801, 11837

Plastocyanin (*cf.* also Electron transport chain) 9210, 9270, 9871-2, 10140, 10401, 10564, 10660, 10952, *10960, 10961, 10976, 11434, 11608

Plastoquinone see Quinones in photosynthesis

Pollution of air, effect on photosynthesis 9427, 9459, 10180-1; 10418, 10878, 11027, *11199, 11340, 11982

Pollution of air, effect on pigments 9172, *10770, 10877, 11708

Porometer (*cf.* also Stomata; Resistance to CO_2 transfer, stomatal) 9116, 9726, *10313, 10404, 10797, 10917-8

Porometer, diffusion (water vapour, nitrous oxide) 9726, *10313, 10917-8

Porometer, viscous 9116, *10313, 10404, 10797

Prediction of canopy productivity see Productivity calculation and prediction, canopy

Pressure, effect on photosynthesis see Altitude and pressure, effect on photosynthesis

Production of dry matter measurement, methods of (*cf.* also Gravimetric methods; Photosynthesis and dry matter production) 9107, 9152, *9211, 9295, 9350, *9541, 9840, *9900-1, 9902-3, 10174, 10246, 10560, *10590-3*, 10751, *B10872*, *11006, 11089, *11091*, 11606, *11662, 11663, 11913, 11920, 11972

Production, primary see Photosynthesis and dry matter production

Productivity calculation and prediction, canopy (*cf.* also Model of leaf and canopy photosynthesis) 9647,

Productivity of algae cultures 9740, 9750

Tissue culture, pigments in 9642, 9754-5, 9787, 9790, 10403, 11443, 11462-3

Tissue culture, photosynthesis in   10112, *10217-8, 10588

Translocation of photosynthates see Photosynthates, translocation, exudation and dis-
    tribution

Transpiration and photosynthesis (*cf.* also Stomata; Stomata and photosynthesis; Resis-
    tance to CO$_2$ transfer, stomatal; Water supply and photosynthesis) 9121-2, 9138,
    9164, 9282, 9469, 9515, 9551, 9624, 9724, 9753, 9896, 9913, 9957, 9983, 10110,
    10158-9, 10161, 10203, *10306, 10316, 10393, 10418-20, 10501, 10545-6, 10548,
    10603-4, 10616, *10819, 10904, *10986, 10987, 10989, 10993, *11081, 11155, 11158,
    11160, 11208, 11483, 11602, 11630, *11670, 11807, *11963

Tricarboxylic acid cycle see Carbon fixation pathways; Photosynthates formation patterns

U

Ubiquinone see Photosynthetic bacteria, electron transport chain; Quinones in photo-
    synthesis

Uncouplers of photosynthesis see Inhibitors and uncouplers ...

V

Vegetation changes in photosynthesis see Photosynthesis, seasonal course

Violaxanthin see Carotenoids ...

Vitamins K see Electron transport chain; Quinones in photosynthesis

Volume of photosynthetic organs see Production of dry matter measurement, methods of

Volume of plant organs, measurement see Leaf area and volume measurement

Volumetric methods see Manometric and volumetric methods

W

Warburg effect (*cf.* also Oxygen, effect on photosynthesis; Photorespiration) 10710

Water and osmotic potential of cell and tissue, effect on photosynthesis 9437-9, 9509,
    9554, 9594, 9714, 9725, 9953, 9956, 10045-6, 10161, 10190-1, 10203, 10258, 10549,
    10636, 11015, 11203, 11955

Water deficit see Water supply ...; Water and osmotic potential ...

Water-plant photosynthesis, methods (*cf.* also Gasometric methods, systems and cir-
    cuits; Infra-red gas analyser ...; CO$_2$ measurement ...) *9188, *9470

Water splitting mechanism see Oxygen evolution mechanism and kinetics

Water supply and pigments *9097, 9430, 9528, 9781, 9935-7, *10205, *10315, 10576,
    10677, *10817, *11192, 11193, *11194, 11195, 12062

Water supply (deficit, soil moisture, drought resistance, xerophytes) and photosyn-
    thesis (*cf.* also Water and osmotic potential of cell and tissue, effect on
    photosynthesis) 9101, 9318, *9322, 9363, 9478, *9522, 9753, *9764, 9942,
    9958, 9996, 10124, 10158-9, 10190-1, 10239, 10241, 10281, 10286, 10290, *10306,
    10393, 10404, 10518, 10562, 10569, 10605, 10639-40, 10678, 10700, 10897, 10902,
    10993, 11015, 11017, 11063-5, 11126, 11160, 11258, 11314, 11347, 11351, 11398,
    11482, 11537, 11603, 11616, 11629-30, *11670, 11672, 11791, 11800, 11890, 11991,
    12018-9, 12062

Wind (air-flow rate), effect on photosynthesis 10271, 10613, 11139, 11441

X

Xanthophyll see Carotenoids ...

Xerophytes see Water supply ...

This index contains a selection of plant genera and types interesting as experimental material for ecophysiological, ecological and agricultural studies. In general,
mainly those plant names have been included which are given in the title of the respective papers or in abstracts. The common English plant names are the main items which
present the reference numbers.

A

Abies see Fir

Acer see Maple

Alder *9541, *10999, 11950

Alfalfa 9303, 9518, 9862, 9986, 10054, 10487, 11001, *11156, 11191-2, 11209-10, 11285,
11290, *11327, *11670, *B11799, 11978, 11997-8, 12021

Algae (*cf.* also *Chlamydomonas, Chlorella,* Diatoms, *Euglena, Scenedesmus*)
9128, *9144, 9162, *9188-9, 9230, 9254, 9257, 9278, *9321, 9345-7, 9352, 9395,
9397, *9406, 9465, 9470, *9493, 9494, *9496, 9502, 9557, 9589, *9650-1, 9692-4,
*9710, 9715, *9728, 9793, 9864, 9889, 9905, 9912, 9920, 9927, 9944, 9965, 9985,
10042, *10043, 10053, 10098, *10122, 10197, *10236, 10344, 10357-8, 10362, 10523-
4, 10551, 10581, 10583-4, *10625-6, 10631, 10643, 10652, 10787, 10795, *10808,
10828, 10851, 10856, *10870, 10909, 10931, *10977, *11037, 11053, 11123, 11131,
11159, 11170, 11181, 11280, 11346, 11348-50, 11396, 11452-3, 11456, 11475, 11493,
11496, 11519, 11607, 11609, 11626, *11627, 11632, 11685-6, *11687, 11694, 11719,
11723, 11758, 11766, 11768, 11774, 11795, 11801, 11837, 11871, 11875, 11931-4,
11980, 12025, 12066

Algae, blue-green 9134, 9298, 9309-10, *9426, 9440, 9443, 9475, 9558, 9575, 9643,
9677, 9681, 9687-8, 9763, 9885, 9891, 9907, 9984, 10036, *10087, 10090, 10098,
10104, 10172, 10178, 10195, 10216, 10254, 10376, *10580, 10632, 10738-9, 10802,
10842, 10884, 10909, 10923, 10931, *10966, 11011, 11046, 11054, 11096, 11124-5,
11261, 11453, 11474, 11494-5, *11595, 11619, 11642, 11656, 11765, 11772, 11819,
11825-6, 11864, 11879, 11976, *12020

Algae, brown 9353, 9476, 9488, 9565, 9610, 9905, *10109, 10433, 10916, 11011, 11699,
12031

Algae, green (*cf.* also *Chlamydomonas, Chlorella, Scenedesmus*) 9125, 9134, *9195,
9433, 9448, 9476, 9481, 9494, 9536, 9547-8, 9609, 9631, 9668, 9684-5, 9716-7,
9858, 9865, 9975, 10036, 10098, 10139, 10153, 10165, 10178, 10209, 10277, 10638,
10730, *10784, 10809, *10876, 10916, 10941, *10974, 11010, 11095, 11099, 11149,
11162, 11244, 11261, 11290, 11294, *11296-7, 11298, 11300-1, 11337-8, 11453,
11472-3, 11504, 11528, 11617, 11643, 11684, 11730, 11764, 11838, 11902-3, 11955-
6, 11976, 11979

Algae, red 9134, 9476, 9501, 9582, 9851, 10433, 10636, 10667, 10811, 10885, 10894-5,
11011, 11058, 11781-2, 12032

Allium cepa see Onion

Allium sativum see Garlic

Alnus see Alder

Amaranthus 10734-5, 11852

Ananas see Pineapple

Apple 9161, *9222, 9253, *9377, *9637, 9708-9, 9737, 10111, 10134, *10232, 10511,
*10555, 10579, 10589, *11028, *11116, 11117, 11708, *11737-8, *11740, 11978

Apricot 10134

Aquatic macrophytes 9231-2, 9276, 9371, 9435, *9447, 9739-41, 9953, 10039, 10060,
10837, 10852, 11089-90, 11231, 11649, *11687, 11688, *11904, 11978

Arabidopsis 9092-3, *9215, 9216-20, 9272, 9898, 9969-70, 10204, 10396-8, *10399, 10530-
1, 10983, 11710, 11849, *11850, 12043

Arachis see Groundnut, peanut

Armeniaca see Apricot

Ash *10446, 10539, *10999, *11199, 11290

Aspen 9274, 10453

Atriplex 9364-6, 10840-1, 11603-4, 11818

Avena see Oat

B

Bacteria, photosynthetic 9132, 9224, 9241-2, *9249, 9256, *9258, 9259, 9262-4, 9346,
    *9403, *9418-9, 9481, 9533, 9581, 9618-9, 9626-7, *9710, 9712, 9727, 9729-30,
    9751, 9788, *9799, 9816, *9821-2, 9829, 9849-50, 9854-6, 9880-1, 9886, 9893, 9912,
    9925, 9929, 9941, *9945-6, *9959, 9995, 10015, *10058, 10063-5, 10076, *10079,
    10083, 10091, 10127, *10200, 10201, 10223-4, 10273-4, *10294, 10301, 10356, 10384,
    10386, 10409, *10414, 10415-7, 10423-4, 10470-2, 10484-5, *10504, *10506-7,
    10508-9, 10585, 10621, 10663, 10844, 10858-9, 10915, 10991, 11022, *11031-2, 11033,
    *11034-5, 11049-50, 11061, 11086, 11111, 11115, 11154, 11166, 11171, 11173, 11182,
    11202, 11304, 11315, 11320, 11368-9, 11395, 11431, 11444-6, 11457-8, 11460, *11470,
    11471, *11477-8, 11485, 11498-502, 11548, 11575, 11588, 11599, 11626, 11628, 11659,
    11721-2, 11725, 11773-4, 11785, 11835, 11851, 11941, 11949, 11973, 11993, 12059-60

Banana 11916

Barley 9089, 9110, 9197, 9319, 9323-4, 9433, 9503, 9621, 9787, 9862, 9951, 10133-4,
    *10162, 10163, *10226, 10323, 1o345, 10360, *10372, 10373, 10442, 10469, 10633,
    *10647, 10658, 10666, *10810, 10967, *10986, 10987, 10990, 11002, 11073, *11081-
    3, 11085, 11121, 11167, *11189, 11221, 11352, 11357, 11359, 11388-9, 11430, 11543,
    11551, 11566, 11665, 11671, 11709, 11729, 11739, 11756, 11791, 11804, 11868,
    11964, 11978

Bean 9104, 9136, 9140, 9166, 9196, 9214, *9343, 9344, 9442, 9515, 9566, *9568, 9686,
    9697, 9771, 9976, *9998, 10029, 10048, 10050-1, 10072, 10149, 10163, 10177,
    10185, 10203, 10207, 10219, 10266, 10331, 10411-2, 10458-60, 10493, 10613, 10641,
    10673, 10676, 10747-8, *10760, 10766, *10819, 10908, 10931, 10993, 11051, 11071,
    11074, *11080-4, *11187, 11230, 11265-6, *11274, 11290, 11356, *11376, 11394,
    11401, 11410, 11489, 11538, 11648, 11651, 11741-2, 11776-8, *B11799, 11896, *11904,
    11908, 11966, 11974, 11977-8, 12038, 12048-9, 12061-2

Beech 10616, 10629, 10657, *10679, *10999, 11198, 11978

Beta see Sugar beet, beet, mangold, spinach beet

Betula see Birch

Birch 9787, 10453, 10684, *10999, 11027, 11198, *11199, 11708, 11978

Blackberry see Raspberry

Blueberry 11198

Brassica see Cabbage

Brassica cauliflora simplex see Broccoli

Brassica napus var. *chinensis* see Chinese cabbage

Brassica oleracea var. *acephala* see Kale

Brassica oleracea var. *botrytis* see Cauliflower

Brassica oleracea var. *gemmifera* see Brussels sprouts

Brassica oleracea var. *gongylodes* see Kohlrabi

Brassica rapa var. *rapa* see Turnip

Broadbean *9265, *9404, 9517, 9738, 10348, 10438, 10478, 10740, 10742, 10871, 11168,
    *11172, 11241, 11384, 11401, 11741-2, 11798, *B11799, 11832, 12061-2

Broccoli 9773

Brussels sprouts *9666, 11957

Buckwheat *9279, 9943, 10407, 11205

C

Cabbage   12062

Canabis see Hemp

Capsicum see Pepper

Carpinus see Hornbeam

Carrot   9754-5, 10067, 10112, 11029, 11084, 11285, 11668, 11842

Castor bean   9465, 10103, 10562

Cauliflower   10954

Cerasus see Cherry

Cereals (*cf.* also Grasses, *etc.*)   *9351, *9423, 10237, 10630, 11283, 11403

Cherry   *9377, 10634, *11028, 12056

Chinese cabbage   11965, 11978

Chlamydomonas   9169, 9205, 9425, 9495, 9575, 9798, *10019, 10020-3, 10039, 10143,
    *10595, 10650, 10814, *10974, 11009, 11057, 11066, 11290, 11424-5, *11514,
    11677-8, 11706, 11940

Chlorella   9181-2, 9285-6, 9297, 9371, 9405, 9428, 9495, 9569, *9577, 9578, 9633,
    *9635, 9636, *9640, 9658, 9777, 9861, 9891-2, 9929, 9932, 9940, 9971, *9980, 9982,
    *9998, *10004, 10005, 10036, 10039, 10041, 10044, 10060, *10108-9, 10131, 10264,
    10367, 10428, *10693, 10695, 10698, 10716, 10732, 10742, *10771-3, 10774, 10782,
    10821, 10928-30, 10943, 10962-3, 10965, *10974, 11052, 11180, 11290, 11370, 11411,
    11421, 11567, 11580, 11584, 11634, 11644, 11656, 11679, 11701, 11762, 11803,
    11872-3, 11899, 12014, *12042, *12063, 12067

Citrullus see Watermelon

Citrus   *9246, *9934, 9990, 10368, 10613, 11155, 11198, 11800, 11839-41

Clover   9108, 9955, 10228, *10229-30, 10560, 10798, *11098, *11339, 11542, 11978

Cocoa   11331, 11978

Coffea see Coffee tree

Coffee tree   9523

Coniferous plants   9103, 9171, *9378, 9469, 9742, 9994, 10439, 10931, 11429, 11478,
    11805, 12010

Corchorus see Jute

Corn see Maize

Cotton   9098, 9119, 9121, 9198, 9781, *9797, *9933, 9935-6, 10173, 10203, 10275, 10299,
    *10313, 10434, 10498, 10752-3, 10758, 10780, 10927, 10945, 10984, 11015, 11078,
    11248-9, 11282-3, 11290, 11367, 11503, 11650, *12068

Cottonwood see Poplar, cottonwood

Cucumber   9120, 9130, 9711, 9776, 9857, 10120, *10278, 10283, 10488-9, 10671, 10701,
    *11308, 11311, *11312-3, 11323, 11410, 11507, *11510, *11555, 11612, *B11799,
    *11855, 11865

Cucumis sativus see Cucumber

Cucurbita see Squash

Currant   10720, 10877, 12056

D

Daucus see Carrot

Dewberry see Raspberry

Diatoms   9494, 9750, 9905, 10156, 10178, 10343, 10838, 11012, 11041, 11244, *11255,
    11261, 11330, 11453, 11656-7, 11699

Hornbeam 10629

Horsetail *10077, 10098, 11967

I

Ipomea batatas see Sweet-potato

J

Juglans see Walnut

Jute *10028

K

Kale 9682, 9862, 10325, 10490

Kohlrabi 9902-3, 10604, 12062

L

Lactuca see Lettuce

Larch 9826, 9977, *10999, 11198

Larix see Larch

Leguminous plants (*cf.* also Bean, Broadbean, Lupine, Pea, Soybean, *etc.*) *9604, *9824,
    *10250, 10708, 10711, 11283, 11800

Lemon see *Citrus*

Lettuce 9559, 11003, *B11799, 11965, 12029

Lichens 9101-2, 9194, 9199, 9480, 9509, *9831-3, 10124, 10364, 10427

Lilac 10453, 10501, 10877, 11027

Linden 9115, 9804, 10539

Linum see Flax

Liverworts 10049, 11290

Lucerne see Alfalfa

Lupine 10701, 10853-4, 11247

Lupinus see Lupine

Lycopersicon see Tomato

M

Maize 9129, 9145, 9197, 9206, 9238, *9253, 9269, 9284, *9322, 9325-6, 9344, 9349,
    9354, 9356, 9398-9, 9410, 9431-2, 9451, 9485, 9528, 9538, 9553, 9571, 9573-4,
    9601, 9614, 9647, 9690, 9701, 9726, *9765, 9870, 9921, *9923, 9968, 9972-3, 9976,
    *9987, *9989, 9993, 9996, 10029, 10039, *10047, 10062, 10081, 10113, *10114,
    10150, 10152, 10225, 10255, *10265, 10304-5, *10329, 10330, 10339-40, 10347,
    10359, 10385, 10393, 10426, *10449, 10483, *10500, 10516-8, *10529, *10586, 10598,
    10602, 10613-4, 10617, 10622-3, 10646, 10648, 10651, 10681, 10704, 10719, 10731,
    10734, *10819, 10860, 10893, 10901, 10922, 10969, *11080, 11082, 11093, 11100,
    *11227, 11228-9, 11290, 11325, 11347, 11354, 11378-9, 11401, 11407, *11518, 11532-
    3, 11541, 11549, 11562, 11569, 11571, 11587, 11589, 11597, 11606, 11607, 11640,
    11648, 11674, 11720, 11731, 11761, 11798, *B11799, 11800, 11817, *11861, 11862,
    11899, *11904, 11905, *11926, 11935, 11947, 11978, 12016, 12026-7, 12041, 12062

Malus see Apple

Picea see Spruce

Pine 9240, 9420, 9427, *9473, 9665, 9932, 10121, 10125, 10161, 10645, 10670, 10683-4,
     10699, 10710, 10726, *10770, *10999, 11005, 11068, 11070, *11075-7, 11078, 11198,
     11476, 11482, 11488, 11616, *11662, 11663, 11708, 11808, 11890, 11893, 11917,
     11982, 12015, 12050-1, 12058

Pineapple 9591

Pinus see Pine

Pirus see Pear

Pisum see Pea

Poplar, cottonwood 9646, 9673, 9787, 10166, *10446, *10679, 10839, *10999, 11004,
     11027, 11140, 11290, 11437, 11708, 11978

Poppy 11742

Populus see Poplar, cottonwood

Populus tremula see Aspen

Potato 9095-6, 9228-9, *9243, 9244, *9421, 9422, *9719, 9743, 9818, 9917, *9918, 9919,
     10296, 10328, *10599, *10702, 10717, 10897, *10900, *10996, 11043, *11067, *11080,
     *11082, 11150-1, 11197, 11520-1, 11668, 11800, 11846, 11906, 11978, 11987, *12045,
     12062

Prunus see Apricot, Cherry, Peach

Pumpkin see Squash

Q

Quercus see Oak

R

Radish 10382, *10746, 10979, 11047-8, 11285, 11552, 11742, 11965, 12062

Raphanus see Radish

Raspberry 10242

Ribes see Currant

Rice *9212, 9213, *9543, 9815, 10116, *10141, 10390-1, 10700, *10816-7, 10833, *10958-
     9, 10978, 11291, *11506, 11668, 11727, 11732-5, 12006, *12028

Ricinus see Castor bean

Rubus see Raspberry

Rye 9862, 9988, 11382, *11828

S

Saccharum see Sugar cane

Salix see Willow

Scenedesmus 9359-61, 9452, 9521, 9740, 9891-2, 10098, 10322, 10716, *10974, 11200,
     11290, 11292, 11324, 11371, 11435, 11695-7, *11700, 11751-2, 11783

Secale see Rye

Sempervirent plants *9378, 9456, 9725, 10095, 10857, *11164, 11198, 11488, 11639,
     11808

Service-tree *10267, 10268

Sinapis see Mustard

Solanum tuberosum see Potato

Sorbus see Service-tree

Sorghum 9233, 9326, 9354, 9356, 9528, 9699-700, 9707, 9890, *10010, 10039, 10319,
     10685, 11219, 11290, 11399, 11537, 11587, 11641, 11744, 11830

Sorgum see Sorghum

Soybean 9164, 9228, *9279, 9335-6, 9380, 9431-2, *9508, 9528, 9554, 9695, 9746, 9830,
     9956-8, 9986, 10085, 10115, 10118, 10569, 10602, 10729, *10777, 10786, 10893,
     10936, 11055, 11219, 11290, 11541, 11554, 11587, 11648, 11664, 11666-7, 11704,
     11744, 11771, *11828, 11978
Spinach 9158, 9190-2, 9225, *9265, 9267, 9288, 9312, 9400, 9428, 9433, 9446, 9449-50,
     9455, 9472, 9476-7, 9482, 9506, 9518, 9520, 9587, 9597, 9649, 9722, 9762, 9780, 9795,
     9859, 9871-5, 9888, 9891-2, 9897, 9932, 10016, 10018, 10066, 10068, *10069, 10070,
     10098-9, *10100, 10101, 10104, 10144, 10147, 10155, 10215, 10260-1, 10284-5, 10303,
     10317, *10321, 10322-3, 10342, 10345, 10374, 10402, 10464-5, 10481, 10527, 10622-
     3, 10749, 10762, 10782, *10813, 10831, 10861, 10880, *10913, 10914, 10931, *10932-
     3, 19034-5, 10944, 10946-52, 10956, 11008, 11021, 11066, 11129, 11165, *11172,
     11203, 11211, 11213, 11289, 11290, 11292, 11334, *11341, 11366, *11380, 11386,
     11392, 11412-3, 11417, 11436, 11467, 11512, 11530, 11558, 11583, 11591, 11690-1,
     *11702, 11703, 11728, 11750, 11761, 11780, 11788, 11823, 11827, 11831, 11988-90,
     12046

Spinach beet see Sugar beet, beet, mangold, spinach beet

Spinacia see Spinach

Spruce 9160, *9170, 9665, 10316, 10419, 10613, 10684, *10999, 11198, 11290, 11488,
     11978

Squash 9201, 9783, 10012, 10441, 10532, 10865, 11329, 11434, *B11799, 12062

Strawberry *9747, 10272, 11238

Submersed plants see Aquatic macrophytes

Succulents 9208, 10207, 10338, 10476-9, 11290, 11672, 11786

Sugar beet, beet, mangold, spinach beet *9131, *9299, 9328, 9478, 9932, 9968, *10014,
     10066, *10081, 10110, 10170, 10196, 10454, *10510, *10556, 10671, 11064, *11080,
     *11082-3, 11132, 11205, 11290, 11343, 11354, *11620, 11753-4, *B11799, 11965, 11978

Sugar cane 9237, 9485, 9489, 9756-7, 9759, 9762, 9983, 10186, 10269-71, 10514, 10968,
     11288, 11290, 11761, 11978

Sunflower 9195, 9204, 9350, 9394, 9437-9, *9683, 9753, 9771, 9809, 9813, 9861, 10039,
     10222, 10234, *10257, 10715, *10733, *11080-3, *11172, 11193, 11217, 11222, 11290, 11432,
     11432, 11646, 11648, 11905, 11918, 11975

Sweet-potato 9932, 11635, 11860, *11861, 11965

Syringa see Lilac

T

Tapioca 11285

Tea 9638, 9661, 11146, 11790

Thea see Tea

Theobroma see Cocoa

Tilia see Linden

Tobacco 9239, *9279, 9453, *9532, 9738, *9758, 9787, 9790, 9932, 10045-6, 10074,
     10084, 10151, 10218, 10234, 10318, *10353-5, 10366, 10394, 10403, 10405-6, 10464,
     10542, 10553, 10567, 10588, 10594, 10596, 11216, 11290, 11293, 11455, 11462-3,
     11505, 11509, 11800, 11830, 11883, 11892, 11908, 11915, 11974, 12017, 12039
Tomato 9142, 9253, 9247-8, *9379, 9454, 9572, 9724, 9900, *9950, 10243, 10255, *10278-
     9, 10280-3, *10522, 10744-5, 10970, 10998, 11143, 11218, 11290, 11305, 11343,
     11507-8, *11510, 11645, 11761, *B11799, *11820, 11842, *11874, 11974, 12057,
     12061

Trees, deciduous (*cf.* also Alder, Ash, Aspen, Beech, Birch *etc.*) 9281, 9679, 9714,
     9731-2, 9742, 9868-9, 10159, 10335, 10835-6, *11199, 11672

Trifolium see Clover

Triticum see Wheat

Turnip 11126, 11465, 11965

U

Ulmus see Elm

V

Vaccinium see Blueberry

Vegetables (*cf.* also Cabbage, Cauliflower, *etc.*) *10220-1, *10278-9, 10280-3, 11306,
     11965, 12057

Vicia faba see Broadbean

Vitis see Grape-vine

W

Walnut 9529, *9696, *10446, 10863

Watermelon 9339, 10349, 11887

Weeds 11367

Wheat *9193, *9260, 9320, *9404, 9407, 9444, *9522, 9592-3, *9623, 9735, 9745, *9769-
     70, 9784, 9738, 9862, 9924, 9932, 9988, *10007, *10047, *10088, 10207, 10235,
     10330, *10341, 10369, 10404, 10440, 10466, *10499, 10521; *10529, 10557, *10575,
     *10718, 10756, 10767, *10791, 10789, *10822, 10881, 10887-8, 10912, *10960, 10961,
     11103-6, *11141, 11153, *11188, 11190, *11194, 11195, *11196, 11201, 11214-5,
     11251-3, 11290, 11303, 11354, 11382, 11400, 11454, 11579, 11587, 11636, 11711,
     11716, 11741, 11748, 11791, 11824, *11828, 11946, 11978, 11995, *12044

Willow *10446, *10999, 11198

Woody plants (*cf.* also Fir, Forest... , Larch, Pine, Spruce, *etc.*) 10086, *10293,
     *10679, *10999, 11027, 11672, 11978, 11983-4, 12011

Z

Zea see Maize